New Perspectives on Applied Industrial Ergonomics

Arturo Realyvásquez Vargas ·
Jorge Luis García-Alcaraz · Emigdio Z-Flores
Editors

New Perspectives on Applied Industrial Ergonomics

Springer

Editors
Arturo Realyvásquez Vargas
Department of Industrial Engineering
Tecnológico Nacional de México/I.T. Tijuana
Tijuana, Baja California, Mexico

Emigdio Z-Flores
Department of Industrial Engineering
Tecnológico Nacional de México/I.T. Tijuana
Tijuana, Baja California, Mexico

Jorge Luis García-Alcaraz
Industrial Engineering & Manufacturing
Universidad Autónoma de Ciudad Juárez
Ciudad Juárez, Chihuahua, Mexico

I.T. Ciudad Juárez
Tecnológico Nacional de México
Ciudad Juárez, Chihuahua, Mexico

ISBN 978-3-030-73467-1 ISBN 978-3-030-73468-8 (eBook)
https://doi.org/10.1007/978-3-030-73468-8

This Springer imprint is published by the registered company Springer Nature Switzerland AG
The registered company address is: Gewerbestrasse 11, 6330 Cham, Switzerland

Preface

Ergonomics is a multidisciplinary science and profession relatively new since it has its origins as scientific discipline in 1949. According to the International Ergonomics Association (IEA), there are three main areas in Ergonomics: Physical Ergonomics, Cognitive Ergonomics, and Organizational Ergonomics. The objective of Ergonomics is to provide safety, health, and comfort to humans when they perform an activity, either physical or mental, and so improve the human performance at short term, and organizational performance at long term. To achieve this, Ergonomics studies the capabilities and limitations of human beings when they interact with other elements in a work system.

Such elements may include environmental conditions (e.g., temperature, noise, lighting, vibration, ventilation, among others), design of tools and machines, design of task and workstations, social relationships, and work schedules, to mention few. When these elements are not adapted to the capacities and limitations of human beings, they cause risk factors for their health, safety, and comfort, and the individual and organizational performance decrease. Such risk factors may include uncomfortable body postures, repetitive movements, manual material handling, mental workload, among others.

An ergonomic intervention basically consists of three stages: detection, evaluation, and control. In the stage of detection, the ergonomist must detect the possible risk factors at which humans can be exposed. In the second stage, evaluation, the ergonomists apply ergonomic methods to evaluate the risk factors and determine the risk level of each of them. Finally, and according to the results obtained in the second stage, in the stage of control, the ergonomists implement improvement changes to decrease the risk level of the evaluated risk factors.

As Ergonomics is a relatively new scientific discipline and profession, researchers in this area continue develop new theoretical and practical knowledge, such as literature reviews, design of new ergonomic evaluation approach, and statistical analyses of ergonomic factors (both risk and success factors) and their effects. Of course, Ergonomists continue to conduct research and interventions that apply popular ergonomic methods, such as the Ovako Working Analysis System (OWAS), Rapid Upper Limbs Assessment (RULA), and Individual Risk Assessment (in Spanish Evaluación de Riesgo Individual, ERIN), to mention few.

This book is divided into three parts. Part I, called New theoretical and practical knowledge, comprises Chapters 1–3. In Chapter 1, a comparative study of nine rehabilitation devices for people with Carpal Tunnel Syndrome is presented. Moreover, the devices that were manufactured with traditional technology and those that were produced using 3D additive manufacturing, as well as their properties and characteristics are mentioned. In Chapter 2, a research review of scientific articles concerning the application of infrared thermography (IRT) in the mensuration of diagnostic studies of carpal tunnel syndrome (CTS) is presented. In a similar research, Chapter 3 presents a literature review to provide a comprehensive literature analysis of mental workload evaluation and management for sustainable processes in the manufacturing industry, from a social perspective. Finally, Chapter 4 presents a dynamic simulation model useful for determining the degree to which ergonomic exposure metrics are impacted by changes in various operational decisions such as scheduled overtime, cycle time, make/model mix.

In Part II, Chapters 5–7 present different statistical analyses to test relationships of ergonomic factors and their effects. Chapter 5 compares the self-reported fatigue of workers from 7-hour to 12-hour and its relationship with those two-night work schedules. Similarly, Chapter 6 analyzes the differences and similarities of the Maslach Burnout Inventory (MBI), and the Shirom-Melamed Burnout Measure (SMBM) scales to evaluate Burnout, and study how these scales relate to Job Strain and Emotional Intelligence. By last, Chapter 7 offers an analysis that uses structural equations to identify those critical success factors as well as the way they relate to financial benefits.

Finally, Part III, called Applied Physical Ergonomics, includes Chapters 8–13 and presents ergonomic interventions in tasks corresponding to different labor sectors. For instance, in Chapter 8, a hierarchical task analysis (HTA) is applied to the exhibition booths set-up process to know and record the tasks and stages in which said process is divided and to know if any of these tasks and stages could cause ergonomic risks to assemblers due to physical efforts in this type of work. Similarly, in Chapter 9, an ergonomic analysis has been made to review the status and discomfort levels of craftsmen manufacturing the agricultural appliances in India. Moreover, a systematic methodology to detect the main ergonomic interventions required in a distribution process for automatic vending machines is presented in Chapter 10. Regard to Chapter 11, it presents an ergonomic evaluation of working conditions in a textile company by means of the Quality Improvement Through Ergonomics (QUITE) method to avoid the presence of workers with discomfort or pain and to improve the quality of the manufactured products. In a similar study, Chapter 12 presents an ergonomic intervention process carried out in a Colombian meat processing plant, using the ERIN method. Finally, in Chapter 13, an integration of postural and fatigue analyses for the ergonomic design and improvement of workstations as a strategy to increase sustainability of manufacturing companies is presented.

Tijuana, Mexico — Arturo Realyvásquez Vargas
Ciudad Juárez, Mexico — Jorge Luis García-Alcaraz
Tijuana, Mexico — Emigdio Z-Flores

Acknowledgements

The editors want to acknowledge to all the authors for their valuable contributions in the development of the book. It is known that, during all the editorial process, and due to the COVID-19 pandemics, the authors had diverse difficulties to attend the comments that reviewers and editors asked them. However, all authors always provide the necessary information at time. Therefore, the editors appreciate and recognize their effort.

Also, the editors wish to thank all the reviewers that, with their knowledge, helped to improve the quality of the chapters with their valuable. A review process is time demanding, so, the editors appreciate all the time the reviewers invested.

Finally, the editors want to acknowledge to their families since without their support during all the editorial process, the book would not be finished at time and quality. For the editors, it is a pride to have the unconditional support of their families in all the projects they undertake.

Editors
Arturo Realyvásquez Vargas (Mexico)
Jorge Luis García-Alcaraz (Mexico)
Emigdio Z-Flores (Mexico)

Contents

Editors and Contributors

About the Editors

Arturo Realyvásquez Vargas is a full-time Professor from the Department of Industrial Engineering at Tecnológico Nacional de México/Instituto Tecnológico de Tijuana in Mexico. He received a master's degree in Industrial Engineering and a Ph.D. in Engineering Sciences from the Autonomous University of Ciudad Juárez in Mexico, and a Ph.D. in Innovation in Product Engineering and Industrial Process at the University of La Rioja (Spain). Specifically, his main research areas are related to optimization of industrial processes, lean manufacturing, and ergonomics. Also, he is an active member of the Society of Ergonomists of Mexico Civil Association (Sociedad de Ergonomistas de México, SEMAC) and the Network of Optimization of Industrial Processes (Red de Optimización de Procesos Industriales, ROPRIN). Currently, Dr. Realyvásquez is a National Researcher recognized by the National Council of Science & Technology of Mexico (CONACYT) as Candidate. Dr. Realyvásquez is an Author/Coauthor in around 12 papers published in journals indexed in the Journal Citation Reports, also, he has attended to international conferences and congress in Mexico as well as in the United States. Nowadays, Dr. Realyvásquez has supervised more than 20 bachelor theses and 5 master theses. Dr. Realyvásquez is Author of one book published by the international publisher Springer, related to ergonomics. Also, Dr. Realyvásquez has edited two books in IGI Global, all of them related to ergonomics.

ORCID: https://orcid.org/0000-0003-2825-2595.

Scopus Author ID: 56167726800.

Jorge Luis García-Alcaraz is a full-time Professor with adscription to Department of Industrial Engineering at the Autonomous University of Ciudad Juárez and full-time Professor at Division of Research and Postgraduate Studies in Tecnológico Nacional de México/I.T Ciudad Juárez. He received a Bachelor degree and a M.Sc. in Industrial Engineering from the Instituto Tecnológico de Colima (Mexico), a Ph.D. in Industrial Engineering from Instituto Tecnológico de Ciudad Juárez (Mexico), a

Ph.D. in Innovation in Product Engineering and Industrial Process from University of La Rioja (Spain), a Ph.D. in Engineering and industrial technologies and a Postdoc in Manufacturing Process from University of La Rioja (Spain). His main research areas are related to Multicriteria decision making applied to lean manufacturing, production process modeling and statistical inference. He is founding member of the Mexican Society of Operation Research and active member in the Mexican Academy of Industrial Engineering. Currently, Dr. García is a National Researcher level III recognized by the National Council of Science & Technology of Mexico (CONACYT) and is working with research groups from Colombia, Spain, Chile, and Dominican Republic. Dr. García is Author/Coauthor of around 150 papers publisher in journals indexed in the *Journal Citation Reports*, more than 200 international conferences and congress around the world. Dr. García is Author of 12 books published by international publishers as Springer and IGI Global, all then related to lean manufacturing, the main tools, and techniques.

ORCID: http://orcid.org/0000-0002-7092-6963.
Scopus Author ID: 55616966800.
Department of Industrial Engineering and Manufacturing
Autonomous University of Ciudad Juárez
Av. del Charro 450 Norte, Ciudad Juárez, Chihuahua, Mexico. Z.P. 32310.
Phone: +52 656 6884843 ext. 5433 and 4249 Fax: +52 656 6884841
e-mail: jorge.garcia@uacj.mx

Emigdio Z-Flores received a Bachelor degree as electrical engineer and a Doctoral degree in Engineering Sciences at Instituto Tecnológico de Tijuana, and his Master degree in Digital Systems at Centro de Investigación y Desarrollo de Tecnología Digital. He worked in the industry sector for nearly 8 years as product designer, developing several projects in different key areas. He currently works as full-time Professor in Tecnológico Nacional de México, campus Tijuana, at the Industrial Engineering department, both in under-graduate and graduate programs. He is Author/Coauthor of several publications in indexed journals and international conferences and holds one patent. His current interests involve applied statistics, computer science, neuroscience, and industrial real-world applications.

ORCID: https://orcid.org/0000-0002-1442-5320.
Industrial Engineering Department
Tecnológico Nacional de México, Campus Tijuana
Calzada Del Tecnológico S/N, Fraccionamiento Tomas Aquino. Tijuana, Baja California, Mexico, 22414
e-mail: emigdio.zflores@tectijuana.edu.mx

Contributors

José Francisco Alatorre-Ávila Campus Ciudad Cuauhtémoc, Tecnológico Nacional de México, Ciudad Cuauhtémoc, Mexico

Cesar Roberto Alferez-Padrón Industrial Engineering and Manufacturing Department, Autonomous University of Ciudad Juárez, Ciudad Juárez, Chihuahua, Mexico

Nancy Ivette Arana-De las Casas Campus Ciudad Cuauhtémoc, Tecnológico Nacional de México, Ciudad Cuauhtémoc, Mexico

Claudia Yohana Arias-Portela Faculty of Engineering, Panamericana University, Benito Juárez, CDMX, México

Oziely Daniela Armenta-Hernández Electrical and Computing Department, Autonomous University of Ciudad Juárez, Ciudad Juárez, Chihuahua, Mexico

Verónica Arredondo Robledo Departamento de Ingeniería Industrial, Universidad Autónoma de Baja California, Mexicali, BS, Mexico

Cesar Omar Balderrama-Armendáriz Industrial Design Department, Autonomous University of Ciudad Juárez, Ciudad Juárez, Chihuahua, Mexico

Anjaneya Bandekar Department of Industrial & Systems Engineering, Auburn University, Auburn, AL, USA

Manuel Alejandro Barajas-Bustillos Electrical and Computing Department, Autonomous University of Ciudad Juárez, Ciudad Juárez, Chihuahua, Mexico

Claudia Camargo-Wilson Faculty of Engineering, Architecture and Design, Autonomous University of Baja California, Ensenada, Mexico

Marlen Castellanos Uralde Instituto de Arquitectura Diseño y Arte, Universidad Autónoma de Ciudad Juárez, Ciudad Juárez, Chihuahua, Mexico

Melissa Airem Cázares-Manríquez Faculty of Engineering, Architecture and Design, Autonomous University of Baja California, Ensenada, Mexico

David Cortés Sáenz Instituto de Arquitectura Diseño y Arte, Universidad Autónoma de Ciudad Juárez, Ciudad Juárez, Chihuahua, Mexico

Raquel Costa Department of Psychobiology, University of Valencia, Valencia, Spain

Jorge De La Riva-Rodríguez Campus Ciudad Juárez, Tecnológico Nacional de México, Ciudad Juárez, Mexico

Yasmina El Arbi Department of Psychobiology, University of Valencia, Valencia, Spain

Pedro Alberto Escárcega Zepeda Departamento de Ingeniería Industrial, Tecnológico Nacional de México, Instituto Tecnológico de Mexicali, Mexicali, BS, Mexico

Jorge Luis García-Alcaraz Department of Industrial and Manufacturing Engineering, Autonomous University of Ciudad Juárez, Ciudad Juárez, Chihuahua, Mexico;

I.T. Ciudad Juárez, Tecnológico Nacional de México, Ciudad Juárez, Chihuahua, Mexico

Enrique García-Grajeda Campus Ciudad Cuauhtémoc, Tecnológico Nacional de México, Ciudad Cuauhtémoc, Mexico

Blanca Rosa García-Rivera Faculty of Administrative and Social Sciences, Autonomous University of Baja California, Tijuana, Mexico

Murray Gibson Department of Industrial & Systems Engineering, Auburn University, Auburn, AL, USA

Juan Luis Hernández-Arellano Instituto de Arquitectura Diseño y Arte, Universidad Autónoma de Ciudad Juárez, Ciudad Juárez, Chihuahua, Mexico

Pablo López Calle Facultad de Ciencias Políticas y Sociología, Universidad Complutense de Madrid, Pozuelo de Alarcón, Madrid, Spain

Juan Gabriel López Hernández Departamento de Ingeniería Industrial, Universidad Autónoma de Baja California, Mexicali, BS, Mexico

Juan Andrés López-Barreras Faculty of Chemical Sciences and Engineering, Autonomous University of Baja California, Tijuana, Mexico

Aide Aracely Maldonado-Macías Department of Industrial and Manufacturing Engineering, Autonomous University of Ciudad Juárez, Ciudad Juárez, Chihuahua, Mexico

Román E. Méndez Department of Technology and Production, CyAD (Sciences and Arts for Design), Universidad Autónoma Metropolitana – Xochimilco, Mexico City, Mexico

Ismael Mendoza Muñoz Departamento de Ingeniería Industrial, Universidad Autónoma de Baja California, Mexicali, BS, Mexico

Debesh Mishra SME, KIIT Deemed to Be University, Bhubaneswar, Odisha, India

Carlos Raúl Navarro González Departamento de Ingeniería Industrial, Universidad Autónoma de Baja California, Mexicali, BS, Mexico

Jesús Everardo Olguín-Tiznado Faculty of Engineering, Architecture and Design, Autonomous University of Baja California, Ensenada, Mexico

Wendy Pereira Acosta Instituto de Arquitectura Diseño y Arte, Universidad Autónoma de Ciudad Juárez, Ciudad Juárez, Chihuahua, Mexico

Elizabeth Pérez School of Industrial Engineering, Universidad Pontificia Bolivariana, Medellín, Colombia

Arturo Realyvásquez Vargas Department of Industrial Engineering, Tecnológico Nacional de México/I.T. Tijuana, Tijuana, Baja California, Mexico

Yordán Rodríguez National School of Public Health, Universidad de Antioquia, Medellín, Colombia

Elizabeth Romero Samaniego Departamento de Ingeniería Industrial, Tecnológico Nacional de México, Instituto Tecnológico de Ensenada, Ensenada, BS, Mexico

David Sáenz-Zamarrón Campus Ciudad Cuauhtémoc, Tecnológico Nacional de México, Ciudad Cuauhtémoc, Mexico

Maria Camila Salazar-Marín School of Industrial Engineering, Universidad Pontificia Bolivariana, Medellín, Colombia

Ana Laura Sánchez Corona Departamento de Ingeniería Industrial, Universidad Autónoma de Baja California, Mexicali, BS, Mexico

Suchismita Satapathy SME, KIIT Deemed to Be University, Bhubaneswar, Odisha, India

Miguel A. Serrano Department of Psychobiology, University of Valencia, Valencia, Spain

Bob Sesek Department of Industrial & Systems Engineering, Auburn University, Auburn, AL, USA

Patricia Eugenia Sortillón-González Department of Manufacturing Industrial Engineering, Universidad Estatal de Sonora, Hermosillo, Sonora, México

Maria Alejandra Trujillo-Murillo School of Industrial Engineering, Universidad Pontificia Bolivariana, Medellín, Colombia

Ricardo Vardasca Faculdade de Engenharia, Universidade do Porto, Porto, Portugal;
INEGI, Universidade do Porto, Porto, Portugal;
ISLA Santarém, Santarém, Portugal

Yanet Villareal González Departamento de Ingeniería Industrial, Tecnológico Nacional de México, Instituto Tecnológico de Mexicali, Mexicali, BS, Mexico

Ann Godelieve Wellens Departamento de Sistemas, Universidad Nacional Autónoma de México, Coyoacán, CDMX, México

List of Figures

List of Tables

Part I
New Theorical and Practical

Chapter 1
Two Groups of Patented Devices for Carpal Tunnel Syndrome Rehabilitation: Comparative Study Between Traditional and Additive Manufacturing

Wendy Pereira Acosta, Juan Luis Hernández-Arellano, David Cortés Sáenz, Marlen Castellanos Uralde, and Pablo López Calle

Abstract This chapter presents a comparative study of two rehabilitation devices groups for people with carpal tunnel syndrome that were registered in different intellectual property offices. The objective was to compare the rehabilitation devices produced using traditional technology and 3D additive manufacturing, to show the advantages and disadvantages of both production processes. The methodology used in this research was structured into three stages. In the first stage, a systematic technological research of existing patents was developed using the search engines PatBase and Google Patent. The second stage describes the main characteristics and functionality of these devices. The last stage consisted of developing a comparative matrix of the properties and characteristics considered to define the relevance of each device. As a result, nine rehabilitation devices were selected describing and comparing their main characteristics. This research can serve as a reference for future studies related to carpal tunnel syndrome.

Keywords Additive manufacturing · Fused filament fabrication · Manpower · Machine · Materials · Methods · Management

1.1 Introduction

Carpal tunnel syndrome (CTS) is the most common musculoskeletal disorder (MSD) with a high incidence in the working guild (Hoyos 2010), being classified as work-related musculoskeletal disorder (WMSDs), especially when a neuropathy occurs in

W. Pereira Acosta (✉) · J. L. Hernández-Arellano · D. Cortés Sáenz · M. Castellanos Uralde
Instituto de Arquitectura Diseño y Arte, Universidad Autónoma de Ciudad Juárez, Av. Del Charro 450, Colonia Partido Romero, Ciudad Juárez, Chihuahua C.P. 32310, Mexico
e-mail: al199273@alumnos.uacj.mx

P. López Calle
Facultad de Ciencias Políticas y Sociología, Universidad Complutense de Madrid, Campus de Somosaguas, Pozuelo de Alarcón, Madrid C.P. 28223, Spain

A. Realyvásquez Vargas et al. (eds.), *New Perspectives on Applied Industrial Ergonomics*,
https://doi.org/10.1007/978-3-030-73468-8_1

the work context (Nunes and McCauley Bush 2012). CTS is identified as a painful and progressive condition that affects the upper limb, caused by the compression of one of the hand peripheral nerves (the median nerve) when passing through the carpal tunnel located in the wrist area (Pardal-Fernández et al. 2004; Cristina et al. 2009; Doughty and Bowley 2019).

According to Armenteros Pedrero et al. (2000), conservative treatment for CTS is the use of immobilization with wrist splints to alleviate the condition symptoms. Splints are orthopedic devices that are located in certain areas of the human body to provide support to injured joints and limit their movement, helping to restore decreased organic functions. These devices must be light, durable, and washable (Ocello and Lovotti 2015); it can be plaster pieces placed in hospital centers by specialists, or commercial static splints fixed with velcro straps, as well as can be purchased in pharmacies or online stores (Herrera Gil 2019). Although it has been used conventionally, these devices usually present certain disadvantages, such as the accumulation of bacteria due to the environmental humidity and the patient's sweat, and they can cause itching, if are not waterproof and are hard to sanitize, among other adversities.

On the other hand, with the technological advance and the development of additive technology, the design and commercialization of other immobilization devices types have increased in the last 10 years (Segnini et al. 2017). For example, splints have been produced using additive manufacturing (also known as 3D printing) with morphologically different designs respecting to the traditionally known. This new kind of production enables the adjustment of the device shape, thanks to the technology used for manufacturing it. Additive manufacturing is also able to adapt the design to the hand anthropometry and proposes a new way to produce and distribute the splints for the rehabilitation of the CTS disease.

Therefore, the diversity of these splints can be divided into two groups according to their manufacturing process, those clusters are conventional devices manufactured with traditional technology, and the novel ones manufactured using additive manufacturing or 3D printing. In the first group are, for example, rigid braces made with hard materials, such as thermoplastics or resins, and wrist brace made of semi-rigid elastic fabric that incorporates an aluminum plate or combination of hard polyurethane and an elastic cotton mesh or thermoplastic tape. Considering this further information, this chapter aims to construct a comparative study of two rehabilitation devices groups for people with carpal tunnel syndrome, based on an extensive technology review of splints registered in different intellectual property offices around the world. Besides, the pros and cons of each of these devices will be described.

1.2 Background

According to Davis et al. (1998), who reviewed more than 200 articles related to the CTS rehabilitation process, one of the most frequent recovery treatments used by physicians is the wrist immobilization splints (WIS) (see Fig. 1.1). In the first three

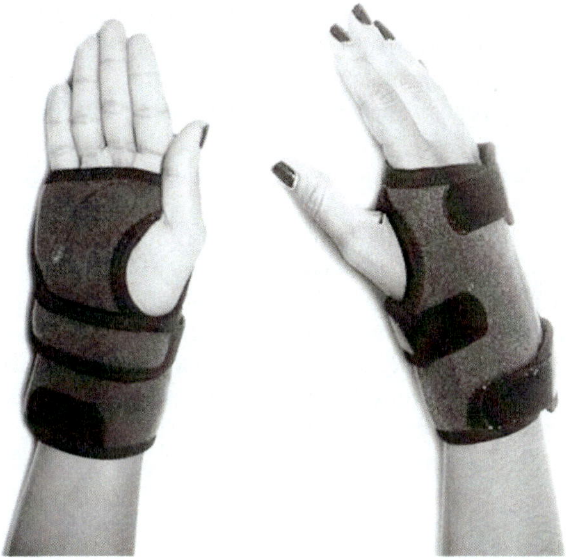

Fig. 1.1 Example of immobilization splints

months of the onset of the symptoms, the target is to decrease the pressure in the carpus and to reduce inflammation of the median nerve (Burke et al. 1994) by keeping the join in a neutral position and preventing flexion and extension movements, or ulnar and radial deviation of the wrist. The WIS should be removed twice per day to keep the arcs of movement.

"Rehabilitation in peripheral nerve injuries of the forearm and hands. Guide of evidence-based clinical practices." Hern et al. (2013) states that there are no perfectly adapted splints for each person. Consequently, they should conform to the anthropometric measurements of hands, the needs, customs, and work of each patient. The adaptation must be optimal to ensure that no major long-term injuries can be incurred, due to immobilization in an incorrect neutral position or due to the lack of ergonomic adjustment of the splint in the hand. According to the specifications of conditions for each patient's injury, the types of adaptation vary between rigid, soft, flying, or dorsal.

In medical consultations, either the patient has been usually provided with a splint, prepared by specialist technicians who, under the doctor's instructions, prepare and place the patient with a plaster prepared to the hand measure; otherwise, they deliver a traditional commercial sale splint available in pharmacies or online stores and adjust it by hand indicating the correct use of it. The second option generally has atypical design, which although they vary between different models is very similar to the first case; differentiation between them depends on its production process, requirements, and use specifications.

On the other hand, the technological advance and development of additive manufacturing (also known as 3D printing) caused the development of the design and commercialization of other types of immobilization devices. For example, splints produced using additive manufacturing with different designs morphologically compared to the traditionally known. These new designs allow us to adjust the device shape to the technology they have manufactured adapting to the hands' anthropometry to propose a new way of producing and distributing the rehabilitation splints for CTS.

Therefore, the splints can be divided into two groups according to their manufacturing process: conventional devices manufactured with a traditional technology (i.e., thermoplastics or resins, semi-rigid elastic fabric that incorporates aluminum or hard polyurethane, elastic cotton mesh, or thermoplastic tape) and the novel ones manufactured using additive manufacturing. Considering this, the chapter has three objectives: to conduct an extensive technology review of splints patented in different intellectual property offices, to develop a comparative study between the rehabilitation devices for CTS produced traditionally or by additive manufacturing, and to describe the devices' advantages and disadvantages.

1.3 Methodology

For the comparative study, the methodology proposed by Aguilar-Duque et al. (2018) was implemented. They propose a three-stage methodology to define the relevance of anthropometric devices. In the first stage, the search criteria are explained; the second stage describes the characteristics and functionality of the selected devices; in the third stage, a comparative matrix table is developed. The three stages are described in detail in the next sections.

1.3.1 Stage 1

To select and analyze a patent sample from each group, a technological search of existing patents about orthopedic rehabilitation devices was carried out using the professional search engines: PatBase and Google Patent.

To integrate the list of devices, the following search engine words were used:

- Rehabilitation, orthopedic, device, and splint;
- Carpal tunnel syndrome and skeletal muscle disorders;
- Superior member, hand and wrist;
- 3D printing.

The list of devices identified was filtered according to the next three phases.

- Phase 1. The general results are filtered considering that all devices are used in the upper limb. Thereby, a list containing the title of each patent for each search engine is established. From this process, the devices meeting these requirements are shown in the first column of Appendix 1.1.

- Phase 2: Results of phase 1 are filtered considering the rehabilitation devices of the upper limb intended for the treatment of CTS and to immobilize the wrist area. From this process, the devices meeting these requirements are shown in the second column of Appendix 1.1.
- Phase 3: In the last filtering, the formal and functional characteristics of patents and their manufacturing process are reviewed and analyzed. The final list is divided into two groups: devices produced by traditional manufacturing and devices produced by additive manufacturing. Finally, designs meeting all criteria and the previous parameters are shown in column 3 of Appendix 1.1.

1.3.2 Stage 2

From the devices selected in the previous stage, the characteristics and functionality of selected patents of each group were described, specifying the title, patent number, inventor, date, and registration country. The devices were divided into two groups according to their manufacturing technology: traditional manufacturing devices and additive manufacturing devices. This information allows conforming the comparative matrix of CTS rehabilitation devices' characteristics.

1.3.3 Stage 3

A comparative matrix was integrated including the different patents evaluated based on the parameters to define the relevance of each one. To demonstrate the advantages and disadvantages of the nine rehabilitation devices, Table 1.1 shows the rubric to evaluate the patents. In the rubric, the parameters used are described and rated in three levels: Low (L), Medium (M), and High (H).

1.4 Results

For the research, analysis, and description of the compared devices in this study, the results were divided into three stages proposed in the methodology.

1.4.1 Stage 1

From the documentary scrutiny in the search engines PatBase and Google Patent, 313 patents were analyzed and submitted to three phases of filtering and analysis

Table 1.1 Criteria for measuring the qualifications of the comparative matrix table

Parameters	Grade measurement criteria		
	Low (L)	Medium (M)	High (H)
Ergonomics	The product does NOT have any adaptation that allows it to be used in an obvious and optimal way	The product has adaptations that allow it to be used clearly but not optimally	The product has adaptations that allow it to be used in an obvious and optimal way
Adaptation anthropometric	The product does NOT allow adaptation to the different user percentiles	The product shows the possibility of adapting to different percentiles; however, they are still insufficient	The product shows the possibility of adapting to different percentiles, simply and effectively
Adjustment	The product does NOT allow an obvious and simple adjustment of its dimensions, parts, and elements	The product allows an adjustment of its dimensions, parts, and elements, but this is neither obvious nor simple	The product allows an evident and simple adjustment of its dimensions, parts, and elements
Placement	Placing the product in the hand is complex and impressive	The placement of the product in the hand is little evident and little evident for the user	Placing the product in your hand is simple and very evident
Daily activities	The product prevents the user from carrying out the rest of their daily activities (it must be removed after office work)	The product allows the user to carry out only some of their daily activities, but not all	The product allows the user to carry out the rest of their daily activities (without removing it from the hand)
Comfort	The product is rigid and uncomfortable for hand movements	The product is semi-rigid and medium-comfortable to carry out the movements of the hand	The product is flexible and comfortable to carry out all movements of the hand
Ventilation	The product prevents ventilation of the hand, generating uncomfortable accumulations of sweat and affecting use	The product prevents optimal ventilation of the hand, generating unnecessary accumulations of sweat but without affecting its use	The product allows adequate ventilation of the hand inside

(continued)

obtaining 242, 64, and 9 patents in each phase. Table 1.2 shows the nine patents selected, the manufacturing process, the inventor, and the country where they were registered.

Table 1.1 (continued)

Parameters	Grade measurement criteria		
	Low (L)	Medium (M)	High (H)
Sanitization	The product is not washable and deteriorates with periodic cleaning; the fabrication material loses properties (elasticity, shape, color, etc)	The product is washable, but with this activity, it easily loses its properties (elasticity, shape, color, etc.)	The product is washable and does not deteriorate with periodic cleaning
Production costs	Production is not expensive	Production is very expensive	Production is expensive
Manufacturing time	Its manufacturing time is simple, and it takes little time to complete	Its manufacturing time is moderately complex and takes a large amount of time (days)	Its manufacturing time is complex, and it takes a large amount of time (days)

Table 1.2 Final patent list

	Tittle	Inventor/patent number	Country
Patents produced by traditional technology			
1	Remedial device for treatment of carpal tunnel syndrome	Aziz (1989)	US
2	Universal wrist splint with a removable dorsal stay	Slautterback et al. (2004)	US
3	Hand brace	Meunchen and Durkin (1991)	US
4	Wrist support	Peters (1992)	US
5	Flexible wrist splint	Elsey (1989)	US
Patents produced by additive manufacturing technology			
6	Post-traumatic immobilization device and production method thereof	Rivero and Ceide (2018)	Spain
7	Customizable fitted apparatus	Hall (2016)	US
8	Methods for integrating sensors and effectors in a custom three-dimensional orthosis	Karasahin (2017)	Turkey
9	3D printed splint and cast	Rivlin et al. (2017)	US

1.4.2 Stage 2

The nine devices selected in the previous stage were divided into two groups according to their manufacturing technology: traditional manufacturing devices with five patents and additive manufacturing devices with four patents.

1.4.2.1 Patents Produced by Traditional Manufacturing

Tables 1.3, 1.4, 1.5, 1.6 and 1.7 show the characteristics (title, patent number, inventor, date, and country where the patent was registered) of the four patents produced by traditional technology.

Table 1.3 Device 1: remedial device for treatment of carpal tunnel syndrome

Information: No. US4883073/Inventor: Farooq Aziz/Filed: July 3, 1989/Country: US
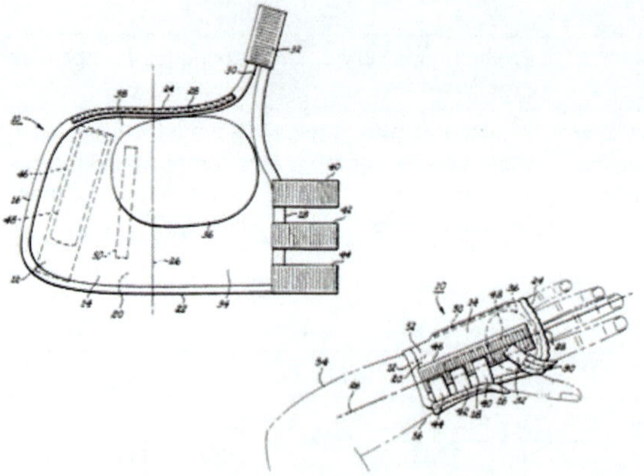
Description: Orthopedic support to improve the medical treatment of carpal tunnel syndrome
Features • The releasable fastening means to ensure easy donning to the wearer's wrist, using copper straps with hook and loop fasteners at lateral locations • Flexible substrate sized to fit the user's forearm • Parts joined together integrally, as the releasable fastening means and the cushioning pad are sewn directly to the fabric substrate
Functionality • Restricts wrist movements in the vertical and horizontal directions of the forearm to a maximum of 10° • Easy placement and adjustment • Relieves pressure on the carpus when the user manipulates the injured wrist • Material with sufficient porosity to provide some ventilation through it • Interchangeable in the right or left hand • Allows most of the usual work tasks to be performed without aggravating STC conditions • Comfortable and with therapeutic benefit

Source Aziz, F. (1989). *Remedial device for treatment of carpal tunnel syndrome*. No. US4883073. Ohio: United States Patent

Table 1.4 Device 2: universal wrist splint with a removable dorsal stay

Information: No. US20040049141A1/Inventor: E. Gerald Slautterback, Rhonda Machin, Weston, Jody Brown/Filed: November 14, 2002/Country: US

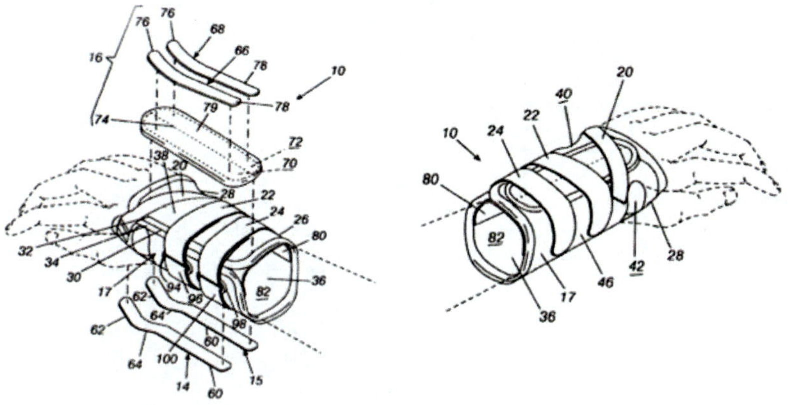

Description: Universal wrist support device to prevent injuries and to rehabilitate various wrist conditions, as well as alleviate the symptoms of carpal tunnel syndrome

Features

- Soft and flexible closure to wrap around the user's wrist
- Pair of palmar supports to hold the wrist at an angle of approximately 15°
- Removable back support structure to immobilize the wrist and straps to tighten and secure the device

Functionality

- Prevents injuries and rehabilitation of various wrist conditions, so multiple splints are not required
- Immobilizer support for wrist injured by various conditions
- Facilitates the rehabilitation of the wrist in different stages
- Injury prevention and treatment of carpal tunnel syndrome
- Keeps the wrist and hand in a neutral position, raising 15°, from the horizontal
- Avoids flexion and extension movements of the wrist
- Allows freedom of movement of the fingers
- For use in any hand, regardless of whether it is in the left or right hand
- Size adjustable to different sizes of user's hands
- Comfortable to use

Source Slautterback, E.G., Machin, R., Weston, & Brown, J. (2002). *Universal wrist splint with removable dorsal stay. No. US20040049141A1.* Florida: United States Patent

1.4.2.2 Patents Produced by Additive Manufacturing

Tables 1.8, 1.9, 1.10 and 1.11 show the characteristics (title, patent number, inventor, date, and country where the patent was registered) of the four patents produced by additive manufacturing technology.

Table 1.5 Device 3: hand brace

Information: No. US5014689A/Inventor: Paul K. Meunchen, Edward T. Durkin/Filed: February 21, 1990/Country: US

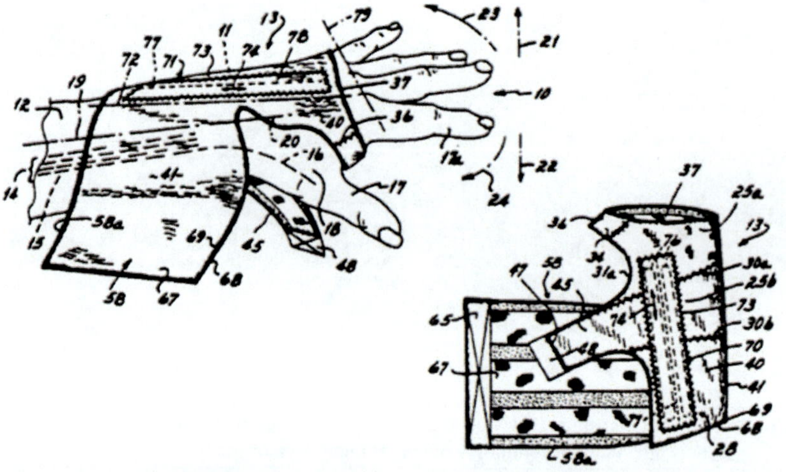

Description: A hand brace to help reduce the impact of carpal tunnel syndrome and De Quervain's condition

Features

- Dorsal and palmar sections of the wrist integrated into the distal and proximal side
- Finger hole and strap-delimited thumb area connecting the palmar section to the dorsal section when the device is assembled
- Structural components of a manual clamp
- The wrist strap covers the palmar side of the user's wrist

Functionality

- Controls or minimizes cumulative trauma to the wrist and helps reduce the impact of carpal tunnel syndrome
- Helps remedy De Quervain's condition
- Limits flexion and extension movements, and ulnar and radial deviation of the hand
- Limits the movement of the thumb concerning the hand
- Easy installation of the brace in the hand of a user

Source Meunchen, P.K., Durkin, E.T. (1990). *Hand brace. No. US5014689A*. Boston: United States Patent

1.4.3 Stage 3

The nine patents were evaluated at three levels: Low (L), Medium (M), and High (H), using the rubric and the results are shown in Table 1.12.

After the analysis of these nine devices, all of them fulfilled the function of immobilizing and limiting the movement of flexion and extension, and ulnar and radial deviation of the wrist. Furthermore, they relieve the pressure on the carpal

Table 1.6 Device 4: wrist support

Information: No. US5160314/Inventor: Helena Peters/Filed: April 26, 1991/Country: US
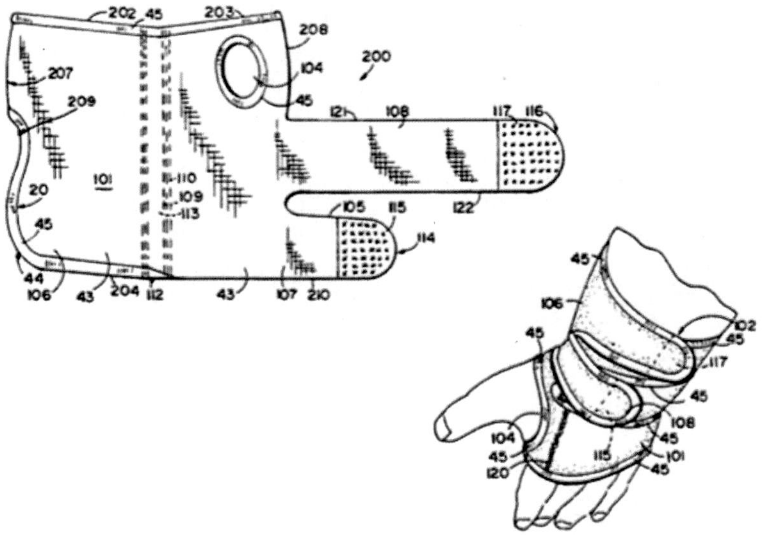
Description: An orthopedic device that provides pain relief and stabilization in the carpal area
Features • Unitary body support adapted to be placed over the hand and wrist area suitable to anatomically adapt to the shape of the wrist in the carpal area, the lower or distal forearm, and the proximal palm area • The cover is constructed from a stretch fabric that includes an outer layer that has a brushed loop texture • Provided with an opening for the thumb and means to receive palmar stabilization support, with the detachable support • Integral compression strap that extends around the wrist • Hook attachment means for releasable hooking with external fabric loops • Adjustable closure at the forearm opening end
Functionality • Limits flexion and extension movements, and ulnar and radial deviation of the hand • Easy fitting and adjustment to the wrist

Source Peters, H. (1991). *Wrist support. No. US5160314.* Michigan: United States Patent

tunnel when the user has an injured wrist, the compose materials are compatible with the skin, and they are easy to place and adjust.

In the first group of devices corresponding to those manufactured by traditional technology (devices 1–5), the most important advantages found are described. They are adjustable and adaptable to different hand sizes. They were made with textile materials and pads to minimize skin irritability, especially when the device needs to be used for long periods and increase patient comfort. They are available for purchase in online stores or pharmacies. Some of the devices are compatible with the right

Table 1.7 Device 5: flexible wrist splint

Information: No. US4854309/Inventor: Denise M. Elsey/Filed: May 6, 1988/Country: US

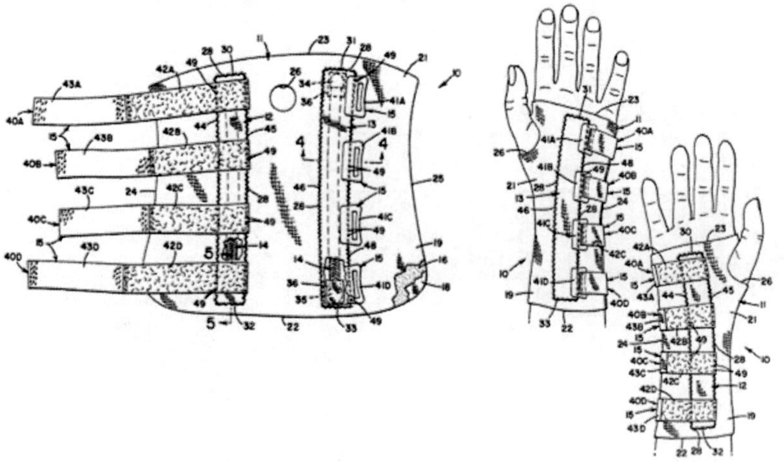

Description: A flexible wrist splint for the treatment of cumulative trauma disorder

Features

- Flexible panel adapted to surround the wrist area
- Attachment straps to secure flexible panel around the wrist area
- Flexible and elastic straps inside the pockets
- Opening for the insertion of a thumb between the flexible panels

Functionality

- Immobilizes the wrist to reduce inflammation of damaged nerves or tendons
- Allows some mobility of the fingers and thumb
- It can be placed on either hand, either right or left
- Usable in the work environment

Source Elsey, D.M. (1988). *Flexible wrist splint. No. US4854309.* Ohio: United States Patent

and left hand (i.e., devices 1, 2, and 5) and allow them to accomplish most of the daily and work tasks (devices 1 and 5). In the case of device 3, it helps to treat the De Quervain's condition.

In addition, these devices have disadvantages that are described. They accumulate moisture and dirt by absorbing moisture and sweat. They trap heat against the patient's skin. They do not adapt correctly to the anatomy of each hand, and as a result, they do not allow proper ventilation of the skin. They are not waterproof, therefore, when getting wet requires removal of the device from the hand until it dries completely, and as a result, they are not easy to sanitize. They depend on market availability, distribution patterns, and transportation time. Finally, various materials and manufacturing processes are required to produce the device.

The devices produced by additive manufacturing technology (devices 6–9), belonging to the second group, included a greater number of advantages concerning the design of the product itself being devices personalized to each patient. They

Table 1.8 Device 6: post-traumatic immobilization device and production method thereof

Information: No. US2018357348A/Inventor: Ricardo Veiga Rivero, Jordi Tura Ceide/Filed: March 15, 2017/Country: Spain

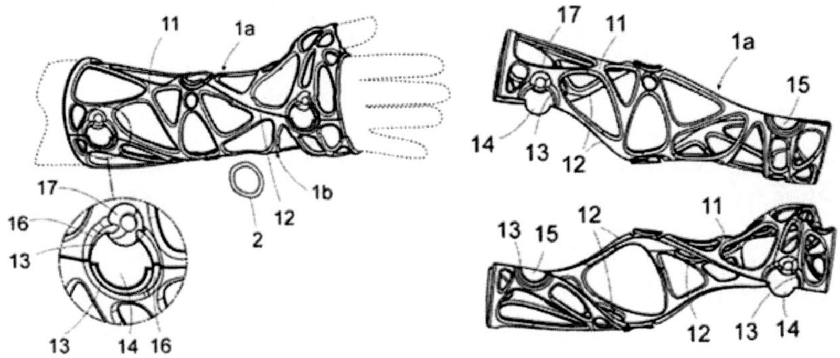

Description: Posttraumatic immobilization device

Features

• Dimensions of the splint from the scan of the patient's arm
• The geometry of triangular motifs with rounded corners
• Composed of two independent pieces which shape only establishing a joining position when both parts are joined together and then form the ferrule itself
• Union of pieces using elastic O-rings placed on projections of each piece, pressing one against the other
• Biocompatible and non-porous material, suitable for contact with the skin

Functionality

• Limits displacement of the bone or injured joint and keeps immobile the affected area while the device is being used
• Aligns the bone segments to facilitate their consolidation, rounded openings to avoid sharp edges
• Allows visualization of wounds and their healing
• Easy sanitization and placement on the arm. Quick and easy fixing of the 2 independent pieces, and easy union of the 2 independent pieces
• Elastic rings that allow some stretching in case of inflammation of the affected area
• It has 0020a manufacturing procedure

Source Rivero, R.V., Ceide, J.T. (2017). *Post-traumatic immobilization device and production method thereof. No. US2018357348A*. Barcelona: United States Patent

have a better anthropometric adaptation to the hand of each patient. They allow the breathability of the skin to be formed by meshed reticles and easy to sanitize as they are made of waterproof plastic. The time to produce a device is shorter than traditional technology because a maximum of two materials are used to produce a device. They can be produced where a 3D printer and the material are available. They allow visual inspection of healing and daily activities and sports, including swimming (devices 6 and 9). They have more medical benefits; for example, the device number 8 detects and monitors wrist movements with sensors that collect, store, and analyze the data to improve the clinical treatment. In the case of device 9, the design can be customized with varying sizes and shapes, including a corporate

Table 1.9 Device 7: customizable fitted apparatus

Information: No. US2016074203A/Inventor: Diana Hall/Filed: September 12, 2015/Country: US
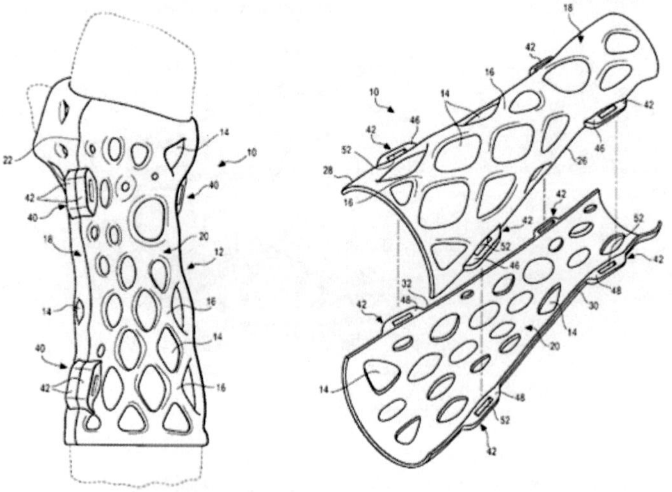
Description: Customizable adjusted device to immobilize injuries
Features • The closed sidewall is divided into one or more housing sections, which can be attached using retaining clips and form one or more longitudinal seams within the sidewall of the device • One or more defined openings in the sidewall of each section of the device while maintaining stiffness in the sidewall • Rigid sidewall, except for a part of the device that has enough flexibility for placement • Structural protrusions in the sidewall that maintains structural rigidity • Secondary openings within the sidewall to accommodate thumbs, fingers, or another part of the body • Includes a 3D wrist image scanning process, modifying the device design to precisely match the contours of the 3D image scan, and using a 3D printer to build the device according to design
Functionality • Immobilizes and stabilizes the wrist • Ease of installation • Reusable • Lightweight • Breathable •Raincoat • Structurally stable in multiple directions

Source Hall, D. (2015). *Customizable fitted apparatus. No. US2016074203A*. Colorado: United States Patent

image, commercial name, sports team logos, instructions for use and care, reminders, patient´s name, among others.

Despite their numerous advantages, they also have disadvantages. The most important is the physical presence of the patient for scanning the affected hand and the subsequent fabrication of the splint and ensuring high fit to the anatomy

Table 1.10 Device 8: methods for integrating sensors and effectors in a custom three-dimensional orthosis

Information: No. US2017224520A/Inventor: Deniz Karasahin/Filed: April 25, 2017/Country: Turkey

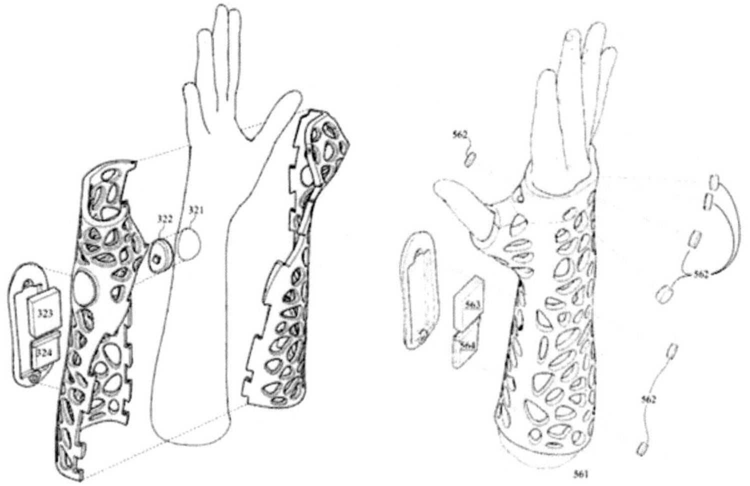

Description: Body splint incorporating wrist injury sensors and treatment elements

Features

- The body structure is composed of a three-dimensional network that is placed on the wrist
- Contains one or more sensors located on the structure
- Sensors are placed near a body joint to detect movement of the body joint
- Combines computer-aided design, software analysis, digital, sensory fabrication, digital data collection, and analysis technologies for custom splint fabrication with 3D printed monitoring and compliance capabilities

Functionality

- Detects movement of the body's joint
- Monitors body movements (flexion, extension, rotation, pronation, and supination) in real time with sensors (inclination, stretching, force, pressure, speed, pulse, electromyography, sensory ultrasound, accelerators, gyroscopes, magnetometers, among others)
- Incorporates diagnostic and/or therapeutic capabilities
- Adjustable to the patient's anatomy
- Digitizes, collects, stores, and analyzes the data obtained by the sensors

Source Karasahin, D. (2017). *Methods for integrating sensors and effectors in custom three-dimensional orthosis. No. US2017224520A*. Turkey: United States Patent

of the hand. This type of production cannot be made for mass productions because it would not comply with the anthropometric adjustments of each patient and the manufacturing times with 3D printers increase overly.

Table 1.11 Device 9: 3D printed splint and cast

Information: No. US20170216078A1/Inventor: Michael Rivlin, Cynthia Watkins, Alexander R. Vaccaro, Mary Grace Maggiano, Pedro Beredjiklian, Michael J. Sileski/Filed: January 27, 2017/Country: US
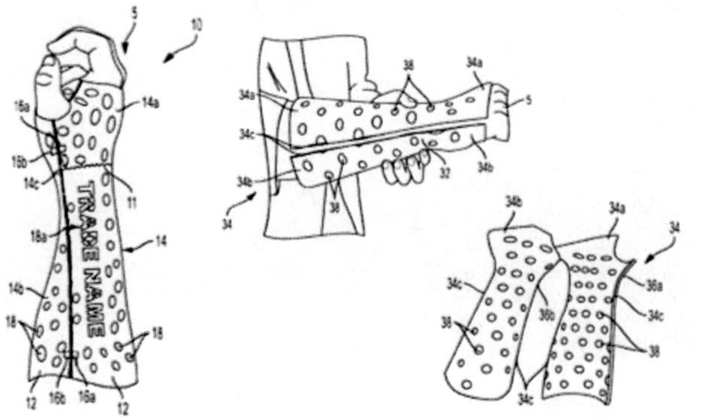
Description: 3D printed splint to immobilize the wrist
Features • 2 shell portions configured to join with a clamping mechanism (straps) simulating the silhouette of the patient's wrist and forearm • The rigid base material is adaptable to 3D printing with an inert polymeric coating adhered to its external surface • Device production consists of scanning the injured area, digitizing, adjusting the device design to the anthropometric measurements of the patient using custom imaging software, and 3D printing
Functionality • Immobilizes and limits wrist movement • Raincoat • Configured to fit and adapt to the hand • Personalized design for each patient • Openings facilitate visual inspection for the healing and breathability of the patient's skin • The coating does not interact chemically with open skin wounds and associated scars • Affordable material • Allows daily activities and sports, including swimming • Fast production since it is manufactured in minutes

Source: Rivlin, M., Watkins, C., Vaccaro, A.R., Maggiano M.G., Beredjiklian, P., Sileski, M.J. (2017). *3D printed splint and cast. No. US20170216078A1*. Philadelphia: United States Patent

1.5 Conclusions

Starting from the technological research for rehabilitation devices for STC registered in different international patent offices, many orthopedic rehabilitation devices of the upper limb were listed and analyzed. Designs focused on the treatment of STC by immobilization from the wrist area were identified, paying special attention to designs that are similar to conventional splints that are sold in pharmacies or online stores, or their manufacturing technology is by additive manufacturing.

Table 1.12 Comparative matrix of characteristics of CTS rehabilitation devices

Parameters	Device								
	1	2	3	4	5	6	7	8	9
Ergonomics	M	M	M	M	M	H	H	H	H
Anthropometric adaptation	M	M	M	H	M	H	H	H	H
Adjustment	H	H	M	H	M	H	H	H	H
Placement	H	M	H	H	M	H	H	L	H
Daily activities	H	M	M	M	H	H	H	M	H
Comfort	M	M	H	H	H	H	H	H	H
Ventilation	M	L	L	L	L	H	H	H	H
Hygiene	L	L	L	L	L	H	H	M	H
Production costs	H	H	M	H	M	L	L	M	L
Manufacturing time	H	H	H	H	H	L	L	L	L

Once the formal and functional features of the patents and their manufacturing were examined, a comparison between the selected devices was established, and some of their characteristics corresponding to the areas of rehabilitation, ergonomics, anthropometry, biomechanics, motion analysis, physiotherapy, industrial design, manufacturing, and costs were evaluated, concluding with their respective advantages and disadvantages.

The obtained results show that the devices manufactured with additive manufacturing facilitate the anthropometric adaptation of the product to the patient's hand since they require a previous scanning process of the affected area. Because of the variability in the anthropometric measures of patients, the standardization of the design and sizes is not possible. Besides, additive manufacturing reduces production costs and facilitates access to the devices since they can be manufactured in any facility that has a 3D printer, changing the way of marketing and distribution of these products. This scheme of production could be applied in other areas of the healthcare sciences, for example, rapid prototyping to facilitate surgery design, 3D printing orthopedic implants, or 3D scaffolds for tissue engineering.

This research facilitates the process of finding similar and homologous devices for the prevention and rehabilitation of occupational diseases of the upper limb, such as CTS, contributing to the development of products that solve existing and future needs in an affected area of the population. Finally, it is necessary to conduct more studies to expand the range of orthopedic devices analyzed.

Appendix 1.1

See Table 1.13.

Table 1.13 Patents selected in the three phases

PatBase Search Engine		
Selected from Phase 1 (Tittle)	Selected from Phase 2	Selected from Phase 3
Orthosis		
Wrist support device		
Upper extremity hand orthosis and method of use		
Wrist supporting device for bowlers		
Bowler's glove		
Football glove		
Orthotic device for assisting with bending and extending movements of the fingers of a hand, for patients suffering from paralysis of the brachial plexus	x	
Reciprocating brace		
Therapeutic device that provides stimulation to an immobilized extremity		
Modeling apparatus for assisting in the recovery of hand, wrist, and foot movement		
Splint	x	
Orthopedic glove		
Thumb orthosis		
Therapy device to increase flexibility and range of motion to the wrist, fingers, and thumb		
Handheld orthosis having a flexible enclosure and method of utilization		
Wrist protector	x	
Rigid foam polyurethane hand splint		
Glove and hand exerciser		
Splint assembly for positioning of the hand		
Brace	x	
Dynamic resting hand splint		
Inflatable device for arthritic therapy		
Retainer for glove		
Fingertip flexor glove		
Splint/therapeutic device	x	
Glove with thumb restraint element		
Low-profile, radial nerve splint with interchangeable resilient digit extensor elements	x	

(continued)

Table 1.13 (continued)

PatBase Search Engine		
Selected from Phase 1 (Tittle)	Selected from Phase 2	Selected from Phase 3
Golf training aid		
A method for producing a brace, the brace as such, and a method to fix the position of a broken bone in a limb		
Epicondylitis clasp		
Clench arresting glove	x	
Orthopedic support apparatus and method of use	x	
Orthopedic device		
Customizable fitted apparatus	x	x
Remedial device for treatment of carpal tunnel syndrome	x	x
Hand splint for stroke patients		
Wrist support device	x	
Handorthese		
Universal articulated splint		
Wrist protector for a sport glove		
Hand splint		
Device for supporting a limb and associated extremity		
Post-traumatic immobilization device and production method thereof	x	x
Wrist brace	x	
Wrist support apparatus	x	
Orthosis for treating muscle disorders in the elbow	x	
Means for actuating artificial or disabled arm members	x	
Handgelenk-orthese	x	
Dynamic mp joint extension splint		
Splint assembly for positioning of a disabled diseased, or injured hand and wrist		
Arm/leg board		
Open-cell mesh cast material	x	
Combination arm splint and finger support means	x	
Hard-grip glove		

(continued)

Table 1.13 (continued)

PatBase Search Engine		
Selected from Phase 1 (Tittle)	Selected from Phase 2	Selected from Phase 3
Arterial wrist support		
Orthopedic splint	x	
Surgical finger and fence splints		
Gloved splint		
Digit apparatus for typing and texting		
Arm protecting device		
Methods for integrating sensors and effectors in a custom three-dimensional orthosis	x	x
Exercise glove incorporating rods which offer resistance to movement of fingers, hands, or wrists		
Golf swing training device and method		
Brace for protecting a region of a limb		
Brace	x	
Splint/therapeutic device		
Molded orthopedic devices		
Dynamic gauntlet and related method of use		
Geometrically aperture protective and/or splint device comprising a re-moldable thermoplastic material		
Hand and digit immobilizer for pulse oximeter		
Swimming glove		
Hand splint		
Digital retractor		
Dynamic hand splints		
Therapeutic exercise glove		
Carpometacarpal joint orthosis of thumb		
Glove		
Orthotic device		
Device for preventing hand contraction, particularly during sleep		
Flexible support brace		
Bowling hand and wrist support device		
Carpal/hand immobilizer		
Ulnar deviation splint		
Finger exercising appliance		

(continued)

Table 1.13 (continued)

PatBase Search Engine		
Selected from Phase 1 (Tittle)	Selected from Phase 2	Selected from Phase 3
Hand and wrist brace and kit		
Rubber glove		
Brace		
Hand immobilizing and positioning apparatus for X-ray examinations		
Method and apparatus for an ulnar collateral ligament thumb support	x	
Wrist brace	x	
Strengthening glove		
Method of providing centralized splint production		
Orthopedic support apparatus with a brace-receiving pocket	x	
Semi-rigid counterforce brace		
Gloved splint for an arthritic hand		
Dynamic hand splints		
Positioning device for first carpometacarpal		
Wrist support	x	
Wrist brace		
Antispasticity aid device and related accessories		
Google Patent Search Engine		
Remedial prophylaxis for carpal tunnel syndrome	x	
Devices for treatment of carpal tunnel syndrome	x	
Adjustable wrist splint	x	
Flexible wrist splint for carpal tunnel syndrome treatment	x	
Diagnostic apparatus and method for evaluation of carpal tunnel syndrome		
Fatigue monitoring device and method		
Computer input device having heating and/or vibrating elements		
Methods and devices to treat compressive neuropathy and other diseases		
Therapeutic hand exercise device		

(continued)

Table 1.13 (continued)

PatBase Search Engine		
Selected from Phase 1 (Tittle)	Selected from Phase 2	Selected from Phase 3
System for projecting content to a display surface having user-controlled size, shape, and location direction and apparatus and methods useful in conjunction therewith		
Therapeutic device		
User-friendly keyboard		
Ergonomic computer mouse		
Ergonomic dual-section computer-pointing device		
Safe and handy pointing device		
Platform for a computer input device		
Computer mouse that prevents or treats carpal tunnel syndrome and methods of use		
Multi-functional ergonomic interface		
Hand and wrist support		
Computer mouse		
Portable wrist rest system		
Inflatable wearable traction device		
Isometric wrist exercise device		
Appliance and method for treating carpal tunnel syndrome		
Device for exercising and strengthening the hand, wrist, and arm		
Carpal tunnel protector		
Carpal tunnel device and method		
Carpal tunnel syndrome traction system		
Combination finger and wrist splint		
Systems for the prevention or treatment of carpal tunnel syndrome		
Molded orthopedic devices		
Low-profile, radial nerve splint with interchangeable resilient digit extensor elements and supination adjustment means		
Wrist brace and method for alleviating and preventing wrist pain		
Handheld hand, wrist, and arm exercise and rehabilitation device		

(continued)

Table 1.13 (continued)

PatBase Search Engine		
Selected from Phase 1 (Tittle)	Selected from Phase 2	Selected from Phase 3
Safety device for prevention of carpal tunnel syndrome		
Digit-supporting therapeutic device for the hand		
Low-profile, radial nerve splint with interchangeable resilient digit extensor elements		
Device for treating carpal tunnel syndrome		
Elastic exercise device		
Wrist support brace		
Support device for administration of cpr		
Method and apparatus for treating carpal tunnel syndrome		
Method and device to alleviate carpal tunnel syndrome and dysfunctions of other soft tissues		
Removable wrist-band for carpal tunnel syndrome		
Digit-supporting therapeutic device for the hand		
System for testing hand, wrist, and forearm strength		
Diagnosing carpal tunnel syndrome		
Surgical hand support apparatus		
Surgical hand support apparatus		
Dynamic support to correct/prevent carpal tunnel syndrome	x	
Neuropathy relief vacuum traction assist system for carpal tunnel relief		
Wrist-band for the prevention and the treatment of the carpal tunnel syndrome and its positioning operating mode	x	
Dynamic splint for carpal tunnel syndrome treatment	x	
Universal wrist splint with removable dorsal stay	x	x
Flexible wrist splint	x	x
Free sliding hand rest		

(continued)

Table 1.13 (continued)

PatBase Search Engine		
Selected from Phase 1 (Tittle)	Selected from Phase 2	Selected from Phase 3
Wrist support device for keyboards		
Wrist rest for keyboard		
Hand and wrist support for computer mouse		
Palm rest for use with computer-pointing devices		
Wrist pad		
Method and apparatus for supporting various parts of person's body		
Wrist/hand support device		
Ergonomic data entry device		
Ergonomic hand support for use with a computer-pointing device		
Forearm support for computer interface device		
Wrist and arm support		
Wrist rest assembly		
Wrist/hand support device		
Hand splint forearm support with optional support of mp joints and thumb and pager assists	x	
Wrist support	x	x
Wrist and hand support device	x	
Non-invasive method for treating carpal tunnel syndrome	x	
Exercise glove incorporating rods which offer resistance to movement of fingers, hands, or wrists	x	
Wrist guard	x	
Dynamic support to correct/prevent carpal tunnel syndrome	x	
Wrist support device	x	
Golf glove and method of forming same		
Gripping ad		
Therapeutic device for wrist injuries		
Movable hand/wrist support for use with a computer mouse		
Hand positioner device for computer mouse		
Wrist support device		

(continued)

Table 1.13 (continued)

PatBase Search Engine		
Selected from Phase 1 (Tittle)	Selected from Phase 2	Selected from Phase 3
Arm support and carpal nerve protection device		
Universal ergonomic handle		
Padded athletic gloves		
Thumb actuated x-y input device		
Handle/grip and method for designing the like		
Yoga wrist saver device		
Computer mouse on glove		
Ergonomic handle and handle sizing method		
Wearable data input device		
Padded cycling glove that reduces nerve injury		
Protective device for boxers		
Wrist brace		
Protective device for the hand	x	
Protective glove		
Protective glove for maximized tactile gnosis		
Glove-like dynamic splint and method of using same		
Combined splint holder and wrist support	x	
Work glove with insert		
Soft-goods type, formable orthopedic cast	x	
Orthopedic device having gel pad with phase change material	x	
Soft-goods type, formable orthopedic cast		
Hand brace	x	x
Adjustable wrist and hand splint		
Hand protector		
Push hand covering with removal assist		
Unitary orthopedic brace	x	
Pivotal brace for prosthesis		
Therapeutic elastic body support		
Non-compressive, distracting wrist brace		
Hockey glove with ventilation holes		
Orthotic device for a joint of the human body		
Carpal tunnel brace	x	

(continued)

Table 1.13 (continued)

PatBase Search Engine		
Selected from Phase 1 (Tittle)	Selected from Phase 2	Selected from Phase 3
Wrap		
Molded orthopedic devices		
Protection aid for protecting the hands and wrists of skaters		
Wrist support with strap	x	
Sealed edge orthopedic casting technique		
Cmc joint splint		
Universal wrist splint	x	
Wrist brace	x	
Low-profile metacarpal fracture brace	x	
Protective glove for boxers		
Batting glove with internal padding		
Wrist brace with adjustable thumb closure	x	
Portable wrist rest system		
Splint or support with quick location technique		
Reversible wrist and thumb support		
Splint	x	
Fracture brace	x	
Orthopedic system for immobilizing and supporting body parts	x	
3D printed splint and cast	x	x
Orthopedic support for immobilizing the thumb	x	
Orthosis for treating rhizarthrosis		
Orthosis		
Wrist brace	x	
Ambidextrous, combination wrist and thumb brace		
Universal wrist brace with enhanced lacing	x	
Moldable wrist brace	x	
Adjustable wrist brace	x	

References

Aguilar-Duque JI, Amaya G, Juarez VM et al (2018) Anthropometry devices: a comparative study. In: Handbook of Research on Ergonomics and Product Design, pp 153–171. https://doi.org/10.4018/978-1-5225-5234-5.ch010

Armenteros Pedrero J, Justo García G, Rey Pita ML et al (2000) Conservative treatment of carpal tunnel syndrome using splint immobilization. Rehabil (Madr, Ed impr) 34(44):313–319 (In Spanish)

Aziz F (1989) Remedial device for treatment of carpal tunnel syndrome (Patent No. US4883073A). U.S. Patent and Trademark Office, Washington, DC

Burke DT, Burke MMH, Stewart GW, Cambré A (1994) Splinting for carpal tunnel syndrome: in search of the optimal angle. Arch Phys Med Rehabil 75:1241–1244. https://doi.org/10.1016/0003-9993(94)90012-4

Cristina G, Parra G, Fernando A et al (2009) Review and update. Carpal tunnel syndrome. Morfolia 1(3):11–23. https://revistas.unal.edu.co/index.php/morfolia/article/view/10857 (In Spanish)

Davis PT, Hulbert JR, Kassak KM, Meyer JJ (1998) Comparative efficacy of conservative medical and chiropractic treatments for carpal tunnel syndrome: a randomized clinical trail. J Manipulative Physiol Ther 21:317–326

Doughty CT, Bowley MP (2019) Entrapment Neuropathies of the upper extremity. Med Clin North Am 103:357–370. https://doi.org/10.1016/j.mcna.2018.10.012

Elsey DM (1989) Flexible wrist splint (Patent No. US4854309A). U.S. Patent and Trademark Office, Washington, DC

Hall D (2016) Customizable fitted apparatus (Patent No. US20160074203A1). Washington, DC: U.S. Patent and Trademark Office

Hern D, Hoyos N, Coordinador A et al (2013) Rehabilitation in hand and forearm peripheral nerve injuries (evidence-based clinical practice guidelines). ISS-Ascofame. http://www.sld.cu/galerias/pdf/sitios/rehabilitacion/nervios_perifericos.pdf (In Spanish)

Herrera Gil L (2019) Design and development of a 3D scanned and printed upper limb splint. Universidad de Valladolid. https://uvadoc.uva.es/bitstream/handle/10324/37756/TFG-I-1230.pdf?sequence=1&isAllowed=y (In Spanish)

Hoyos DH (2010) Rehabilitation in injuries of peripheral nerves of the forearm and hands. ASCOFAME. http://www.sld.cu/galerias/pdf/sitios/rehabilitacion/nervios_perifericos.pdf (In Spanish)

Karasahin D (2017) Methods for integrating sensors and effectors in custom three-dimensional orthosis (Patent No. US20170224520A1). U.S. Patent and Trademark Office, Washington, DC

Meunchen PK, Durkin ET (1991) Hand brace Hand brace (Patent No. US5014689A). U.S. Patent and Trademark Office, Washington, DC

Nunes IL, McCauley Bush P (2012) Work-related musculoskeletal disorders assessment and prevention. In: Ergonomics—a systems approach. InTech, pp 1–30

Ocello M, Lovotti V (2015) Orthotics and prosthetics. 1a ed.-Santa Fe: Ediciones UNL, 2015. https://bibliotecavirtual.unl.edu.ar:8443/bitstream/handle/11185/5534/ortesisyprotesis.pdf (In Spanish)

Pardal-Fernández JM, Martín-Garrido MJ, García-Reboiro G et al (2004) Carpal tunnel syndrome diagnosis. Clinical and neurophysiological evaluation. Rehabilitación 38(3):137–147. https://doi.org/10.1016/s0048-7120(04)73446-5 (In Spanish)

Peters H (1992) Wrist support (Patent No. US5160314A). U.S. Patent and Trademark Office, Washington, DC

Rivero RV, Ceide JT (2018) Post-traumatic immobilization device and production method thereof (Patent No. US20180357348A1). U.S. Patent and Trademark Office, Washington, DC

Rivlin M, Beredjiklian P, Vaccaro AR et al (2017) 3D printed splint and cast (Patent No. US10758396B2). U.S. Patent and Trademark Office, Washington, DC

Segnini JM, Vergara MJ, Provenzano S (2017) Prospective for the design and manufacture of a 3D printed orthosis. In: Vergara Paredes MJ, Díaz Rodríguez MÁ, Rivas Echeverría F, Restrepo Moná M (eds) Design of devices for rehabilitation and orthotics. Universidad de los Andes, Venezuela, 1st ed., pp 153–175 (In Spanish)

Slautterback E, Machin R, Brown J (2004) Universal wrist splint with removable dorsal stay (Patent No. US20040049141A1). U.S. Patent and Trademark Office, Washington, DC

Chapter 2
A Review on Infrared Thermal Imaging as a Tool in Carpal Tunnel Syndrome

Melissa Airem Cázares-Manríquez, Claudia Camargo-Wilson, Ricardo Vardasca, Jorge Luis García-Alcaraz, Jesús Everardo Olguín-Tiznado, Juan Andrés López-Barreras, and Blanca Rosa García-Rivera

Abstract This research reviews 18 scientific articles concerning the application of infrared thermography (IRT) in the mensuration of diagnostic studies of carpal tunnel syndrome (CTS). In addition, the proposed future challenges in this research area are identified. A review of articles is performed in databases such as PubMed, Scopus, EBSCO, ELSEVIER, Springer, and Oxford Academic using the keywords: carpal tunnel syndrome and (thermography OR infrared image OR thermal image). Its contents, journals publishing the topic, and the year of publication are reviewed, and graphs and cross tables are constructed. Using databases such as PubMed, Scopus, EBSCO, ELSEVIER, Springer, and Oxford Academic, 937 articles are identified, 37 of which were duplicates. The titles and abstracts of the remaining articles were reviewed, and 855 articles were deleted due to exclusion criteria. Eighteen articles were found written in foreign language, five were removed for not covering the topic (three reviews and two on liquid crystal thermography), and four were not available

M. A. Cázares-Manríquez · C. Camargo-Wilson · J. E. Olguín-Tiznado
Faculty of Engineering, Architecture and Design, Autonomous University of Baja California, Ensenada BC 22860, Mexico

R. Vardasca
Faculdade de Engenharia, Universidade do Porto, 4200-465 Porto, Portugal

INEGI, Universidade do Porto, 4200-465 Porto, Portugal

ISLA Santarém, 2000-241 Santarém, Portugal

J. L. García-Alcaraz (✉)
Department of Industrial Engineering and Manufacturing, Autonomous University of Ciudad Juárez, Ciudad Juárez CHIH 32310, Mexico
e-mail: jorge.garcia@uacj.mx

I.T. Ciudad Juárez, Tecnológico Nacional de México, Ciudad Juárez, Chihuahua, Mexico

J. A. López-Barreras
Faculty of Chemical Sciences and Engineering, Autonomous University of Baja California, Tijuana BC 22390, Mexico

B. R. García-Rivera
Faculty of Administrative and Social Sciences, Autonomous University of Baja California, Tijuana BC 22390, Mexico

A. Realyvásquez Vargas et al. (eds.), *New Perspectives on Applied Industrial Ergonomics*,
https://doi.org/10.1007/978-3-030-73468-8_2

online. Finally, eighteen articles were selected for the full text review, from which 13 articles meet the CTS diagnostic classification and 5 consider the CTS studies. IRT is a reliable method in the diagnosis of CTS, mainly in the first stage. To improve diagnostic accuracy, it is recommended nerve conduction studies.

Keywords Carpal tunnel syndrome · Cumulative trauma disorders · Infrared thermal imaging · Medical diagnosis

2.1 Introduction

Musculoskeletal disorders of the upper extremity, such as tendonitis or nerve entrapments, may be due to the repetitive and forceful use of hands and arms, common among certain occupations. These injuries are the result of small, accumulated tissue damage that is sustained by performing repetitive tasks and are collectively known as cumulative traumatic disorders (CTD). They may also be referred to as overuse syndromes, regional musculoskeletal disorders, cervical and brachial disorders, repetitive stress injuries, or repetitive movement disorders. According to data published by the Bureau of Labor Statistics of the United States of America, the incidence of CTD has increased dramatically in recent years, and since 1989, these injuries have accounted for more than 50% of all occupational diseases reported in that country (Rempel et al. 1992).

Carpal tunnel syndrome (CTS) is one of the most conventional compressive and canalicular neuropathies of the upper limbs, which causes hand discomfort and impaired function. CTSs are the result of compression or injury to the median nerve of the wrist (Fig. 2.1) within the limits of the carpal tunnel. CTS patients often

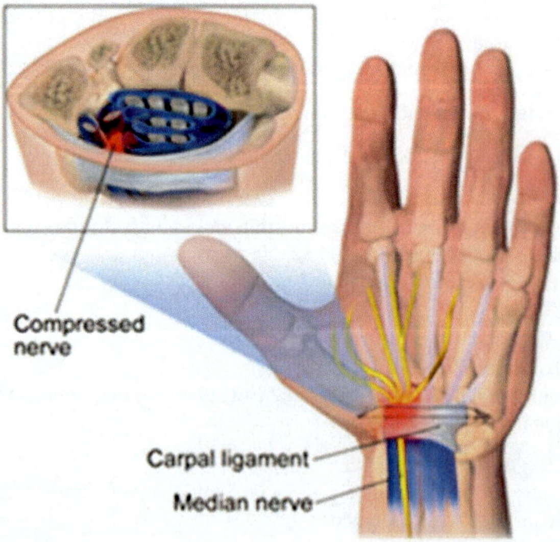

Fig. 2.1 Nerves and ligaments involved in carpal tunnel syndrome

experience pain, numbness, tingling, and a feeling of swelling in the median nerve distribution area of the hand. Among other characteristic symptoms of this pathology, it consists of waking up suddenly at night due to numbness and pain in the hand. Both pain and numbness can extend to the upper forearm or even the shoulder (Papež et al. 2008).

According to Ghasemi-Rad et al. (2014), 1 in 5 patients who present the previously mentioned symptoms in the hands is diagnosed with CTS, based on clinical examination and electrophysiological tests. It is estimated that 3.8% of the general population suffers from CTS, with an incidence rate of 276:100,000 per year. This pathology is more frequent in women than in men. The female sex has a prevalence rate of 9.2%, while the male sex has a rate of 6%. The most frequent age ranges for CTS are in a maximum age of 40–60 years; however, there are cases of CTS in young people of twenty years old and on the other hand, in patients older than eighty-six years.

Specifically, in Mexico, the incidence of CTS is 99 per 100,000 people per year and the prevalence is approximately 3.4% in women and 0.6% in men.

Costs caused by this pathology are diverse in nature, as derived from health care, surgery, and rehabilitation. It is estimated that due to this, the United States of America spends approximately 1000 million dollars per year, resulting in the loss of productivity from the affected worker, the economic compensation of the companies, and the days of absence from work (Roel-Valdés et al. 2006).

In Europe, the estimated documented incidence of CTS is 3.5% of the active population between the ages of 16 and 65; however, CTS is not yet recognized as an occupational CTD in all the European Union member countries (Redzwan Habib 2017).

Early detection of CTS increases the probability of treatment success (Alcan et al. 2018). Wright and Atkinson (2019) confirm that the diagnosis begins with clinical history. This is followed by a physical examination and confirmed by an electrodiagnostic evaluation (Kanafi et al. 2018).

According to Schols et al. (2018), there is currently no universal diagnostic method for CTS. There are several tests to detect this pathology, which are palpation tests, such as the Phalen's test and the Tinel's test, and electromyography (EMG), especially nerve conduction studies, which allows the diagnosis of CTS within a certain range, according to Baic et al. (2017).

However, EMG is invasive, so it is uncomfortable for patients, and not totally reliable. Several studies show that routine electrodiagnostic tests have limited sensitivity and specificity for mild CTS cases. Therefore, an expensive and uncomfortable test with uncertain results is usually not convenient, according to Maxel et al. (2014) and Ming et al. (2005)

Electrodiagnostic studies have a sensitivity of 56–85% and a specificity of 94–99% for CTS. These studies should be reserved for atypical cases of CTS, since results may be normal in one-third of patients with mild CTS. It is recommended that electrodiagnostic studies be performed prior to surgery to confirm the diagnosis, since patients with severe CTS are less likely to recover after surgery (Wipperman and Goerl 2016).

There are other techniques such as magnetic resonance imaging (MRI) and ultrasonic detection, which reveal the morphologic changes of carpal tunnel and its contents. However, their results are not reliable (Ming et al. 2005).

Among other tools used in the study and diagnosis of CTS is infrared thermal imaging (IRT), which has been adopted in medicine as a method of monitoring physiology in real time and can be used to document vascular conditions of the autonomic nervous system and musculoskeletal; CTS is one of the conditions in which the use of the IRT image can improve medical diagnosis (Ring and Ammer 2012).

Changes in temperature gradient (decrease and increase) on the surface of the skin or in the center of the body are indicators of disease, and it is possible to evaluate changes in metabolism and local blood flow, mostly in the superficial layer of the skin, according to Boerner et al. (2015) and Cholewka et al. (2010).

Various studies indicate that the symmetry of the extremities and the trunk will not have a temperature difference on the two sides along a dermatome or thermatoma by more than 0.30 °C and in the forearms no more than 0.90 °C (Uematsu and Long 1976).

The diagnosis of neuromuscular pathology by infrared thermography (IRT) is based on the existence of thermal symmetry and asymmetry between normal and abnormal sites (Fischer 1986).

IRT works by measuring the temperature distribution of a surface, which offers several advantages, as it is non-invasive and non-contact, non-radioactive, painless, with easy reproducibility of results (thermal images) and low cost of operation, according to Živčák et al. (2011).

Due to the advantages of IRT in the provision and treatment of CTS, the purpose of this study is to identify and discuss the application of this technique in studies related to the diagnosis of CTS, as well as its support in conducting medical research studies related to this pathology, and how to identify future challenges that may arise in this research area.

2.2 Methodology

To achieve the above objective, a search strategy is first developed in scientific databases, and then, a review and eligibility of articles are carried out and classified, and their contributions are analyzed.

2.2.1 Stage 1. Search Strategy

Initially, the bibliographic search was performed during the period from August 2018 to July 2019, through the PubMed, Scopus, EBSCO, ELSEVIER, Springer, and Oxford Academic databases, where the following keywords were used: carpal tunnel syndrome AND (thermography OR Infrared imaging OR thermal imaging).

The Boolean operator OR has been used in the present investigation because the word “thermography” presents as synonymous words “Infrared imaging” AND “thermal imaging.” It is worth mentioning that the same word structure was used to guarantee consistency in the search through the information sources. Subsequently, duplicate articles were identified in the databases consulted, where a total of 37 articles were identified.

The principles of exclusion and inclusion of articles are as follows:

- Contain in the title or abstract search keywords.
- They must be written in English.
- Liquid crystal thermography or review articles are not included.

2.2.2 *Stage 2. Review and Eligibility Results*

A review of the results obtained in each database is performed, selecting the research papers whose title and abstract contain the previously determined keywords, that is, the use of infrared thermography (IRT) in relation to the CTS, thus complying with the first selection criterion. Once the initial filter has been approved and the duplicates have been located among the sources of information, the second criterion is evaluated, in which the articles in a language other than English (eighteen of them) are eliminated. The third criterion is to exclude papers on the use of liquid crystal thermography since this article focuses only on the use of IRT. Five articles were excluded because two had liquid crystal as their central topic and three were reviews. Furthermore, six articles were not available, of which five have been requested and only two of these have been provided. Therefore, a total of eighteen articles make up the development of this work.

During the development of this work, the Preferred Reporting Items for Systematic Reviews and Meta-Analysis PRISMA method for reviews was used, whose flow diagram is shown in Fig. 2.2.

2.3 Results

Through the literary search among the databases, a total of 937 articles were found, of which 33 correspond to the PubMed database, 66 to Scopus, 36 to EBSCO, 268 ELSEVIER, 361 to Springer, and 173 to Oxford Academic.

Among the data sources used, 37 duplicate articles were identified, and after reviewing the titles and abstracts of the remaining articles, 855 were eliminated and classified as off-topic according to the first established criterion. Furthermore, 18 articles written in languages other than English were identified (second criterion), and then, 6 articles were not available, so that 5 of these were requested from the

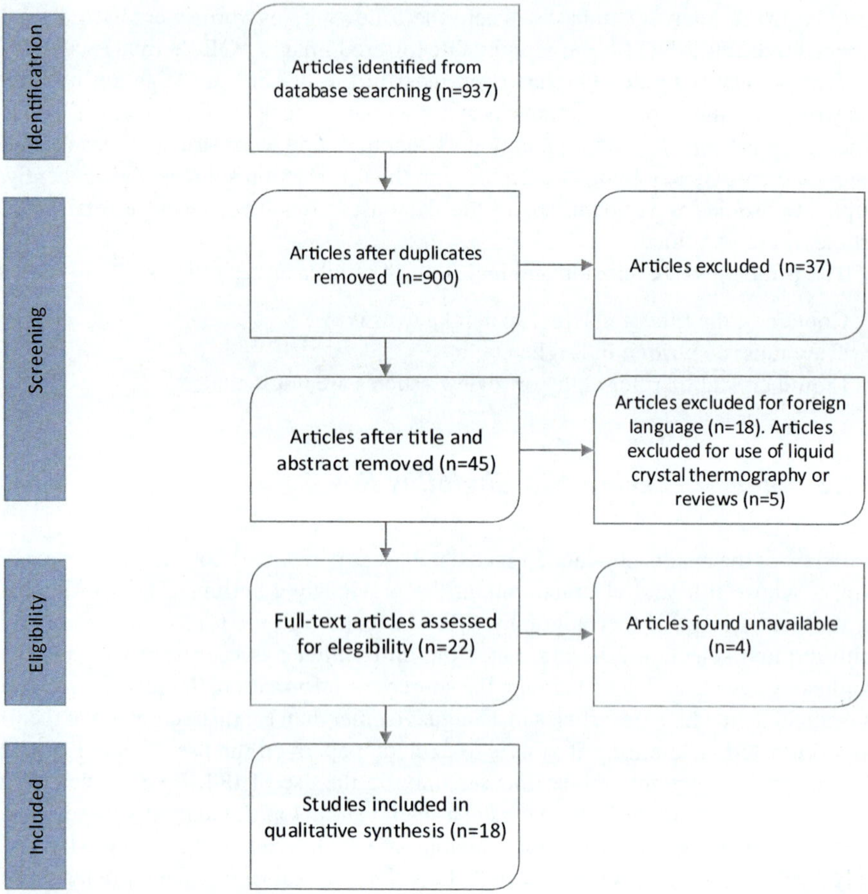

Fig. 2.2 PRISMA flowchart

corresponding authors but only two provided the articles (same researcher). The missing article has not been obtained since emails are not available and neither are the authors' pages.

In addition, 5 articles were rejected due to the third and fourth criterion; that is, 3 reviews and 2 articles on liquid crystal thermography were identified. Therefore, 18 articles form the present analysis, whose methodology search was carried out by two reviewers to guarantee a reliable selection process. Table 2.1 summarizes the results found in the selection process for the development of this bibliographic research, whose results are classified into two sections: *Application of infrared thermography for the diagnosis of CTS (13 articles) and Application of infrared thermography for CTS studies (5 articles)* listed below.

Figure 2.3 illustrates the Journals that publish topics related to the application of IRT in CTS-related studies. In International Symposium of Computer-based Medical

Table 2.1 Articles analyzed and their sources

Database	Found articles	Duplicates	Off-topic articles (1st criterion)	Articles in different language to English (2nd. criterion)	Articles not available	Requested articles	Articles received	Deleted articles (3rd criterion)	Valid articles
PubMed	33		8	9	2	2		3	11
Scopus	66	18	32	7	3	3	2	2	6
EBSCO	36	13	20	2					1
ELSEVIER	268	4	263		1				
Springer	361	2	359						
Oxford Academic	173		173						
Totals	937	37	855	18	6	5	2	5	18

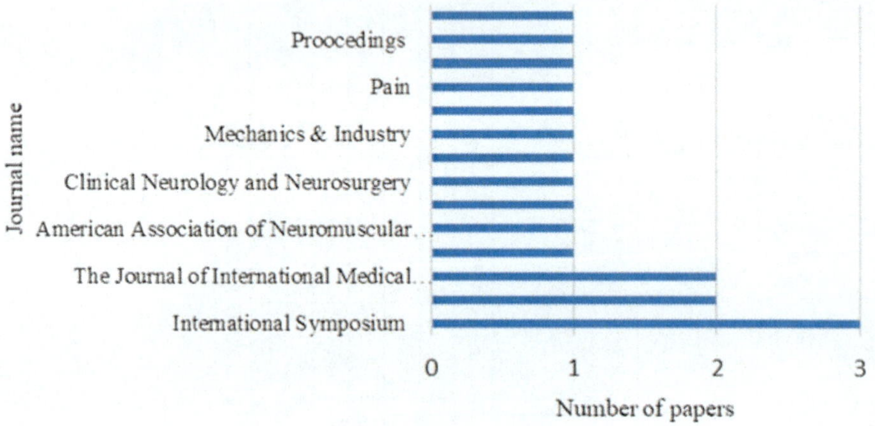

Fig. 2.3 Journals published about the application of IRT in CTS-related studies

Systems, Computational Intelligence, and Computer Science, 3 papers have been published in total, and in the Journal of International Medical Research and The Journal of Hand Surgery, two papers have been published in each one. The rest of the journals only have one publication on the subject.

Figure 2.4 illustrates the number of articles analyzed per year. It can be seen that from 1987 to 2017, there is only one publication per year, except for the years 1995, 2005, 2007, and 2008, which had 2 articles.

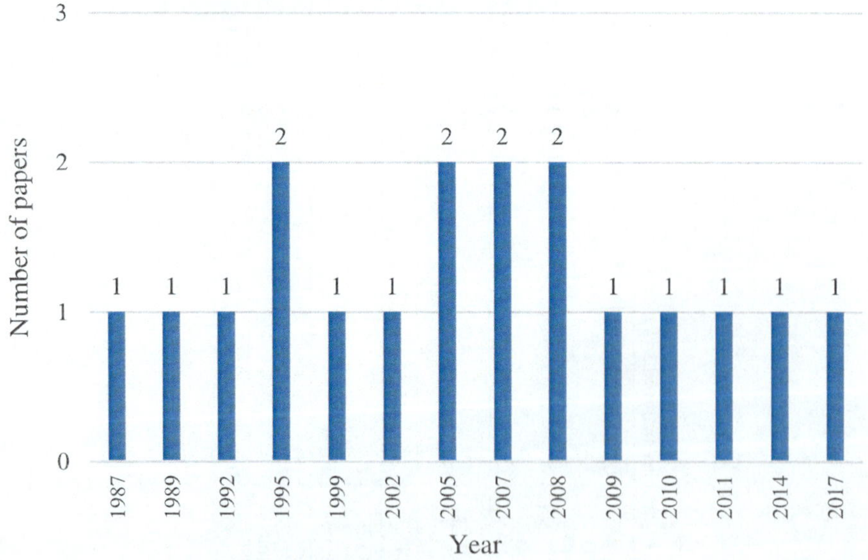

Fig. 2.4 IRT as a tool in CTS publications by year

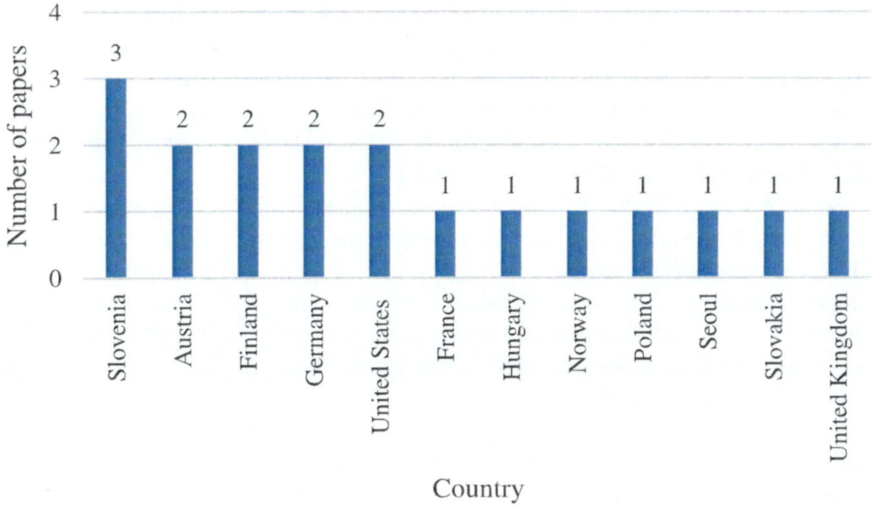

Fig. 2.5 List of publications by country

Figure 2.5 illustrates the countries that have developed research related to the use of IRT as a tool for CTS. It can be seen that Slovenia presents the highest number of articles, that is, 3 publications, followed by Austria, Finland, Germany, and the United States with 2 publications, while France, Hungary, Norway, Poland, Seoul, Slovakia, and the United Kingdom present only one publication each, regarding IRT application issues in CTS.

The departments with greater commitment regarding the use of IRT in CTS studies are associated with neurology, neurosurgery, pathophysiology, rheumatology, among others.

2.3.1 Application of Infrared Thermography for CTS Diagnosis

Below is a series of results from various research works focused on determining if IRT is actually a useful tool for diagnosing CTS.

Five out of eighteen studies focus on determining the effectiveness of IRT as a diagnostic method of CTS, by analyzing thermal images of hands on their dorsal and palmar parts, between healthy people and patients with CTS. These authors agree that the dorsal side of the hand provides more successful results when diagnosing CTS and therefore is more important than the palmar area of the hand. Indeed, thermography is a potential method in the diagnosis of CTS (Papež et al. 2008, 2009; Tchou et al. 1992; Tkáčová et al. 2010; Živčák et al. 2011).

Papež et al. (2009) establish that IRT allows correct classification of 72.2% of the hands, between healthy and pathological, based on the dorsal part of the hand, while >80% when evaluating severely affected hands with healthy hands. On the other hand, they obtained success rates for the classification of healthy hands with CTS pathologies close to or greater than 80% (Palfy and Papež 2007; Papež et al. 2008; Tchou et al. 1992; Tkáčová et al. 2010).

While Živčák et al. (2011) were analyzed a database of 268 thermograms, they catalogued 120 healthy hands and 14 unhealthy hands (with clinically diagnosed CTS). The temperature distribution was observed, and the partial temperatures were calculated in five regions: center point of the carpophores (D1), center point of the metacarpals (D2) and the tips of the third finger from the proximal phalanges (D3) and the intermediate phalanges (D4) to the distal phalanges (D5), and the median nerve index (MI = D1–D5).

In addition, artificial intelligence systems have been used to improve the diagnosis of CTS (Palfy and Papež 2007; Papež et al. 2008, 2009; Tkáčová et al. 2010).

In one of these studies, the reliability of the method was tested through artificial neural networks for data analysis, using 112 thermogram shots (26 healthy hands and 30 pathological ones), and specifically divided the hand into 12 areas of interest: the fingers (five segments), metacarpus (five segments), and carpus with wrist (two segments). In addition, they determined that the dorsal side of the hand provides greater success results when diagnosing CTS, and therefore, it is more important than the palmar (Papež et al. 2008).

Baic et al. (2017) verified the usefulness of IRT as a diagnostic method of CTS and analyzed its possible use for monitoring the healing process in CTS surgery. Images were obtained on both hands of palms and back to obtain temperature gradients. CTS surgery patients were examined before and after surgery to verify treatment effects. Their study involved 15 patients with CTS, 15 patients gone through surgery for CTS, and 15 healthy volunteers. The thermal images analyzed showed that healthy controls and patients, as well as the manual thermal image performed before and after surgery, differ completely. Thermal analysis shows that the thumb recovers faster than the rest of the fingers, which could be due to thenar eminence, a very strong muscle that can protect nerves and blood vessels. In fact, fingers, heart, and index are exposed to the effects of the CTS more than the thumb.

Herrick and Herrick (1987) used thermography to obtain images of the cervical spine, shoulders, forearms, and hands of a total of 90 patients, of whom 55 had CTS. Their studies to diagnose CTS by thermography establish that the results were effective and sensitive in differentiating the diagnosis of CTS from other peripheral compressive neuropathies, including cervical radiculitis, thoracic outlet syndrome, ulnar tunnel syndrome, and Guyon canal. The results reached an overall specificity of the thermographic studies of 80% with sensitivity of 96%, while when considering only the CTS patients, the diagnostic success rose to 97% with 100% of sensitivity.

Palfy and Papež (2007) used 44 thermograms of healthy and pathological hands (23 pathological and 21 healthy) to which he measured their temperatures and divided the hand into the following areas of interest: fingers (5 segments), metacarpus (middle part of the hand: 5 segments), and the carpus with the wrist (2 segments) to determine the effectiveness of IRT as a method of diagnosing CTS. They reached success rates for the classification of healthy hands with CTS pathologies close to or greater than 80%.

Papež et al. (2009) segmented the hand region as previously described and used a database of 502 thermal images including 132 healthy and 119 pathological hands to test the effectiveness of IRT in diagnosing CTS. They obtained success rates from the classification of healthy hands and CTS pathology of 72.2% based on images of the back and 80% of the hands when they were only seriously affected and considering healthy hands. Additionally, they found that the dorsal side of the hand provides more successful results when diagnosing CTS, and therefore, is more important than the palmar area of the hand.

Tkáčová et al. (2010) recorded 14 thermal images (7 healthy and 7 affected) and measured the temperature of each hand divided into 5 segments to determine the level of effectiveness of IRT to diagnose CTS. The success rates found for classifying healthy and pathology CTS hands approached 80%. In addition, they identified that the dorsal side of the hand provides greater success results when diagnosing CTS, and therefore, is more important than the palmar area of the hand.

Unlike the works previously analyzed, they studied the thermal responses in 6 patients with CTS and 5 healthy people, so that the patients' hands were excited. The procedure consisted of asking the participant to wash and dry his hands and then immediately place his hands on a plate to obtain his thermal images. The patient was then asked to place the palms of both hands in ice cubes for 60 s. This was done to induce a sympathetic nerve stimulation. At the end of the stimulation phase, the participant placed his hands on the analysis table, and then, thermal images were gathered. Using infrared views, they demonstrated that the hands with CTS give a thermal response. Thus, the difference increases if the patient has CTS, because the thermal pattern of the hand is different. The temperature variation is weak for healthy people; instead, it rises for patients. It also shows that the temperature variation must be above or below 0 °C, which corresponds to a vasoconstriction or the vasodilation of the veins. Also, this study proved that CTS can cause vasomotion problems in the thumb and index finger, where paresthesia is felt. From this, it follows that the IRT method is sensitive to median nerve variation (Maxel et al. 2014).

Ming et al. (2005) implemented a research concerning sympathetic pathology in CTS and the benefits of IRT in the diagnosis of this pathology. Temperatures were measured at the tips of the fingers, central thenar point, and eminences in 38 pathological hands and 41 healthy hands. They calculated the thermal differences between the fingertips. The results showed highly significant differences in the temperatures of the median nerve distribution area in the hands among the CTS and the control group. The differences between the distribution area of the median and ulnar nerve were also highly significant in the hands with CTS. The susceptibility and specificity of infrared thermography were 84% and 91% accordingly.

Orlin et al. (2005) included 22 patients with CTS symptoms and 16 healthy subjects, who underwent two-hand exercises as follows: (a) passive dorsiflexion in the radiocarpal joint (range 0–75°) and (b) hand exercise with hand booster (one compression per second) with wrist in neutral position for a maximum period of 90 s (if no fatigue occurs before). Before and after the manual exercise, the temperature was measured at the fingertips. The findings in thermography showed interrelationship with the deterioration of symptoms after manual exercise. A remarkable reduction in fingertip temperature was determined after the exercises performed by the patients. However, no changes were identified in healthy subjects. Low-grade CTS patients had a reduction in skin temperature caused by increased sweating.

In the same context, Tchou et al. (1992) recorded 122 thermograms, where 61 patients presented CTS and 40 volunteers healthy patients. They measured the temperatures of the thumb, index, and middle fingers of each hand from the phalangeal metacarpus of the junction to the tip of the fingers and toward the width of the fingers in order to test the effectiveness of IRT in the diagnosis of CTS. They obtained success rates for the classification of healthy hands with CTS pathologies close to or greater than 80%. Furthermore, they determined that the dorsal side of the hand provides more successful results when diagnosing CTS, and therefore, it is more important than the palmar area of the hand.

Park et al. (2008) determined that IRT does not contribute to the detection of CTS. In their research work, the feasibility of said method to diagnose unilateral CTS was evaluated. They evaluated a population of 28 patients with this pathology, where 19 of 28 patients with unilateral CTS manifested relevant differences in the temperature of the thumb tip, index, and middle finger. Meanwhile, 13 of these 19 patients had significant differences in at least one of the regions not inverted by the median nerve passing through the carpal tunnel. Of the 28 patients with unilateral CTS, only 4 had significant thumb, index, or middle finger tip temperature and only 2 had significant temperature in at least one of the regions that are not innervated by the median nerve passing through the carpal tunnel.

Lang et al. (1995) conducted a more in-depth study to assess the functions of the thick and thin nerve fibers in CTS, in order to deduce whether the data of the thin nerve fibers can contribute to the diagnosis of CTS. Twenty-two patients and 16 participants were studied, examining motor and sensory nerve conduction, vibration tests, thresholds of hot and cold sensations, upper pain threshold magnitude and burst response, as well as sympathetic reflexes before and after median nerve decompression. Thermography and photoplethysmography of the injured hand were recorded before and 1, 3, 6, and 12 months after median nerve decompression. Time comparisons were made using univariate analysis of variance in a repeated measures design and post hoc comparisons with the Newman-Keuls test or the Wilcoxon-Wilcox test. The T test was used to compare patients and control subjects and finally the Pearson correlation or the Spearman rank correlation.

The preoperative heat sensation (5.59 ± 0.62 °C) and cold (7.11 ± 0.9 °C) sensation thresholds were found to increase significantly in the index finger compared to control subject's data. After median nerve decompression, the cold and hot sensation thresholds enhanced significantly only in the index finger ($p < 0.05$). The highest rating of pain intensity by harmful mechanical stimulation was ($30 \pm 15\%$) in patients. The reflective decrease in skin temperature on the palm side of the third and fifth fingers tended to be deeper ($p < 0.01$) in patients than in healthy subjects. However, no significant differences were reported between the third and fifth fingers.

Regarding the response of the dilator vessel locally around the stimulation zone, no significance was detected among patients (0.71 ± 0.13 °C) and control subjects (0.88 ± 0.18 °C), even after median nerve decompression.

The data on the functions of the thick and thin nerve fibers were not significantly correlated at any time during surgical treatment, i.e., before or after. As far as thermography and plethysmography studies are concerned, they showed no significant relationship. Therefore, it can be inferred that these methods do not contribute to the diagnosis of CTS.

Reilly et al. (1989) found the characteristic thermal patterns in the CTS in 23 patients with a clinical history that suggested suffering from this pathology. Thermography and median nerve conduction tests (NCTs) were compared. Six patients presented normal NCTs, where two were thermographically normal. On the other hand, 17 patients presented non-normal NCTs, 12 unilaterally and 5 bilaterally. Sixteen participants showed non-normal thermal results. Six of eleven patients with right CTS presented hot right wrist and only one with hot left wrist. Regarding the Tinel´s test, all obtained negative results (present in 8 of 17 with CTS). Similarly, the Phalen´s test did not show a positive result in 11 of 17 patients. Therefore, it correlated significantly with thermal abnormalities. However, despite having found thermal irregularity in wrists and fingers, no clear diagnostic pattern was found. Similarly, no implication found among clinical tests and thermography or conduction tests. Thus, nerve conduction studies offer the most accurate diagnosis of CTS.

2.3.2 Infrared Thermography Application for CTS Studies

Lang et al. (1994) assessed the functional profile of pain-related middle nerve fibers due to CTS. The objective was to predict pain in CTS by analyzing the functional deficits of certain nerve fibers. The frequency and intensity of pain were followed for 14 days in 23 patients with CTS and 16 volunteers. Measurements were made of distal motor latency (DML) of the median nerve. Through thermotest, cold, heat, and heat pain perception tests were performed. Thermograms and photoplethysmograms were also obtained on the hand stipulated for surgical treatment. Once all the tests were performed, data were analyzed with the Kolmogorov-Smirnov test. It was determined that CTS-induced pain was situated in fingers 1–4 and palm. Intensity and frequency of pain attacks were significantly correlated ($p < 0.01$). The same was true for pain intensity and area of pain ($p < 0.05$). Greater significance was observed in the median nerve DML in patients (6.0 ± 1.4 ms.), compared to volunteers (3.6 ± 0.5 ms.).

Patients showed significant increase in heat and cold perception thresholds when compared to control subjects. After clinical intervention of the median nerve, pain levels were significantly reduced in the index and small fingers. There was no significant correlation between pain intensity and other neurophysiological parameters. However, significance was obtained between the area of pain and DML ($p < 0.05$).

Pain intensity due to CTS was predicted with $R = 0.72$ ($p < 0.001$, $n = 23$). The inclusion of more parameters as independent variables did not contribute to the prediction of pain due to CTS.

Ming et al. (2007) studied the recovery of patients after CTS surgery. Thermograms were obtained from the hands before and after CTS release between 22 patients and 41 volunteers. Subsequently, they obtained the temperatures of the fingertips from digit 1 to digit 5, the central point of the thenar and hypothenar eminences, the mean nervous index, and the temperature difference among the median and the area of distribution of the ulnar nerve. Based on the outcomes of IRT, it is determined that regulation of blood flow in CTS is abnormal, which is probably due to the altered regulation of the sympathetic motor vessels, and that circulation is normalized along with the relief of other CTS symptoms documented 6 months post-surgery.

Ammer et al. (2002) focused on his research in establishing a normal temperature range in the finger joints. For this purpose, the hands of 140 participants were thermographically evaluated, of which 37 patients had symptoms of painful osteoarthritis, 21 were diagnosed with arthritis, and 22 with CTS. Cold water tests were then performed to observe the pattern of temperature recovery. To establish the normal temperature ranges, a standard deviation value was defined, as well as the median and interquartile range. The results showed that, in the case of non-symptomatic joints, the highest temperatures were found in the thumb joints and the lowest in the little finger. However, the temperature of the interphalangeal joints of the ring finger and the little finger was lower than the non-symptomatic joints. The temperature of the colored joints in the walls and certain woofer joints recovered more quickly than other joints after a slight cold spell.

Ammer (1999) studied 154 hand thermographs to identify patients with thoracic outlet syndrome or CTS from healthy subjects. This is through the presence of thermal asymmetry between the index and little fingers, defined by the temperature difference of 0.5 °C between those fingers. According to neurography and thermographic criteria, the hands were classified into four groups: healthy controls, CTS, thoracic outlet syndrome, and the combination of the syndromes. Subsequently, a discriminant analysis of the asymmetries in the groups was carried out. As a result, a correct classification of 44.8% of the cases was obtained.

Once the syndrome combination has been passed to the thoracic outlet, the number of correct classifications increased to 63.3%. Values of 71.6 and 42.9% were obtained, for sensitivity and specificity of pathological temperature differences for the diagnosis of thoracic outlet syndrome. The sensitivity was 11.9% and the specificity 42.9% for the comparison controls and CTS.

Therefore, it is possible to identify patients with thoracic outlet syndrome by means of temperature distribution in the hand, but not for CTS cases. Table 2.2 is shown below, summarizing the results previously presented.

Table 2.2 Summary of results found

Reference Ref.	Country	Journal	Type	Participants	Method	Temperature measurement regions	Results
Baic et al. (2017)	Poland	Medicine	CTS	15 patients with CTS	Obtaining temperature gradients by palm and back thermograms of each hand	Palm and back of each hand	IRT may be useful in diagnosing CTS
			Diagnostic	15 patients gone through surgery for CTS			
				15 healthy volunteers			
Maxel et al. (2014)	France	Mechanics & Industry	CTS	6 STC patients	Energize hands of patients (hot-cold changes) and analyze their thermal response	Palm and back of each hand	Thermograms show CTS hands give thermal response
			Diagnostic	5 healthy people			
Živčák et al. (2011)	Hungary	International Symposium on Computational Intelligence and Informatics	CTS	14 pathological hands	Analysis of the temperature distribution of the whole hand and calculate partial temperatures	Central point of the carpophores & metacarpus, tips of the third finger from the proximal phalanges, intermediate phalanges to the distal phalanges and the median nerve index	Thermography is a potentially method in the diagnosis of CTS
			Diagnostic	120 healthy people			
Tkáčová et al. (2010)	Slovakia	International Symposium on Applied Machine Intelligence and Informatics	CTS	7 pathological hands	Skin temperature of both hands by computer-assisted IRT	The hand was divided into 5 segments	Classification of healthy hands and with CTS pathologies close to or greater than 80%
			Diagnostic	7 healthy hands			

(continued)

Table 2.2 (continued)

Reference Ref.	Country	Journal	Type	Participants	Method	Temperature measurement regions	Results
Papež et al. (2009)	Slovenia	The Journal of International Medical Research	CTS	502 thermal data	Use of thermal images of palm and back on hands and artificial neural networks for the diagnosis of CTS	Fingers (five segments), metacarpus (five segments), and wrist carpus (two segments)	Classification of 72.2% of the hands, between healthy and pathological, based on the dorsal part of the hand, while > 80% when evaluating severely affected hands with healthy hands
			Diagnostic	132 healthy hands			
				119 pathological hands			
Papež et al. (2008)	Slovenia	The Journal of International Medical Research	CTS Diagnostic	26 healthy hands 30 pathological hands	Compare thermal images (back and palm) with respect to a reference, using an artificial intelligence system	The fingers (five segments), metacarpus (five segments), and carpus with wrist (two segments)	Classification of healthy hands and with CTS pathologies close to or greater than 80%
Park et al. (2008)	Seoul, Republic of Korea	American Association of Neuromuscular & Electrodiagnostic Medicine	Diagnostic STC	350 patients	Measure finger temperature by IRT in patients with unilateral CTS	Tip of thumb, index and middle, thenar area, hypothenar area, back of hand, and forearm	IRT is not useful in diagnosing unilateral CTS

(continued)

Table 2.2 (continued)

Reference Ref.	Country	Journal	Type	Participants	Method	Temperature measurement regions	Results
Ming et al. (2007)	Finland	Clinical Neurology and Neurosurgery	Medical study	22 CTS patients 41 healthy volunteers	Obtain thermograms before and 6 months after CTS release and measure the temperatures of the fingertips	Fingertips from digit 1 to digit 5, central point of thenar and hypothenar eminences, median nerve index, and the temperature difference between the median and the ulnar nerve distribution area	Found abnormal regulation of blood flow in CTS, possibly due to impaired sympathetic motor vessel regulation and that circulation normalizes along with relief of other CTS symptoms recorded 6 months after surgery
Palfy and Papež (2007)	Slovenia	International Symposium on Computer-Based Medical Systems	CTS Diagnostic	23 pathological hands 21healthy hands	Segmentation of hand thermograms and temperature extraction and image analysis to diagnose CTS	Fingers (5 segments), metacarpus (middle part of the hand—5 segments), and the carpus with the wrist (2 segments)	Classification of healthy hands and with CTS pathologies close to or greater than 80%
Ming et al. (2005)	Finland	Pathophysiology	CTS Diagnostic	38 pathological hands 41 healthy hands	Measure temperatures in regions of hands to clarify sympathetic pathology in CTS	Fingertips, central thenar point, and eminences	IRT with 84% sensitivity and 91% specificity
Orlin et al. (2005)	Norway	European Journal of Neurology	CTS Diagnostic	22 patients with symptoms of CTS 16 control subjects	Study microvascular perfusion in fingertip skin and skin temperature during dorsiflexion of the hand before and after manual exercise	Fingertips	Thermography findings correlate well with deterioration of symptoms after the manual exercise test

(continued)

Table 2.2 (continued)

Reference / Ref.	Country	Journal	Type	Participants	Method	Temperature measurement regions	Results
Ammer et al. (2002)	Austria	International Thermology	Medical study	98 patients 42 participants	Cold water test in patients to observe the temperature recovery pattern	Index finger and little finger	It is inferred that normal temperature range of the finger joints by thermography did not prove to be clinically useful
Ammer (1999)	Austria	Proceedings of The First Joint BMES/EMBS Conference	Medical study	77 patients	Thermograms taken during the hyper abduction test and the modified Adson maneuver to measure temperatures and discriminant analysis of the data grouping	Ring fingers, thumb, and little finger	CTS patients and healthy people were classified correctly in 44.8% of cases. By transferring combined CTS to the thoracic outlet the number of correct classifications increased to 63.3%
Lang et al. (1995)	Germany	MUSCLE & NERVE	CTS Diagnostic	22 patients 16 control subjects	Examine sensory and motor nerve conduction, vibration tests, and thresholds for hot and cold sensations, magnitude of upper pain threshold and flare response, as well as sympathetic reflexes before and after median nerve decompression	Second, third, and fifth fingers; thenar and hypothenar and the area around the site of the mechanical stimuli and skin and skin stimulation between the second and third finger	Assessment of the flare and sympathetic response by IRT and photoplethysmography did not contribute to the diagnosis of CTS

(continued)

Table 2.2 (continued)

Reference Ref.	Country	Journal	Type	Participants	Method	Temperature measurement regions	Results
Lang et al. (1994)	Germany	Pain	Medical study	23 patients with CTS 16 volunteers	Obtaining thermograms on painfully stimulated hands and fingers	Fingers 2, 3, and 5, the thenar and hypothenar, the area around the site of the mechanical noxious stimulation of the skin between the index and middle fingers and in a reference area around the skin between the ring finger and the little finger	The sweat motor nerve fibers are not considerably damaged by CTS. The intensity of pain due to CTS depends on disturbances of peripheral and central nerve functions
Tchou et al. (1992)	United States	The Journal of Hand Surgery	CTS Diagnostic	61 patients with unilateral CTS 40 volunteers	Thermographic examinations for classification of healthy and pathological hands	Thumb, index, and middle fingers	Classification of healthy hands and with CTS pathologies close to or greater than 80%
Reilly et al. (1989)	United Kingdom	British Journal of Rheumatology	CTS Diagnostic	23 patients with CTS	Evaluation of CTS thermal patterns, comparing tests of mean nerve conduction and thermography	Hands and wrists	No clear diagnostic pattern of CTS was found by IRT
Herrick and Herrick (1987)	United States	The Journal of hand surgery	CTS	90 patients	Obtaining thermograms to get diagnoses and stress series Studies were conducted to differentiate CTS from peripheral neurovascular injuries	Cervical spine, shoulders, forearms, and hands	General specificity of 80% and sensitivity of 96%. When considering only CTS patients, the diagnostic success was 97% with 100% sensitivity

2.4 Discussion

It has been determined that IRT is useful in the diagnosis of CTS (Baic et al. 2017; Herrick and Herrick 1987). In addition, Baic et al. (2017) support its application to monitor the recovery process, and Herrick and Herrick (1987) consider its application in treatments and preventive measures of the CTS, which could eliminate the high cost of the loss of labor and medical care.

Employing IRT in conjunction with intelligent computing systems improves the diagnosis of CTS (Palfy and Papež 2007; Papež et al. 2009; Tkáčová et al. 2010). On the other hand, several investigations agreed on their findings about IRT, as effective in the diagnosis of CTS, particularly in the first stage (Lang et al. 1995; Maxel et al. 2014; Ming et al. 2005; Orlin et al. 2005).

However, IRT is not recommended when CTS is severe (Papež et al. 2009). On the other hand, they supported that IRT is used in the diagnosis of unilateral CTS, but limited in bilateral CTS (Tchou et al. 1992). Conversely, the use of IRT in the diagnosis of unilateral CTS is not recommended (Park et al. 2008).

The measurements made by IRT demonstrate blood flow regulation abnormality in CTS. This is probably due to the sympathetic vasomotor disturbance and the simultaneous normalization of the relief of CTS symptoms and the circulation (Ming et al. 2007). Ammer et al. (2002) determined that the definition of the normal range for temperature readings from thermal images of the finger joints has not been shown to be clinically useful. Whereas Ammer (1999) determined that patients suffering from CTS, it is indistinguishable from normal patients in thermal imaging by applying tests for thoracic outlet syndrome.

Lang et al. (1994) established that the intensity of pain caused by CTS relies on the disturbances of the peripheral and central nervous functions. Finally, the absence of a clear CTS diagnostic pattern is pointed out, since no association has been found between clinical tests and thermography or conduction tests (Reilly et al. 1989).

2.5 Conclusion

IRT imaging is a useful technique for diagnosing CTS, especially at an early stage. However, its precision is limited in bilateral disorders, while it is very useful in unilateral CTS. An attempt has been made to find a diagnostic guideline for CTS, but has not yet been achieved. It is considered that to achieve a completely reliable diagnosis of CTS, it is necessary to carry out nerve conduction studies.

Overall, IRT imaging provides useful information for prompt detection, treatment, and even preventive measures that can remove the high costs of lost human resources and medical care related to CTS.

2.6 Future Works

Some authors have proposed as future work to study the identification of new unknown interrelationships between skin surface temperatures and nerve entrapment syndromes or even electromyography results (Palfy and Papež 2007; Papež et al. 2008, 2009).

Furthermore, it is proposed that to determine how the recovery process will unfold and how longer it will take, more research that includes measurements after completion and beyond just 4 weeks after surgery is needed, allowing an objective evaluation of the procedures used (Baic et al. 2017). On the other hand, it is recommended that the analyses carried out always be based on statistical analyses (Tkáčová et al. 2010). In addition, a quantitative assessment of neuropathy and the importance of the diagnostic method are suggested (Tkáčová et al. 2010).

Moreover, given the current working conditions, it is considered that the studies have been focused on the population with restricted patients, so it is now required to be generalized. Also, the robustness of the tests should be studied in depth, to understand why the thermal signature of the CTS was generally positive but also negative in particular cases.

It is also suggested to analyze the cause of kinetics in thermal reactions among one patient and another, which allows optimizing the experimental protocol (duration on excitation, type of excitation, duration analysis, type of post-processing, etc.). Additionally, it is important to take into account the exclusion criteria selected (Maxel et al. 2014).

Additional follow-up studies should be considered to explain the advancement of neuropathophysiology to distinguish diverse stages of neural lesions and to investigate the reversibility of functional abnormalities (Ming et al. 2005). Likewise, studies focused on increasing diagnostic susceptibility and specificity are required in cases of chronic unilateral CTS and recurrent postsurgical CTS, since almost 30 years ago (Tchou et al. 1992).

Studies are also needed on the reproducibility of IRT digital recordings. In addition, further studies are suggested on the functional impairment of different types of nerve fibers in relation to nerve compression. Employing digital IRT, nerve conduction studies, and, for example, the tactile sensitivity test could provide information on the responses of the various groups of fibers under such conditions (Ming et al. 2007).

Additional information must be obtained from observing the recovery of temperature after a mild attack of cold. It is possible that the assessment of finger temperatures from the reference database of normal human body thermographs will solve the problem of establishing a normal range of finger joint temperatures in the nearby future (Ammer et al. 2002).

In general, it is recommended to continue with the review of related research between CTS and IRT which will allow to enrich this work, so that the reliability of IRT as a diagnostic method for CTS is firmly demonstrated, as well as a support technique for studies related to this pathology involving prevention, pre-operative and post-operative studies, assessment of short-, medium-, and long-term treatment, among other research that will foster the continuity in this topic given its importance and support for the prevention and treatment of CTS, which contributes to improving the quality of life of patients suffering from it.

References

Alcan V, Zinnuroglu M, Karatas G, Bodofsky E (2018) Comparison of interpolation methods in the diagnosis of carpal tunnel syndrome. Balkan Med J 35:378–383

Ammer K (1999) Diagnosis of nerve entrapment syndromes by thermal imaging proceedings of the first joint BMES/EMBS conference, p 1117

Ammer K, Engelbert B, Kern E (2002) The determination of normal temperature values of finger joints. Thermol Int 12:23–33

Baic A, Kasprzyk T, Rzany M, Stanek A, Sieron K, Suszynski K, Marcol W, Cholewka A (2017) Can we use thermal imaging to evaluate the effects of carpal tunnel syndrome surgical decompression? Med (Baltimore) 96(39): https://doi.org/10.1097/MD.0000000000007982

Boerner E, Bauer J, Kuczkowska M, Podbielska H, Ratajczyk B (2015) Comparison of the skin surface temperature on the front of thigh after application of combined red-IR radiation and diadynamic currents executed in different sequence. J Therm Anal Calorim 120:921–928

Cholewka A, Drzazga Z, Knefel, Kawecki, Nowak (2010) Thermovision diagnostics in chosen spine diseases treated by whole body cryotherapy. J Therm Anal Calorim 102:113–119

Fischer A (1986) The present status of neuromuscular thermography. Postgrad med special report

Ghasemi-Rad M, Nosair E, Vegh A, Mohammadi A, Sayed D, Davarian A, Maleki-Miyandoab T, Hasan A (2014) A handy review of carpal tunnel syndrome: from anatomy to diagnosis and treatment. World J Radiol 6:284–300

Herrick RT, Herrick SK (1987) Thermography in the detection of carpal tunnel syndrome and other compressive neuropathies. J Hand Surg 12(5):943–949. https://doi.org/10.1016/s0363-5023(87)80262-9

Kanafi L, Arianpur A, Gharedaghi M, Rezaei H (2018) Ultrasound as a diagnostic tool in the investigation of patients with carpal tunnel syndrome. Eur J Transl Myol 28:193–197

Lang E, Claus D, Neundörfer B, Handwerker HO (1994) Parameters of thick and thin nerve-fiber functions as predictors of pain in carpal syndrome. Pain 60:295–302

Lang E, Spitzer A, Claus D, Neundörfer B, Pfannmüller D, Handwerker HO (1995) Function of thick and thin nerve fibers in carpal tunnel syndrome before and after surgical treatment. Muscle & Nerve: Official J Am Assoc Electrodiagnostic Med 18(2):207–215

Maxel X, Bodnar JL, Stubbe L (2014) Detection of carpal tunnel syndrome by infrared thermography. Mech Ind 15(5):363–370. https://doi.org/10.1051/meca/2014034

Ming Z, Siivola J, Pietikainen S, Narhi M, Hanninen O (2007) Postoperative relieve of abnormal vasoregulation in carpal tunnel syndrome. Clin Neurol Neurosurg 109(5):413–417. https://doi.org/10.1016/j.clineuro.2007.02.014

Ming Z, Zaproudina N, Siivola J, Nousiainen U, Pietikainen S (2005) Sympathetic pathology evidenced by hand thermal anomalies in carpal tunnel syndrome. Pathophysiology 12(2):137–141. https://doi.org/10.1016/j.pathophys.2005.05.002

Orlin JR, Stranden E, Slagsvold C (2005) Effects of mechanical irritation on the autonomic part of the median nerve. Eur J Neurol 12:144–149

Palfy M, Papež BJ (2007) Diagnosis of carpal tunnel syndrome from thermal images using artificial neural networks international symposium on computer-based medical systems, pp 1–6

Papež B, Jesenšek, Palfy M, Mertik M, Turk Z (2009) Infrared thermography based on artificial intelligence as a screening method for carpal tunnel syndrome diagnosis. J Int Med Res 37:779–790

Papež B, Jesenšek Palfy M, Turk Z (2008) Infrared thermography based on artificial intelligence for carpal tunnel syndrome diagnosis. J Int Med Res 36:1363–1370

Park D, Seung NH, Kim J, Jeong H, Lee S (2008) 66. Infrared thermography in the diagnosis of unilateral carpal tunnel syndrome. Clin Neurophysiol 119(3). https://doi.org/10.1016/j.clinph.2007.11.116

Redzwan Habib K (2017) Estimation of Carpal Tunnel Syndrome (CTS) prevalence in adult population in Western European countries: a systematic review. Eur J Clin Biomed Sci 3(1). https://doi.org/10.11648/j.ejcbs.20170301.13

Reilly P, Clark AK, Ring EF (1989) Thermography in carpal tunnel syndrome. British J Rheumatol, 553–554

Rempel DM, Harrison RJ, Barnhart S (1992) Work-related cumulative trauma. JAMA 267:838–842

Ring EF, Ammer K (2012) Infrared thermal imaging in medicine. Physiol Meas 33(3):R33–R46. https://doi.org/10.1088/0967-3334/33/3/R33

Roel-Valdés J, Arizo-Luque V, Ronda-Pérez E (2006) Epidemiology of occupational carpal tunnel syndrome in Alicante Province, 1996–2004. Spain Public Health Magazine, pp 395–409

Schols A, Beekman R, Cals J, van Alfen N, Ottenheijm R (2018) Carpal tunnel syndrome: a clinical diagnosis. Ned Tijdschr Geneeskd

Tchou S, Costich J, Burgess RC, KY L, Wexler CE, Calif T (1992) Thermographic observations in unilateral carpal. J Hand Surg 17:631–637

Tkáčová M, Foffová P, Hudák R, Švehlík J, Živčák J (2010) Medical thermography application in neuro-vascular diseases diagnostics. International symposium on applied machine intelligence and informatics, pp 293–296

Uematsu S, Long D (1976) Thermography in chronic pain. Medical thermography, theory and clinical applications. Los Angeles, Brentwood

Wipperman J, Goerl K (2016) Carpal tunnel syndrome: diagnosis and management. Am Fam Physician 94:993–999

Wright A, Atkinson R (2019) Carpal tunnel syndrome: an update for the primary care physician. Hawai'i J Health Soc Welf 78:6–10

Živčák J, Madarász L, Hudák R (2011) Application of medical thermography in the diagnostics of carpal tunnel syndrome. International symposium on computational intelligence and informatics, pp 535–539

Chapter 3
Mental Workload Management and Evaluation: A Literature Review for Sustainable Processes and Organizations

Nancy Ivette Arana-De las Casas, Aide Aracely Maldonado-Macias, Jorge De La Riva-Rodríguez, David Sáenz-Zamarrón, José Francisco Alatorre-Ávila, and Enrique García-Grajeda

Abstract Current organizational and manufacturing processes imply high mental workload demands that can be evaluated through the construct of MWL (mental workload). This term is often used in new manufacturing and organizational environments, which have replaced physical tasks with cognitive activities involving a high MWL. By overusing the attentional resources given to the tasks, such work environments are placing high cognitive loads on operators, thus affecting their performance and causing them to experience mental fatigue. A formal evaluation of MWL offers the opportunity to prevent mental disorders and maintain mental health. On the other hand, the lack of evaluation and proper management of MWL in the industry can result in errors that create economic costs, accidents, injuries, or even deadly events. Finally, MWL assessment and management can be a human-oriented strategy designed to improve and sustain the future of an organization. Industries must find competitive advantages in sustainable processes from the economic, environmental, and social view. In this sense, this chapter aims to present a literature review to provide a comprehensive literature analysis of MWL evaluation and management for sustainable processes in the manufacturing industry, from a social perspective.

Keywords Mental workload · Sustainable process · Cognitive ergonomics

N. I. Arana-De las Casas (✉) · D. Sáenz-Zamarrón · J. F. Alatorre-Ávila · E. García-Grajeda
Campus Ciudad Cuauhtémoc, Tecnológico Nacional de México, Ciudad Cuauhtémoc, Mexico
e-mail: narana@itcdcuauhtemoc.edu.mx

A. A. Maldonado-Macias
Industrial Engineering and Manufacturing Department, Universidad Autónoma de Ciudad Juárez, Ciudad Juárez, Mexico

J. De La Riva-Rodríguez
Campus Ciudad Juárez, Tecnológico Nacional de México, Ciudad Juárez, Mexico

A. Realyvásquez Vargas et al. (eds.), *New Perspectives on Applied Industrial Ergonomics*,
https://doi.org/10.1007/978-3-030-73468-8_3

3.1 Introduction

Workload has an impact on any kind of work, especially in performance. Gore (2018) classifies workload into a physical load, mental load, or the load associated with the effort required to meet task demands. Current workplaces in manufacturing organizations show a growing trend in the amount of mental workload (MWL) demand on workers, which has led researchers to focus on different ways of evaluating MWL (Estes 2015; Haji et al. 2015; Park et al. 2018).

3.1.1 Mental Workload Evaluation in Organizations

Several models have been proposed to evaluate MWL in a wide range of contexts. However, most of them have been applied and used in organizations. These models entail diverse techniques for conducting employees' MWL assessment; they can be classified into physiological, subjective, and analytical ones. This section describes its main characteristics, advantages, and disadvantages.

Physiological measures are based on the use of certain cognitive activities associated with physiological variations in people, where variation in the measurements is affected by the increase or decrease in the MWL generated by the working task. The main advantage of this kind of measurement is that they do not interfere with the performance of the primary task. Some disadvantages, however, are that they can be affected by the worker's health and that the equipment necessary for its execution can be expensive and/or difficult to manage.

In contrast, the subjective rating models analyze the worker's perception of the task's complexity and are usually performed once the task is executed. The advantages of these types of techniques are their ease of use, their short administration time, and their low cost. In contrast, they show certain disadvantages when the MWL rate correlates with workers´ performance. Additionally, the outcomes can be affected by respondents' factors such as partiality, response sets, errors, and attitudes of protest (Yan et al. 2017).

The analytical techniques to measure MWL are performed through mathematical modeling and computer simulation. One advantage is that computer systems are quite common and highly potent nowadays, which facilitates the work of math modeling and simulation significantly. A disadvantage, however, is that they are more difficult to understand and apply.

3.1.2 Techniques for Measuring Mental Workload

3.1.2.1 Properties of Techniques Used for Measuring Mental Workload

Wierwille and Eggemeir (1993) compile and determine the properties that any mental workload evaluation technique should comply with, and they are as follows:

- Sensitivity: the degree to which a given workload technique can distinguish differences in the levels of the load imposed on an operator.
- Intrusion: an undesirable feature in which the introduction of the MWL measuring technique causes a change in worker's performance.
- Diagnosticity: defined as the ability to distinguish the cause or type of MWL or the ability to attribute it to an aspect of the task.
- Global Sensitivity: the potential to reflect discrepancies in different types of resource expenditure or factors that influence MWL.
- Transferability: refers to the capability of a technique to be used in various applications.
- Implementation requirements: consist of any instrumentation necessary to present and analyze information or record data, including aspects such as time, data collection procedures, or subject training.

3.1.2.2 Physiological Measurement Techniques

The following physiological techniques to measure MWL can be highlighted:

- Heart rate (HR) is the number of times a person's heart beats per minute. The normal range for adults is 60 to 100 beats per minute but can vary from person to person. Several authors mention that heart rate increases when MWL appears (Fallahi et al. 2016; Yan et al. 2017).
- Heart rate variability (HRV) is a measure that indicates the variation in a person's heartbeat within a specific interval. When the intervals between the heartbeats are relatively constant, the HRV is considered low; if their length varies, the HRV is high. The spectral HRV analysis enables the break down into components associated with different biological mechanisms such as the low-frequency power/high-frequency power (LF/HF) ratio, the men inter-beat (RR), and the standard deviation of normal RR intervals (SDNN) (Yan et al. 2017). HRV decreases concerning MWL (Heine et al. 2017; Mandrick et al. 2016).
- Salivary cortisol: Cortisol (hydrocortisone) is a steroidal or glucocorticoid hormone synthesized by the adrenal gland, released in response to stress and responsible for the reduction of glucocorticoid levels in the blood. Cinaz et al. (2013) state that in some individuals, cortisol levels increase concerning MWL, while in other cortisol levels stay the same or decrease. These responses might be explained by the conclusions that uncontrollable and social-evaluative stressors are associated with the changes in cortisol.

- Blink rhythm and duration: Blinking is a semi-autonomic, rapid closing of the eyelid. Nowadays, researchers think blinking may help humans to disengage their attention as following blink onset, cortical activity decreases in the dorsal network and increases in the default-mode network, which can be associated with internal processing. Thus, eye blinking frequency and duration decrease with MWL (Yung and Wells 2017).
- Ocular fixation duration: This is the period when the eyes remain relatively static. Teo et al. (2015) reference that fixation duration is shorter when MWL is present.
- Pupil dilation (PD): The size of the pupil is controlled by the activities of two muscles: the circumferential sphincter muscle, found in the margin of the iris and innervated by the parasympathetic nervous system; and the iris dilator muscle, running radially from the iris root to the peripheral border of the sphincter. Yan et al. (2017) mention that human pupil dilation may be used as a measure of psychological load because it is related to the amount of cognitive control, attention, and cognitive processing required for a given task.
- Intraocular pressure (IOP) is a measurement involving the magnitude of the force exerted by the aqueous humor on the internal surface area of the anterior eye. Higher intraocular pressure values are associated with high workload levels (Jiménez et al. 2018; Vera et al. 2017).
- Electromyography (EMG) is a diagnostic procedure that evaluates the health condition of muscles and the nerve cells that control them. Fallahi et al. (2016) found a relationship between EMG amplitude increase and MWL.
- Electroencephalography (EEG) measures voltage fluctuations resulting from ionic current within the brain's neurons. EEG activity is often decomposed into frequency bands delta (up to 2 Hz), theta (4–7 Hz), alpha (8–13 Hz), and beta (14–25 Hz). Charles and Nixon (2019) pointed out that brain activity establishes a promising indication of MWL; however, data measurement and analysis remain complex.
- Near-infrared reflectance spectroscopy (NIRS): A non-invasive technique uses light transmission and absorption to measure hemoglobin and mitochondrial oxygenation. Reduced cerebral blood oxygen saturation in the left prefrontal cortex (PFC) is associated with fatigue (Liu et al. 2016).
- Functional near-infrared spectroscopy (fNIRS) is the use of near-infrared spectroscopy with the objective of functional neuroimaging. Brain activity is assessed through hemodynamic reactions associated with neuronal behavior. Banville et al. (2019) point out that several fNIRS features correlate significantly with the temporal demand and performance dimensions of the NASA-Task Load Index (NASA-TLX) methodology to measure MWL.
- Respiration depth (RD) refers to the volume of the breath. Respiration volume decreases as MWL increases (Charles and Nixon 2019).
- Respiration rate (RR) accounts for the number of actions indicative of inspiration and expiration per unit of time. The normal respiratory rate for healthy adults is between 12 and 20 breaths per minute. However, RR increases as MWL increases, but it is also contingent on physical activity (Charles and Nixon 2019).

- Saccade rate (SR): The brain commands sent to the eye muscles result in the eyes making a rapid, step-like rotation following which the eyes remain stationary at their new position. These step movements are known as saccade eye movements, which increase with MWL (Cai et al. 2016).
- Systolic blood pressure (SBP) is the amount of pressure in the arteries during the contraction of the heart muscle. SBP increases with the difficulty of a task (Fairclough and Ewing 2017).
- Diastolic blood pressure (DBP) is the pressure in the arteries when the heart rests between beats. DBP increases with an increment in workload (Klonowicz 2015).
- Facial thermography: This technique detects heat patterns created by the branching of blood vessels and issued from the skin. Marinescu et al. (2016) mention that facial thermography is a good method for non-intrusive MWL measurement as temperature variations on some areas of the face appear to be strongly associated with the changes in MWL.
- Electrodermal activity (EDA) describes the phenomenon in which the skin momentarily becomes a better conductor of electricity when either external or internal physiologically arousing stimuli occur. The disadvantage is that they require the control of several factors such as temperature, humidity, time of day, emotional state, and irregularities in respiration (Charles and Nixon 2019).

3.1.2.3 Subjective Techniques

Concerning subjective techniques to measure MWL, fourteen methods can be mentioned:

- NASA-TLX: It is a subjective evaluation technique designed by Hart and Staveland (1988), divided into six subjective subscales: mental demand, physical demand, temporal demand, performance, effort, and frustration. In the first part of this test, each subscale is to be rated within a 100-point range with 5-point steps. These ratings are then merged to obtain the task load index. The other part of the test proposes to generate an individual weighting of these subscales by letting individuals equate them pairwise based on their perceived weight. NASA-TLX is currently one of the most widely used indices to evaluate subjective MWL (Bommer and Fendley 2018; Cowan et al. 2018; Galy et al. 2018).
- Workload profile (WP) is the label for a multidimensional MWL assessment tool, elaborated by Tsang and Velazquez (1996) and used to determine demand ratings imposed by a task on the eight following dimensions: perceptual processing, response selection and execution, spatial processing, verbal processing, visual processing, auditory processing, manual output, and speech output.
- Subjective rating of mental effort (SRME). It was developed by Paas (1992) as a unidimensional 9-point rating scale, which asks participants to rate how much mental effort they've invested in a task, ranging from (1) very, very low mental effort to (9) very, very high mental effort (Paas 1992).
- Modified Cooper Harper: The Cooper Harper Scale (Cooper and Harper 1969) is a decision-tree rating scale that was originally developed as an aircraft handling

measurement tool. The modified Cooper Harper scale is a unidimensional measure that uses a decision tree to elicit operators' MWL (Wierwille and Casali 1983) based on the assumption that there is a direct relationship between the level of difficulty of aircraft controllability and workload on the pilot (Wierwille and Casali 1983).

- Standardized Copenhagen Psychosocial Questionnaire II (COPSOQ II): It was developed by Pejtersen et al. (2010) because of the first version of the Copenhagen Psychosocial Questionnaire (COPSOQ I) failure to address important aspects relative to work. The COPSOQ II solved these limitations and incorporated aspects arising from the former questionnaire.
- Dundee Stress State Questionnaire (DSSQ): It was developed by Matthews and colleagues in 1999 as a questionnaire based on a factor model distinguishing eleven primary state factors that cohere around three higher-order dimensions: task engagement, distress, and worry (Matthews et al. 1999).
- Short State Stress Questionnaire (SSSQ) is a self-report measure of stress based on the DSSQ and developed by Helton (2004).
- Integrated Workload Scale (IWS): This is a scale for real-time assessment of subjective workload, initially designed to be used with train traffic controllers (Pickup et al. 2005). It consists of nine points that describe the degree of MWL and captures the multidimensionality of workload by integrating elements that reveal time, demand, and effort.
- State-Trait Anxiety Inventory (STAI): An assessment developed by Spielberger et al. (1970) and a Spanish edition the next year (Spielberger et al. 1971); it is used to measure trait and state anxiety through 40 items, 20 for assessing trait anxiety and 20 for state anxiety, all rated on a 4-point scale.
- Instantaneous self-assessment of workload technique (ISA): It is a unidimensional scale developed by the United Kingdom Civil Aviation Authority to provide subjective ratings of workload during air traffic control tasks. It uses five different ratings for perceived workload: excessive, high, comfortable, relaxed, and underutilized; Castle and Leggatt (2002) published the validity and reliability of the method.
- Subjective Workload Assessment Technique (SWAT): Reid and Nygren (1988) developed this tool, based on the thought that MWL is planned to be a multidimensional concept that can be largely explained by three factors: time load, mental effort load, and psychological stress load.
- Simplified Subjective Workload Assessment Technique (S-SWAT): SWAT presented two main problems: The first one is related to its lack of sensitivity for low MWL (Nygren 1991) and the second one to the time-consuming card sorting (Luximon and Goonetilleke 2001). The discrete SWAT subscale and the other four used the continuous SWAT subscale, resulting in the S-SWAT technique being more sensitive than the original SWAT scale (Luximon and Goonetilleke 2001).
- Verbal Online Subjective Opinion scale (VOSO): It consists of unidimensional rating systems, where subjects provide a verbal estimate of MWL on a scale

between 0 and 10. This scale is very sensitive to short periods of mental load (Miller 2001).

- For very particular investigations and in rare cases, researchers choose to develop their own assessment techniques that could be a direct interview with the subjects including very specific questions and recording to analyze emotions and gestures and/or assessments made specifically with the points related to the research on hand; an example is found in the research by Cruz Espinoza (2017).

3.1.2.4 Analytical Techniques

Additionally, five models related to analytical techniques for MWL measurement published from 2015 to 2019 were found:

- Improved Performance Research Integration Tool (IMPRINT): It is a human performance-modeling tool that uses discrete event simulation to predict MWL. The Human Research and Engineering Directorate (HRED) of the United States Army Research Laboratory (ARL) Combat Capabilities Development Center (CCDC). Data and analysis Center developed this tool. It was developed using the NET framework, which operates under the Microsoft Windows operating system to evaluate human-system function allocation, human performance, and MWL (Allender 2000).
- Model Sleep, Activity, Fatigue, and Task Effectiveness (SAFTE): This is a biomathematical fatigue model developed by the US Army Research Lab. It analyzes an array of relevant sleep factors; among them are acute sleep interruptions, cumulative sleep debt, the consistency of sleep onset and wake times, and circadian disruptions that influence a change in cognitive function to accurately predict fatigue. Once this information is provided, the SAFTE Fatigue Model can accurately predict changes in cognitive effectiveness throughout the work hours (US Army Research Lab 2019).
- Adaptive Control of Thought-Rational (ACT-R): This cognitive architecture-based system stemmed from the assumption of a unified theory of mind. Simulations with ACT-R allow for the prediction of typical measures in psychological experiments such as latency, accuracy, and neurological data, which help understand human cognition. It was developed by John Robert Anderson and Christian Lebiere at Carnegie Mellon University (Anderson and Matessa 1998).

3.1.3 Issues, Controversies, Problems

3.1.3.1 International Businesses in the Mexican Manufacturing Sector

Mexico is a country that is offering significant advantages related to manufacturing, such as low labor rates, convenient localization, and transportation infrastructure with cheap transportation costs of raw materials, and the ability to incorporate important sustainable manufacturing strategies such as just-in-time, kaizen, lean, and ergonomics. Additionally, Mexico counts on a highly skilled and motivated workforce at every business level. Additionally, an important factor is the existence of the IMMEX program (maquiladora program) which allows foreign manufacturers to import raw materials and components into Mexico, tax and duty-free, under the condition that 100% of the product will be exported out of Mexico within a government-mandated timeframe. The program aims to modernize the globalization of Mexico's manufacturing infrastructure by bringing new, specialized technologies and knowledge to the region (Stanley 2020).

Recently, in January 2020, the president of the United States signed the new North American Free Trade Agreement call USMCA (United States-Mexico-Canada Agreement) with new laws on intellectual property protection, the Internet, investment, currency, and additional provisions designed to identify and prevent labor violations, particularly in Mexico, all this in pursue of better international business relations.

3.1.3.2 Ergonomics and Humanistic Management for Sustainable Processes

Sustainable development has become a relevant topic. It refers to the development that fulfills the needs of the present, without compromising future generations' capacity to meet their own needs (United Nations 1987). This approach grants importance not only to the environmental aspect, but also to sustainable socio-economic development, where production processes are involved and terms such as humanistic management. In this approach, people turn out to be a priority above other variables to have a sustainable process. On the other hand, Ergonomics is the scientific discipline that undertakes the understanding of interactions between humans and other elements of a system, as well as the profession that applies theory, principles, and design to optimize human well-being and total system performance (IEA 2019). Accordingly, Radjiyev et al. (2015) affirm that ergonomics can be a significant factor in the transition to the sustainable development of production processes having because of sustainable processes.

Accordingly, several studies that have addressed how significant knowledge of ergonomics, such as human behavior, performance, and interactions with technology, can play an important role in the systems of sustainable and socially responsible manufacturing that current trends require (Thatcher et al. 2018; Kujawinska et al. 2015). These human factors are also key in the man-machine systems that new manufacturing environments and advanced manufacturing technology involve. These new working conditions demand a greater mental (cognitive) workload from workers, while the physical load required by work activities has decreased significantly, leading to the search for sustainable processes where the MWL is greater than the physical one (Haslam and Waterson 2013).

This means that new problems regarding workers' health have emerged in the work environment. These are mainly related to mental well-being rather than physical conditions. In cognitive ergonomics, it is critical to understand the influence of MWL over workers' performance, as mentioned by Young et al. (2015) in their work "State of Science: Mental Workload in Ergonomics," and more importantly, the actions that ergonomists can take to lessen the negative effects of MWL.

Human factors, especially cognitive ergonomics, should focus on being an alternative to study and improve the variables in the new international business/sustainable processes that affect the workers' performance in the search of better labor conditions, fewer job-related accidents, continuous improvement, and more productivity.

3.1.4 Cognitive Ergonomics and Sustainability

It is important to mention that MWL is not a variable that occurs only in office work environments or in association with senior management positions; it is also present in all kinds of work activities. That is the reason why Apud (2012) recommends the implementation of ergonomics and its guidelines in all types of workstations. On the other hand, according to Nieto (2014), occupational injuries and diseases have increased with a significant differentiation among the types of work as the conditions currently faced by employees that are related more to the cognitive aspect of the job than to the physical one. Accordingly, the mental demand that new sustainable processes place on their workers is increased to meet quality and customer service expectations.

In fact, the term "psychosocial risks" is becoming more common concerning occupational health. Such risks have been grouped, according to Gil-Monte (2012), in five areas:

1. New forms of contracting distinguished by the emergence of less favorable work contracts in conjunction with phenomena such as production scheduling and outsourcing and insecurity in the workplace.
2. The risk associated with the working population's aging and the delay in retirement age.
3. The intensification of work activities characterized by the need to handle more information and greater workload along with increased pressure from chief executive officers (CEOs).
4. The strong emotional demands at work as well as the increase in psychological harassment in the workplace.
5. The imbalance and conflict between work and personal life.

It can be inferred that these emerging psychosocial risks result in a greater MWL.

In conclusion, if the new international business wants to have a sustainable process, they need to measure MWL in their workers and have changes in their processes to increase productivity and avoid health risks in their job force.

3.2 Methodology

The specific objective in the literature review was to answer the question: What variables are being considered when measuring MWL and how are these variables being measured? To achieve this goal, a systematic literature review was conducted in the following databases: Science Direct, Google Scholar, Taylor & Francis Group, and SAGE journals. Likewise, the following journals were selected as they were considered to be more closely related to the subject of this work: Ergonomics, International Journal of Occupational Safety and Ergonomics (JOSE), Human Factors, the Ergonomics Open Journal, Procedia Manufacturing, Consciousness and Cognition, the American Journal of Engineering Research (AJER) and Medicina y Seguridad del Trabajo. Research articles, literature reviews, and theses that were written in Spanish and English from 2015 to 2019 were reviewed.

The general inclusion-exclusion criteria included the focus on adult subjects, research articles, thesis, and book chapters or review articles published between 2015 and 2019. It was focused on manufacturing, monitoring, and/or controlling processes, and studies in relation to the medical field were excluded because the activities performed are too specific to this field and not to any other; experimental and quasi-experimental studies were accepted with randomized assignation.

To perform the literature review in relation to ergonomics and sustainable production, six (6) articles were found in the last five years, all of them highlighting the relevant impact of ergonomics on sustainable development. Regarding mental workload, 610 articles were found in English, out of which 93 were selected considering these ones are related to manufacturing processes or activities that can be related to manufacturing activities. A considerable number of articles were found in relation to mental workload in medical activities of doctors and nurses and other health professionals; none of these were selected, because they are very specific to this profession. Ninety (90) more were found in Spanish, from which 4 articles and 3 theses/dissertations were selected after applying the same aspects applied in the English ones.

The authors of this work resorted to the ScienceDirect, Taylor & Francis, and Sage Journals' databases to find results in English as a greater number of indexed journals focusing mainly on the area of engineering (ergonomics/neuroergonomics/human factors) and manufacturing are published in this language.

Once the databases to be used were established and based on each of their advanced search characteristics, including only from 2015 to date and the words: ergonomics, sustainable production, and mental workload were chosen that would be the basis on which the searches would be performed.

In the final phase, titles and abstracts were thoroughly analyzed concerning the questions stated before.

When utilizing the ScienceDirect database, the term used was "mental workload." Once articles and book chapters related to medical processes or those dated before 2015 were excluded, 2808 results since 1995 decreased to 925. Then, only those publications related to manufacturing, vigilant, or attention processes

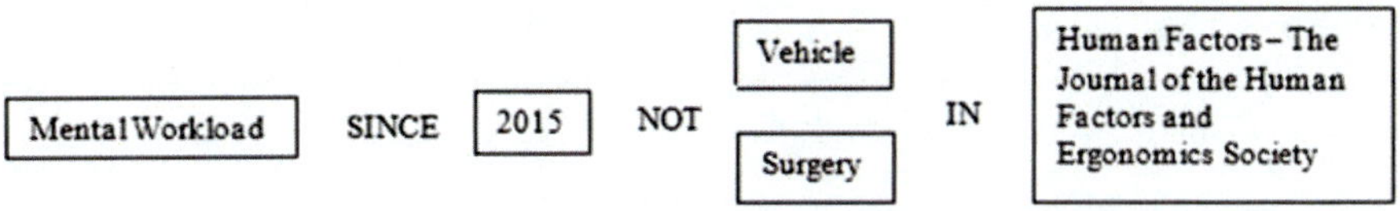

Fig. 3.1 Relationship between keywords and the logical operators for the SAGE database

with high mental workload were selected. We search in the following journals: Applied Ergonomics, Procedia Manufacturing, International Journal of Industrial Ergonomics, Accident Analysis & Prevention, Safety Science, International Journal of Human-Computer Studies, Neuroergonomics and Computer in Human Behavior excluding only IFAC-PapersOnLine and Transportation Research Part F: Traffic Psychology and Behavior, which reduced the number of articles to 308. Next, the 308 titles and their corresponding summaries were read, excluding those that did not seem to focus on the question "which are the variables to consider when measuring MWL, and how are these variables measured?" Eventually, only 56 articles were left to be included in the systematic review.

As for the Taylor & Francis database, 15,863 results were initially obtained using the term "Mental Workload." When excluding publications before 2015, the number lowered to 4,350. Then, the journals of Ergonomics and the International Journal of Occupational Safety and Ergonomics were selected because they focus on manufacturing or related processes and not medical, thus resulting in 247 publications being obtained, which were analyzed regarding the variables exposed before to end up with 27 articles significant to the research.

The SAGE database was searched through an advanced search using the term "mental workload"; 11893 results were found. When the search was refined to include only content published from 2015 to 2019 and to which the authors had full access, the number of results lowered to 3456, excluding the words "vehicle" and "surgery" in the journal "Human Factors—The Journal of the Human Factors and Ergonomics Society." The relationship in which these words and logical operators were used is shown in Fig. 3.1. The content of the remaining 55 articles was analyzed more thoroughly with a focus on the above-stated question, and in the end, only 10 articles remained.

3.3 Results

3.3.1 Summary of the Search in the Databases in English

A summary of the results obtained in the different phases of the systematic review is offered below.

In the first search, a total of 30564 articles were found. Table 3.1 shows the number obtained per database. During the second phase, all items that were not full

Table 3.1 Number of articles found by database in phase 1

Database	Number of articles
ScienceDirect	2808
Taylor & Francis	15863
SAGE Journal	11893
Total	30564

Table 3.2 Number of articles left by database after phase 2

Database	Number of articles
ScienceDirect	308
Taylor & Francis	247
SAGE Journal	55
Total	610

Table 3.3 Number of articles in the final selection

Database	Number of articles
ScienceDirect	57
Taylor & Francis	27
SAGE Journal	10
Total:	94

access articles, or book chapters, or were published before 2015 were eliminated. In addition, depending on the database, some journals were selected, while others were eliminated, after which keywords for further exclusion were used. Table 3.2 shows the results obtained after this process.

The final number of articles selected in each database is shown in Table 3.3.

3.3.2 Results of the Search by Keywords Related to Mental Workload Found in Spanish

When the words "Carga Mental," which is the Spanish term for "mental workload," were searched in Google Scholar, 2850 results were obtained. After filtering them by their title and publication date (2015 to present), the results decreased to 90. Finally, once those publications were analyzed concerning the question of variables and types of techniques related to MWL measurement, three thesis and four research articles remained.

3.3.3 Classifications of MWL Measurement Results

After this selection process, all one hundred and one items were reviewed in detail. As a result, they were classified in relation to their type of MWL measurement as shown in Fig. 3.2.

These articles feature a variety of different MWL measurements and techniques related to subjective measures. Their classification is shown in Table 3.4. In 41% of the studies, the NASA-TLX is used as a subjective measure for MWL; 15% use only subjective measurements; and from those, 53% use NASA-TLX alone. Although it can be concluded that the NASA Task Load Index is the most widely used measurement for subjective MWL in recent years, 35% of the studies use at least one subjective and one physiological method to determine MWL (see Table 3.5).

Eighty percent of the total studies found include at least one physiological measure (Table 3.6). Thirty-eight percent of the studies use only physiological methods to measure MWL. Fifty-two percent of the studies use an electrocardiogram in the subjects with measurements of heart rate (HR), heart rate variability (HRV), or the mean inter-beat-interval (IBI), this being the most widely used type of physiological methodology to find MWL. Finally, thirty-three of the studies used electroencephalography.

In relation to the analytical methodologies used to measure MWL (Table 3.7), of the eight studies found, five use already validated methodologies (IMPRINT, SAFTE, and ACT-R). The remaining three, on the other hand (Fang et al. 2015; Cai et al. 2016; Kostenko et al. 2016), propose new methods that require validation through further studies.

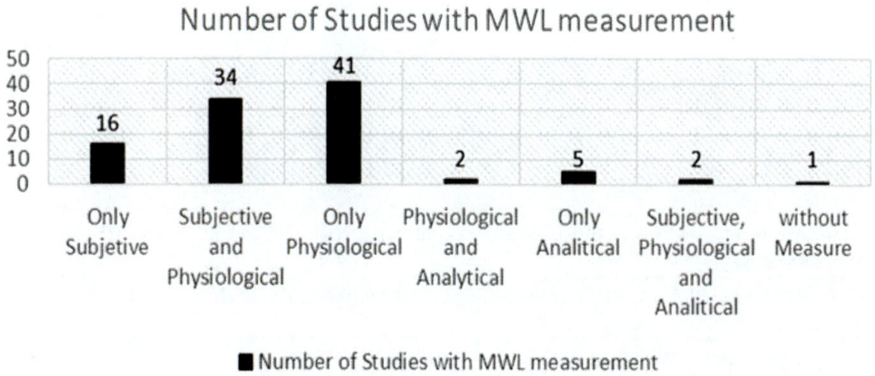

Fig. 3.2 Classification of studies from 2015 to June 2020 in relation to their type of MWL measurement

Table 3.4 Studies which employ subjective measures

Measure	Reference
NASA-Task load Index (NASA-TLX)	Chi et al. (2019), Dudek and Koniarek (2015), Jiménez et al. (2018), Bommer and Fendley (2018), Galy et al. (2018), Atalay et al. (2016), Mansikka et al. (2016), Yan et al. (2017), Alberdi et al. (2016), Claypoole et al. (2019), Ahmed et al. (2016), Banville et al. (2019), Boulhic et al. (2018), Cai et al. (2016), Darvishi and Melmanatabadi (2015), Causse et al. (2015), Fairclough and Ewing (2017), Fallahi et al. 2016, Faure and Benguigui (2016), Fernandes and Olivind Braarud (2015), Foy and Chapman (2018), García-Mas et al. (2016), Grassmann et al. (2017), Heine et al. (2017), Finkbeiner et al. (2016), Huggins and Claudio (2019), Jaquess et al. (2018), Mandrick et al. (2016), Marinescu et al. (2016), Morales et al. (2019), Mun et al. (2017), Ogawa et al. (2016), Orlandi and Brooks (2018), Shakouri et al. (2018), Shaw et al. (2018), Shuggi et al. (2017), Teo et al. (2015), Vera et al. (2017), Wang et al. (2015), Gore (2018), and Chen et al. (2019)
Workload Profile (WP)	Bommer and Fendley (2018)
Subjective mental effort rating (SRME)	Haji et al. (2015)
Modified Cooper Harper	Mansikka et al. (2016)
Standardized Copenhagen Psychosocial Questionnaire II (COPSOQ II)	Wixted et al. (2019)
Dundee Stress State Questionnaire (DSSQ)	Teo et al. (2015), Fairclough and Ewing (2017), and Claypoole et al. (2019)
Short State Stress Questionnaire (SSSQ)	Craig and Klein (2019) and Wixted et al. (2019)
Integrated Workload Scale (IWS)	Balfe et al. (2015)
State-Trait Anxiety Inventory (STAI)	Mandrick et al. (2016)
Instantaneous self-assessment of workload technique (ISA)	Marinescu et al. (2016) and Gore (2018)
Subjective Workload Assessment Technique (SWAT)	Dudek and Koniarek (2015)
Simplified Subjective Workload Assessment Technique (S-SWAT)	Fallahi et al. (2016)
Verbal online subjective opinion scale (VOSO	Marchitto et al. (2016)
Interview, questionnaire, assessment	Cruz Espinoza (2017) and Orsila et al. (2015)

Table 3.5 Studies which employ physiological measures and at least one subjective MWL measure

Measure	Reference
NASA-Task load Index (NASA-TLX)	Jiménez et al. (2018), Bommer and Fendley (2018), Mansikka et al. (2016), Yan et al. (2017), Ahmed et al. (2016), Banville et al. (2019), Cai et al. (2016), Causse et al. (2015), Fairclough and Ewing (2017), Fallahi et al. (2016), Faure and Benguigui (2016), Foy and Chapman (2018), García-Mas et al. (2016), Grassmann et al. (2017), Heine et al. (2017), Huggins and Claudio (2019), Jaquess et al. (2018), Mandrick et al. (2016), Marinescu et al. (2016), Morales et al. (2019), Mun et al. (2017), Ogawa et al. (2016), Orlandi and Brooks (2018), Shakouri et al. (2018), Shaw et al. (2018), Teo et al. (2015), Vera et al. (2017), Wang et al. (2015), Gore (2018), Chen et al. (2019)
Workload Profile (WP)	Bommer and Fendley (2018)
Modified Cooper Harper (MCH)	Mansikka et al. (2016)
Dundee Stress State Questionnaire (DSSQ)	Teo et al. (2015) and Fairclough and Ewing (2017)
Short State Stress Questionnaire (SSSQ)	Craig and Klein (2019)
State-Trait Anxiety Inventory (STAI)	Mandrick et al. (2016)
Instantaneous self-assessment of workload technique (ISA)	Marinescu et al. (2016) and Gore (2018)
Simplified Subjective Workload Assessment Technique (S-SWAT)	Fallahi et al. (2016)
Verbal Online Subjective Opinion scale (VOSO	Marchitto et al. (2016)
Interview, questionnaire, assessment	Orsila et al. (2015)

3.4 Solutions and Recommendations

Interest in mental workload evaluation has increased significantly in recent years. Companies are becoming aware of the diverse problems that result from dealing with high cognitive demands and significant MWL in the new manufacturing systems, which leads the researchers to recommend the inclusion of MWL in the design of the human-machine systems (job workstations) to help improve human performance and workers' well-being.

Table 3.6 Studies which employ physiological measures only

Measure	Reference
Electrocardiogram (HR, HRV, and/or IBI)	Yan et al. (2017), Young et al. (2015), Ahmed et al. (2016), Cai et al. (2016), Foy and Chapman (2018), Grassmann et al. (2017), Huggins and Claudio (2019), Mandrick et al. (2016), Marinescu et al. (2016), Orlandi and Brooks (2018), Mansikka et al. (2016), Morris et al. (2018), Orsila et al. (2015), Zhang et al. (2018b), Boele-Vos et al. (2017), Charles and Nixon (2019), Cui et al. (2016), Gore (2018), Hidalgo-Muñoz et al. (2018), Hsu et al. (2016), Omurtag et al. (2019), Heine et al. (2017), Luque-Casado et al. (2016), Vera et al. (2017), Li et al. (2019), Singh and Bharti (2015), Fallahi et al. (2016), Teo et al. (2015), and Huang et al. (2018)
Salivary cortisol	García-Mas et al. (2016)
Pupil dilation	Yan et al. (2017), Chen et al. (2019), Jiang et al. (2015), Peysakhovich et al. (2015), Omurtag et al. (2019), Marquart et al. (2015), Kim and Yang (2017), Ke et al. (2015), Bernhardt et al. (2019), Kearney et al. (2019), Cai et al. (2016), and Jiménez et al. (2018)
Eye-tracking	Teo et al. (2015), Fallahi et al. (2016), Charbonnier et al. (2016), and Charles and Nixon (2019)
Blink rhythm and duration	Yan et al. (2017), Faure and Benguigui (2016), Ogawa et al. (2016), Yung and Wells (2017), Charles and Nixon (2019), Chen et al. (2019), Marquart et al. (2015), and Ohtsuka et al. (2015)
Ocular fixation duration	Marquart et al. (2015) and Ohtsuka et al. (2015)
Intraocular pressure (IOP)	Vera et al. (2017) and Jiménez et al. ()2018Jiménez et al.
Electromyography (EMG)	Zhang et al. (2018c), Zadry et al. (2016), and Fallahi et al. (2016)
Electroencephalography (EEG)	Cui et al. (2016), Borghetti et al. (2017), Wang et al. (2016, 2017), Roy et al. (2019), Omurtag et al. (2019), Kumar and Kumar (2016), Kosti et al. (2018), Horat et al. (2016), Fairclough et al. (2019), Di Flumeri et al. (2019), Cui et al. (2016), Chen et al. (2016), Charles and Nixon (2019), Charbonnier et al. (2016), Causse et al. (2017), Bernhardt et al. (2019), Zadry et al. (2016), Liu et al. (2016), Teo et al. (2015), Shaw et al. (2018), Orlandi and Brooks (2018), Mun et al. (2017), Morales et al. (2019), Jaquess et al. (2018), Fallahi et al. (2016), Fairclough and Ewing (2017), Causse et al. (2015), Banville et al. (2019), Young et al. (2015), Mun et al. (2017), and Shaw et al. (2018)
Near-infrared reflectance spectroscopy (NIRS)	Craig and Klein (2019), Liu et al. (2016), and Foy and Chapman (2018)
Functional near-infrared spectroscopy (fNIRS)	Roy et al. (2019), Causse et al. (2017), Liu et al. (2016), Li et al. (2019), Teo et al. (2015), Mandrick et al. (2016), and Banville et al. (2019)
Respiration depth (RD)	Cai et al. (2016)

(continued)

Table 3.6 (continued)

Measure	Reference
Respiration rate (RR)	Omurtag et al. (2019), Charles and Nixon (2019), Cai et al. (2016), Foy and Chapman (2018), Grassmann et al. (2017), Huggins and Claudio (2019), Grassmann et al. (2015), and Charles and Nixon (2019)
Saccade rate (SR)	Ohtsuka et al. (2015), Marquart et al. (2015), Biswas and Prabhakar (2018), Cai et al. (2016), Kearney et al. (2019), Biswas and Prabhakar (2018), and Marquart et al. (2015)
Systolic (SBP) or diastolic (DBO) blood pressure	Klonowicz (2015) and Charles and Nixon (2019)
Facial thermography	Marinescu et al. (2016) and Murai et al. (2015)
Electrodermal activity (EDA)	Foy and Chapman (2018), Huggins and Claudio (2019), Li et al. (2019), Ghaderyan and Abbasi (2016), and Ghaderyan et al. (2018)

Table 3.7 Studies which employ analytical methodologies to measure MWL

Measure	Reference
Improved Performance Research Integration Tool (IMPRINT)	Bommer and Fendley (2018), Borghetti et al. (2017), and Rusnock and Borghetti (2018)
Model Sleep, Activity. Fatigue, and Task Effectiveness (SAFTE)	Peng et al. (2018)
Adaptive Control of Thought Rational (ACT-R)	Park et al. (2018)
Observational	Fang et al. (2015)
Model	Cai et al. (2016) and Kostenko et al. (2016)

3.4.1 Recommendations and Strategies for MWL Management

Some research mentions that MWL in organizations can be managed and diminished in diverse ways. Among them are work activities reorganization, workstation, and infrastructure design, physical conditions improvement (noise reduction, proper lighting, and temperature control). Additionally, Technologies of Information and communication (TIC's) usage and more recent, work breaks which can be passive or active, and of different lengths of time, been a very effective way to reduce MWL (Zhang et al. 2018a; Rupp et al. 2017).

3.5 Future Research Directions

So far, there is no evidence of a model of work breaks focusing on the reduction of MWL—one that establishes the time when pauses should be implemented during the workday, as well as their duration, or whether they should be active or passive pauses. The lack of the above leads to excessive mental load in different work activities, which affects the workers' performance and directly influences the company's sustainable production goals.

Moreover, the literature shows that various authors, such as Apud (2012), Duque López (2015), and Nadima (2014), have addressed the issue of how the implementation of active pause models has had an impact on the decrease in physical load. However, the interruption of work with pauses designed to decrease MWL is considered a particularly important field of study, which has not been dealt with in a significant way to this day.

3.6 Conclusions

The objective of this chapter has been accomplished since international businesses face significant challenges to remain competitive in a global market. Thus, sustainability has become relevant in all senses. Global initiatives have considered humanistic management, which involves the quest for human employee's well-being. This aspect is essential to achieve sustainable operations. Actual organizational and manufacturing processes imply high mental work demands that can be evaluated through the construct of mental workload. Nowadays, work environments are placing high cognitive loads on operators, which affect their performance by overusing the attentional resources given to the tasks and causing workers to experience mental fatigue. Lack of evaluation and proper management of MWL in the industry can result in errors that can create economic costs, accidents, injuries, or even fatal events. MWL assessment and management can be a human-focused strategy designed to improve and sustain the future of the organization. Industries must find competitive advantages in sustainable processes from the economic, environmental, and social view.

Acknowledgements The Tecnológico Nacional de México Campus Cd. Juárez, Tecnológico Nacional de México Campus Ciudad Cuauhtémoc and The Universidad Autónoma de Ciudad Juárez supported this research.

References

Ahmed S, Babski-Reeves K, DuBien J, Webb H, Strawderman L (2016) Fatigue differences between Asian and Western populations in prolonged mentally demanding work tasks. Int J Ind Ergon 54:103–112

Alberdi A, Aztiria A, Basarab A (2016) Towards an automatic early stress recognition system for office environments based on multimodal measurements: a review. J Biomed Inform 59:49–75

Allender L (2000). Modeling human performance: impacting system design, performance, and cost. In: Proceedings of the Military, Government and Aerospace Simulation Symposium, 2000 Advanced Simulation Technologies Conference. M. Chinni, Washington, DC, pp 139–144

Anderson J, Matessa M. (1998) The rational analysis of categorization and the ACT-R architecture. In: Oaksford M, Chater N (eds), Rational models of cognition. Oxford University Press, pp 197–217

Apud E (2012) Ergonomics in mining: the chilean experience. Hum Factors 54:901–907

Atalay KD, Can GF, Erdem SR, Muderrisoglu IH (2016) Assessment of mental workload and academic motivation in medical students. J Pak Med Assoc 66:574–578

Balfe N, Sharples S, Wilson JR (2015) Impact of automation: Measurement of performance, workload, and behavior in a complex control environment. Appl Ergon 47:52–64

Banville H, Parent M, Tremblay S, Falk TH (2019) Toward mental workload measurement using multimodal EEG-fNIRS monitoring. Elsevier. In: Neuroergonomics: 245–246. http://dx.doi.org/10.1016/B978-0-12-811926-6.00057-9

Bernhardt KA, Poltavski D, Petros T, Ferraro FR, Jorgenson T, Carlson C, Drechesel P, Iseminger C (2019) The effects of dynamic workload and experience on commercially available EEG cognitive state metrics in a high-fidelity air traffic control environment. Appl Ergon 77:83–91

Biswas P, Prabhakar G (2018) Detecting drivers' cognitive load from saccadic intrusion. Transp Res Part F 54:63–78

Boele-Vos MJ, Commandeur JF, Twisk DM (2017) Effect of physical effort on mental workload of cyclists in real traffic in relation to age and use of pedelecs. Accid Anal Prev 105:84–94

Bommer SC, Fendley M (2018) A theoretical framework for evaluating mental workload resources in human systems design for manufacturing operations. Int J Ind Ergon 63:7–17

Borghetti BJ, Giametta JJ, Rusnock CF (2017) Assessing continuous operator workload with a hybrid scaffolded neuroergonomic. Hum Factors 59:134–146

Boulhic L, Bignon A, Sillone F, Morineau T, Rechard J, Bouillon JF (2018) Effects of color codes used on marine supervision HMI on mental workload and information retrieval: experimentation with novices and experts. Int J Ind Ergon 67:180–191

Cai Z, Wu Q, Huang D, Ding L, Yu B, Law R, Huag J, Fu S (2016) Cognitive state recognition using wavelets singular entropy and ARMA entropy with AFPA optimized GP classification. Neurocomputing 197:29–44

Castle H, Leggatt H (2002) Instantaneous self-assessment (SA) validity & reliability. BAE Systems. Advanced Technology Centre, Internal Report, JS, 14865

Causse M, Chua Z, Peysakhovich V, Del Campo N, Matton N (2017) Mental workload and neural efficiency quantified in the prefrontal cortex using fNIRS. Sci Reports 7:5222. https://doi.org/10.1038/s41598-017-05378-x

Causse M, Fabre E, Glraudet L, Gonzalez M, Peysakhovich V (2015) EEG/ERP as a measure of mental workload in a simple piloting task. Procedia Manufacturing 3/6th International Conference on Applied Human Factors and Ergonomics (AHFE 2015) and the Affiliated Conferences 3:5230–5236

Charbonnier S, Roy RN, Bonnet S, Campagne A (2016) EEG index for control operators' mental fatigue monitoring using interactions between brain regions. Expert Syst Appl 52:91–98

Charles RL, Nixon J (2019) Measuring mental workload using physiological measures: a systematic review. Appl Ergon 74:221–232

Chen Y, Yan S, Tran CC (2019) Comprehensive evaluation method for user interface design in nuclear power plant based on mental workload. Nucl Eng Technol 51:453–462

Chi CF, Cheng CC, Shih YC, Sun IS, Chang TC (2019) Learning rate and subjective mental workload in five truck driving tasks. Ergonomics. https://doi.org/10.1080/00140139.2018.1545054

Cinaz B, Arnrich B, La Marca R (2013) Monitoring of mental workload levels during an everyday life office-work scenario. Pers Ubiquit Comput 17:229–239

Claypoole VL, Dever DA, Denues KL, Szalma JL (2019) The effects of event rate on a cognitive vigilance task. Hum Factors 61:440–450

Cooper GE, Harper RP (1969) The use of pilot rating in the evaluation of aircraft handling qualities. Anes Research Center, Moffett field

Cowan C, Girdner J, Majdoc B, Barrella E, Anderson R, Watson M (2018) Validating the use of B-Alert live electroencephalography in measuring cognitive load with the NASA task load index. ASEE southern section conference. American society for engineering education.

Craig CM, Klein MI (2019) The abbreviated vigilance task and it's attentional contributors. Hum Factors 61:426–439

Cruz Espinoza BS (2017) Model to determine the incidence of mental workload in the work performance of the workers of the CONALVISA company in the City of Riobamba, in 2016 [in Spanish]. Graduate studies Thesis. Riobamba, Ecuador: Chimborazo National University

Cui X, Zhang J, Wang R (2016) Identification of mental workload using imbalanced EEG data and DySMOTE-based neural network approach. IFAC-PapersOnLine 49(19):567–572

Darvishi E, Melmanatabadi M (2015) The rate of subjective mental workload and its correlation with musculoskeletal disorders in bank staff in Kurdistan, Iran. Procedia Manufacturing 3:37–42

Di Flumeri G, Borghini G, Arico P, Sciaraffa N, Lanzi P, Pozzi S, Vignali V, Lantieri C, Bichicchi A, Simone, A, Babilon (2019) EEG-based mental workload assessment during real driving: a taxonomic tool for neuroergonomics in highly automated environments. Elsevier. In Neuroergonomics, pp 121–126. https://doi.org/10.1016/B978-0-12-811926-6.00020-8

Dudek B, Koniarek J (2015, January 8) The subjective rating scales for measurement of mental workload-thurstonian scaling. Retrieved from International Journal of Occupational Safety and Ergonomics. https://www.tandfonline.com/doi/abs/10.1080/10803548.1995.11076308

Duque Lopez V (2015) Las pausas activas como estrategia para el control de la fatiga. QUITO: UNIVERSIDAD CENTRAL DEL ECUADOR

Eastwood Gruginski B, Amarai Gontijo L, Merino E (2015) Frequency of application of mental workload measuring instruments in recent publications in Brazil. Procedia Manuf 3:5134–5138

Estes S (2015) The workload curve: subjective mental workload. Hum Factors 57(7):1174–1187. https://doi.org/10.1177/0018720815592752

Fairclough S, Ewing K (2017) The effect of task demand and incentive on neurophysiological and cardiovascular markers of effort. Int J Psychophysiol 119:58–66

Fairclough, S, Ewing, K, Burns, C, Kreplin, U (2019) Neural efficiency and mental workload: locating the red line. Elsevier. In Neuroergonomics, pp 73–77. https://doi.org/10.1016/B978-0-12-811926-6.00012-9

Fallahi M, Motamedzade M, Heidarimoghadam R, Soltanian AR, Miyake S (2016) Effects of mental workload on physiological and subjective responses during traffic density monitoring: a field study. Appl Ergono 52:95–103

Fang W, Liu Y, Guo B, Zhang Y (2015) OCC Controller workload evaluation model and application. Procedia Manufacturing 3246–3253

Faure V, Benguigui N (2016) The effects of driving environment complexity and dual tasking on drivers' mental workload and eye blink behavior. Transp Res Part F: Traffic Psychol Behav 40:78–90

Fernandes A, Olivind Braarud P (2015) Exploring measures of workload, situation awareness, and task performance in the Main Control Room. Procedia Manuf 3:1281–1288

Finkbeiner KM, Russell PN, Helton WS (2016) Rest improves performance, nature improves happiness: assessment of break periods on the abbreviated vigilance task. Conscious Cogn 42:277–285

Foy HJ, Chapman P (2018) Mental workload is reflected in driver behavior, physiology, eye movements and prefrontal cortex activation. Appl Ergono 73:90–99

Galy E (2018) Consideration of several mental workload categories: perspectives for elaboration of new ergonomic recommendations concerning shiftwork. Theor Issues Ergon Sci 19(4):483–497

Galy E, Paxion J, Berthelon C (2018) Measuring mental workload with the NASA-TLX needs to examine each dimension rather than relying on the global score: an example with driving. Ergonomics 61(4):517–527

García-Mas A, Ortega E, Ponseti J, de Teresa C, Cardenas O (2016) Workload and cortisol levels in helicopter combat pilots during simulated flights. Revista Andal Med Deporte 9(1):7–11

Ghaderyan P, Abbasi A (2016) An efficient automatic workload estimation method based on electrodermal activity using pattern classifier combinations. Int J Psychophysiol 110:91–101

Ghaderyan P, Abbasi A, Ebrahimi A (2018) Time-varying singular value decomposition analysis of electrodermal activity: a novel method of cognitive load estimation. Measurement 126:102–109

Gil-Monte P (2012) Riesgos psicosociales en el trabajo y salud ocupacional. Rev Peru Med Exp Salud Publica 237–241

Gore BF (2018) Chapter 3 - Workload and fatigue. In: Sgobba T, Kanki B, Clervoy J-F, Sandal GM (eds) Space safety and human performance. Butterworth-Heinemann, pp 53–85. https://doi.org/10.1016/B978-0-08-101869-9.00003-0

Grassmann M, Vlemincx E, von Leupoldt A, Van den Bergh O (2015) The role of respiratory measures to assess mental load in pilot selection. Ergonomics 59(6):745–756

Grassmann M, Vlemincx E, von Leupoldt A, Van den Bergh O (2017) Individual differences in Cardiorespiratory measures of mental workload: an investigation of negative affectivity and cognitive avoidant coping in pilot candidates. Appl Ergon 59:274–282

Haji FA, Rojas D, Childs R, de Ribaupierre S, Dubrowski A (2015) Measuring cognitive load: performance, mental effort, and simulation task complexity. Med Educ 49:815–827

Hart SG, Staveland LE (1988) Development of NASA-TLX (Task Load Index): results of empirical and theoretical research. In Hancock PA, Meshkati N (eds), Human mental workload. Advances in psychology, pp 139–183

Haslam R, Waterson P (2013) Ergonomics and sustainability. Ergonomics 56(3):343–347

Heine T, Lenis G, Reichensperger P, Beran T, Doessel O, Deml B (2017) Electrocardiographic features for the measurement of drivers' mental workload. Appl Ergon 61:31–43

Helton WS (2004) Validation of a short stress state questionnaire. Proceedings of the Human Factors and Ergonomics Society Annual Meeting, pp 1238–1242

Hidalgo-Muñoz AR, Mouratille D, Matton N, Causse M, Rouillard Y, El-Yagoubi R (2018) Cardiovascular correlates of emotional state, cognitive workload, and time-on-task effect during a realistic flight simulation. Int J Psychophysiol 128:62–69

Horat SK, Hermann FR, Favre G, Terzis J, Debatisse D et al (2016) Assessment of mental workload: a new electrophysiological method based on intra-block averaging of ERP amplitudes. Neuropsychologia 82:11–17

Hsu FW, Chiuhsiang JL, Lee YH, Chen HJ (2016) Effects of elevation change on mental stress in high voltage transmission tower construction workers. Appl Ergono 56:101–107

Huang S, Li J, Zhang P, Zhang W (2018) Detection of mental fatigue state with wearable ECG devices. Int J Med Informatics 119:39–46

Huggins A, Claudio D (2019) A mental workload-based patient scheduling model for a cancer clinic. Oper Res Health Core 20:56–65

IEA (2019) Definition and domains of rrgonomics, December 17. Retrieved from International Ergonomics Association. https://www.iea.cc/whats/index.html

Jaquess KJ, Lo LC, Oh H, Lu C, Ginsberg A, Tan YY, Gentili RJ (2018) Changes in mental workload and motor performance throughout multiple practice sessions under various levels of task difficulty. Neuroscience 393:305–318

Jiang X, Zheng B, Bednarik R, Atkins MS (2015) Pupil responses to continuous aiming movements. Int J Human Comp Stud 83:1–11

Jiménez R, Cárdenas D, González-Anera R, Jiménez JR, Vera J (2018) Measuring mental workload: ocular astigmatism aberration as a novel objective index. Ergonomics 61(4):506–516

Ke Y, Qi H, Zhang L, Chen S, Jiao X, Zhou P, Zhao X, Wan B, Ming D (2015) Towards an effective cross-task mental workload recognition model using electroencephalography based on feature selection and support vector machine regression. Int J Psychophysiol 98:157–166

Kearney P, Li WC, Yu CS, Braithwaite G (2019) The impact of alerting designs on air traffic controller's eye movement patterns and situation awareness. Ergonomics 62(2):305–318

Kim JH, Yang X (2017) Applying fractal analysis to pupil dilation for measuring complexity in a process monitoring task. Applied Ergonomics, pp 61–69

Klonowicz T (2015) Mental workload and health: a latent threat. Int J Occup Saf Ergono 1(2):130–135

Kosti MV, Georgiadis K, Adamos DA, Laskaris N, Spinellis D, Angelis L (2018) Towards an affordable brain-computer interface for the assessment of programmers' mental workload. Int J Hum Comput Stud 115:52–66

Kostenko A, Rauffet P, Chauvin C, Coppin G (2016) A dynamic closed-looped and multidimensional model for mental workload evaluation. IFAC-PapersOnLine 49(19): 549–554

Kujawinska A, Vogt K, Wachowiak F (2015) Ergonomics as a significant factor of sustainable production. In: Golinska P (ed), Technology management for sustainable production and logistics. Springer, Berlin, pp 193–203. https://doi.org/10.1007/978-3-642-33935-6_10

Kumar N, Kumar J (2016) Measurement of cognitive load in HCI systems using EEG power spectrum: an experimental study. Procedia Comput Sci 84:70–78

Li LP, Liu ZG, Zhu HY, Zhu L, Huang YC (2019) Functional near-infrared spectroscopy in the evaluation of urban rail transit drivers' mental workload under simulated driving conditions. Ergonomics. https://doi.org/10.1080/00140139.2018.1535093

Liu T, Pelowski M, Pang C, Zho Y, Cai J (2016) Near-Infrared spectroscopy as a tool for driving research. Ergonomics 59(3):368–379

Luque-Casado A, Perales JC, Cárdenas D, Sanabria D (2016) Heart rate variability and cognitive processing: The autonomic response to task demands. Biol Psychol 113:83–90

Luximon A, Goonetilleke RS (2001) Simplified subjective workload assessment technique. Ergonomics 44(3):229–243

Mandrick K, Peysakhovich V, Rémy F, Lepron E, Causse M (2016) Neural and psychophysiological correlates of human performance under stress and high mental workload. Biol Psychol 121:62–73

Mansikka H, Simola P, Virtanen K, Harris D, Oksama L (2016) Fighter pilots' heart rate, heart rate variation and performance during instrument approaches. Ergonomics 59(10):1344–1352

Marchitto M, Benedetto S, Baccino T, Cañas JJ (2016) Air traffic control: Ocular metrics reflect cognitive complexity. Int J Ind Ergon 54:120–130

Marinescu AC, Sharples S, Ritchie C, Sánchez López T, McDowell M, Morvan H (2018) Physiological parameter response to variation of mental workload. Hum Factors 60:31–56

Marinescu A, Sharples S, Ritchie AC, Sánchez López T, McDowell M, Morvan H (2016) Exploring the relationship between mental workload, variation in performance and physiological parameters. IFAC-PapersOnLine 49(19):591–596

Marquart G, Cabrall C, de Winter J (2015) Review of eye-related measures of drivers' mental workload. Procedia Manuf 3:2854–2861

Matthews G, Joyner L, Gilliland K, Campbell S, Falconer S, Huggins J (1999) Validation of a comprehensive stress state questionnaire: towards a state 'big three'? In: Mervielde IDI (ed), Personality psychology in Europe 7:335–350

Miller S (2001) Literature review workload measures. National Advanced Driving Simulator, Iowa. http://www.nads-sc.uiowa.edu/publicationStorage/200501251347060.N01-006.pdf. Accessed 27 Oct 2019

Morales JM, Ruiz-Rabelo JF, Diaz-Piedra C, Din Stasi LL (2019) Detecting mental workload in surgical teams using a wearable single-channel electroencephalographic device. J Surg Educ. https://doi.org/10.1016/j.jsurg.2019.01.005

Morris CE, Winchester LJ, Jackson AJ, Tomes AS, Neal WA, Wilcoxen D, Chander H, Arnett SW (2018) Effect of a simulated tactical occupation task on physiological strain index, stress, and inflammation. Int J Occup Saf Ergon. https://doi.org/10.1080/10803548.2018.1482053

Mun A, Whang M, Park MC (2017) Effects of mental workload on involuntary attention: a somatosensory ERP study. Neuropsychologic, pp 7–20

Murai K, Kitamura K, Hayashi Y (2015) Study of a port coordinator's mental workload based on facial temperature. Procedia Comput Sci 60:1668–1675

Nadima H (2014) CD de Monografias. Regimenes de Trabajo y Descanso. Matanzas, Cuba: Universidad de Matanzas "Camilo Cienfuegos"

Nieto J (2014) Occupational diseases, a pandemic that requires prevention [in Spanish]. Medicina y Seguridad del Trabajo 60(234):1–3

Nygren TE (1991) Psychometric properties of subjective workload measurement techniques: implications for their use in the assessment of perceived mental workload. Hum Factors 33(1):17–33

Ogawa T, Takahashi M, Kawashima R (2016) Human cognitive control mode estimation using JINS MEME. IFAC-PapersOnLine 49(19):331–336

Ohtsuka R, Wang J, Chihara T, Yamanaka K, Morishima K, Daimoto H (2015) Estimation of mental workload during motorcycle operation. Procedia Manuf 3:5313–5318

Omurtag A, Roy R, Dehais F, Chatty L, Garbey M (2019) Tracking mental workload by multimodal measurements in the operating room. Elsevier. In Neuroergonomics, pp 99–103. https://doi.org/10.1016/B978-0-12-811926-6.00016-6

Orlandi L, Brooks B (2018) Measuring mental workload and physiological reactions in marine pilots: Building bridges towards redlines of performance. Appl Ergon 69:74–92

Orsila R, Virtanen M, Luukkaala T, Tarvainen M, Karjalainen P, Viik J, Savinainen M, Nygard C (2015) Perceived mental stress and reactions in heart rate variability - a pilot study among employees of an electronics company. Int J Occup Saf Ergon 14(3):275–283

Paas FG (1992) Training strategies for attaining transfer of problem-solving skills in statistics: a cognitive load approach. J Educ Psychol 64:429–434

Park S, Jeong S, Myung R (2018) Modeling of multiple sources of workload and time pressure effect with ACT-R. Int J Ind Ergon 63:37–48

Pejtersen JH, Kristensen TS, Borg V, Bjorner JB (2010) The second version of the Copenhagen psychosocial questionnaire. Scandinavian J Public Health 38:8–24

Peng HT, Bouak F, Wang W, Chow R, Vartanian O (2018) An improved model to predict performance under mental fatigue. Ergonomics 61(7):988–1003

Peysakhovich V, Causse M, Scannella S, Dehais F (2015) Frequency analysis of a task-evoked pupillary response: luminance-independent measure of mental effort. Int J Psychophysiol 97:30–37

Pickup L, Wilson JR, Norris BJ, Mitchell L, Morrisroe G (2005) The Integrated Workload Scale (IWS): a new self-report tool to assess railway signaller workload. Appl Ergon 36:681–693

Radjiyev A, Qiu H, Xiong S, Nam K (2015) Ergonomics and sustainable development in the past two decades (1992-2011): research trends and how ergonomics can contribute to sustainable development. Appl Ergon 46:67–75

Reid GB, Nygren TE (1988) The Subjective workload assessment technique: a scaling procedure for measuring mental workload. Adv Psychol 53:185–218

Rodriguez Erhar R (2006) Mental load assessment of workstation workers in computing with alternative natural and artificial lighting. Hum Environment and Housing Laboratory CRICYT, Mendoza, Colombia

Roy RN, Moly A, Dehais F, Scannella S (2019) EEG and fNIRS connectivity features for mental workload assessment: a preliminary study. Elsevier In Neuroergonomics, pp 327–328. http://dx.doi.org/10.1016/B978-0-12-811926-6.00098-1

Rupp MA, Sweetman R, Sosa AE, Smither JA, McConnell DS (2017) Searching for affective and cognitive restoration: examining the restorative effects of casual video game play. Hum Factors 59:1096–1107

Rusnock C, Borghetti BJ (2018) Workload profiles: a continuous measure of mental workload. Int J Ind Ergon 63:49–64

Shakouri M, Ikuma LH, Aghazadeh F, Nahmens I (2018) Analysis of the sensitivity of heart rate variability and subjective workload measures in a driving simulator: the case of the highway work zone. Int J Ind Ergon 66:136–145

Shaw EP, Rietschel JC, Hendershot BD, Pruziner AL, Miller MW, Hatfield B, Gentili R (2018) Measurement of attentional reserve and mental effort for cognitive workload assessment under various task demands during dual-task walking. Int J Ind Ergon 134:136–145

Shuggi IM, Oh H, Shewokis PA, Gentili RJ (2017) Mental workload and motor performance dynamics during practice of reaching movements under various levels of task difficulty. Neuroscience 360:166–179

Singh B, Bharti N (2015) Software tools for heart rate variability analysis. Int J Recent Sci Res 6(4):3501–3506

Spielberger CD, Gonzalez-Reigosa F, Martinez-Urrutia A, Natalicio LF, Natalicio DS (1971) Development of the Spanish edition of the State-Trait anxiety inventory. Interam J Psycholo 5:145–158

Spielberger CD, Gorsuch RL, Lushene RE (1970) The state-trait anxiety inventory (test manual). Consulting Psychologists Press, Palo Alto, CA

Stanley S (2020) NAPS, April 30. Retrieved from Manufacturing in Mexico. https://napsintl.com/manufacturing-in-mexico/

Teo G, Reinerman-Jones L, Matthews G, Szalma J (2015) Comparison of measures used to assess the workload of monitoring an unmanned system in a simulation mission. Procedia Manuf 3:1006–1013

Thatcher A, Waterson P, Todd A, Moray N (2018) State of science: ergonomics and global issues. Ergonomics 61(2):197–213

Tsang PS, Velazquez V (1996) Diagnosticity and multidimensional subjective workload ratings. Ergonomics 39(3):358–381

United Nations (1987) Report of the World Commission on environment and development: our common future. United Nations, New York, NY

US Army Research Lab (2019). Ready by fatigue science, December 2. Retrieved from https://www.fatiguescience.com/sleep-science-technology/

Vera J, Jiménez R, García JA, Cárdenas D (2017) Intraocular pressure is sensitive to cumulative and instantaneous mental workload. Appl Ergon 60:313–319

Wang D, Chen J, Zhao D, Zheng C, Wu X (2017) Monitoring workers' attention and vigilance in construction activities through a wireless and wearable electroencephalography system. Autom Constr 82:122–137

Wang J, Ohtsuka R, Yamanaka K, Shioda K, Kawakarni M (2015) Relation between mental workload and visual information processing. Procedia Manuf 3:5308–5312

Wang Y, Zhang J, Wang R (2016) Mental workload recognition by combining wavelet packet transform and kernel spectral regression techniques. IFAC-PapersOnLine 49(19):561–566

Wierwille WW, Casali JC (1983) A validated rating scale for global mental workload measurement applications. Proc Hum Factors Soc 27:129–133

Wierwille WW, Eggemeier FT (1993) Recommendations for mental workload measurement in a test and evaluation environment. Hum Factors 35(2):263–281

Wixted F, Shewlin M, O'Sullivan LW (2019) Distress and worry as mediators in the relationship between psychosocial risks and upper body musculoskeletal complaints in highly automated manufacturing. Ergonomics 61(8):1079–1093

Yan S, Tran CC, Wei Y, Habiyaremye JL (2017) Driver's mental workload prediction model bases on physiological indices. Int J Occup Saf Ergon. https://doi.org/10.1080/10803548.2017.1368951

Young M, Brookhuls K, Wickens C, Hancock P (2015) State of science: mental workload in ergonomics. Ergonomics 58(1):1–17

Yung M, Wells R (2017) Responsive upper limb and cognitive fatigue measures during light precision work: an 8-hour simulated micro-pipetting study. Ergonomics 60(7):940–956

Zadry HR, Dawal SZ, Taha Z (2016) Development of statistical models for predicting muscle and mental activities during repetitive precision tasks. Int J Occup Saf Ergon (JOSE) 22(3):374–383

Zhang C (2018) Work and Non-Work Activities in Replenishing Workday Energy: Meetings, Individual work, and Micro Breaks. Michigan: University of Michigan

Zhang N, Fard M, Bhulyan MH, Verhagen D, Azari MF, Robinson SR (2018a) The effects of physical vibration on heart rate variability as a measure of drowsiness. Ergonomics 61(9):1259–1272

Zhang Y, Wang W, Chu Y, Yuan X (2018b) Real-time and user-independent feature classification of forearm using EMG signals. J Soc Inf Display 27:101–108

Chapter 4
Ergonomic Control Panel (ECP): A Proposed Comprehensive Ergonomics Evaluation Tool Using Multi-task Evaluation Models

Murray Gibson, Bob Sesek, and Anjaneya Bandekar

Abstract The authors propose a dynamic simulation model useful for determining the degree to which ergonomic exposure metrics are impacted by changes in various operational decisions such as scheduled overtime, cycle time, and make/model mix. Conventional ergonomics analyses and risk assessment approaches are incapable of rapidly assessing the impact of these types of operational decisions on resultant ergonomic exposure metrics. It is common practice to treat ergonomic exposure metrics as static/constant, when in actuality they change in response to variations in a multitude of operational parameters. The proposed simulation model extends the capabilities of current multi-task ergonomics assessment tools to account for this variation. State-of-the-art multi-task evaluation models such as RCRA (Recommended Cumulative Recovery Allowance) and Fatigue Failure theory-based models can be utilized to recommend appropriate job rotation schedules and modifications in shift duration to limit exposure duration, increase recovery time, and protect employees.

Keywords Ergonomic simulation · Ergonomic evaluation · Multi-task evaluation · RCRA · Fatigue Failure

4.1 Introduction

In the industrial ergonomics profession, the word simulation is typically associated with the use of Digital Human Modeling (DHM) in the initial design phase (Baines and Kay 2002; Paquet and Lin 2003; Baines et al. 2004; Chaffin 2005; Case and Piamonte 2007; Kumar et al. 2013; Eldar and Fisher-Gewirtzman 2019) with modeling software such as Jack™ (Badler et al. 1999) and Santos™ to efficiently gather realistic motion data by primarily simulating human motions (Jung et al. 1995; Ma et al. 2015). Other DHM simulation efforts include assessing workstations (Chaffin 2005; Chang and Wang 2007; Örtengren et al. 2009), spatial conditions and facility layouts (Broberg et al. 2011; Andersen and Broberg 2015), lifting motions

M. Gibson (✉) · B. Sesek · A. Bandekar
Department of Industrial & Systems Engineering, Auburn University, Auburn, AL, USA
e-mail: murray@saturnergonomics.com

A. Realyvásquez Vargas et al. (eds.), *New Perspectives on Applied Industrial Ergonomics*,
https://doi.org/10.1007/978-3-030-73468-8_4

(Song et al. 2016), vision tracking and analysis (Karmakar et al. 2012; Kim et al. 2018), and cognitive workload (Mayer et al. 2012).

Less widespread is the use of simulation as a tool to assess the impact that proposed future changes to operational parameters (working overtime, cycle time, make/model mix, etc.) might have on MSD (musculoskeletal disorder) risk and muscle fatigue metrics (Rychtyckyj and Stephens 2009). There is great value in a visual ergonomic assessment tool that not only analyzes physical risk factors such as postures, loads and repetition, and provides suitable recommendations, but also seamlessly integrates into production planning as well as other aspects of manufacturing (Schaub et al. 2012). An assessment tool should be dynamic and adaptable as a common problem is recurrence of ergonomic concerns with process changes, which can be expensive to abate (Drinkaus et al. 2005; Hendrick 2008).

The primary goal of ergonomics is to create working conditions that improve the safety, well-being, and performance of employees (Kroemer and Kroemer 1994). With that consideration, the authors present a novel approach where a control panel enables even the non-expert to quickly and accurately assess the impact that changing operational parameters can have on resultant ergonomics exposure. The authors have applied this control panel concept utilizing multi-task ergonomic evaluation models such as Fatigue Failure (Gallagher and Schall 2017) and RCRA (Recommended Cumulative Recovery Allowance) (Gibson and Potvin 2016). These multi-task models utilize continuous equations and leverage common input parameters, enabling changes in operational variables to cascade down into model output metrics, allowing the practitioner to observe the impact by line, job/station, and even task/element.

4.2 Single-Task Model Limitations

Ergonomics evaluation methodologies provide an assessment value or rating of the ergonomic risk factors of force, posture, and repetition/duration. Traditional methods assume static or non-changing conditions for operational parameters such as cycle time or line speed, make/model mix, shift duration, and employee rotation path/sequence. Generally, ergonomics risk assessment and evaluation tools are incapable of efficiently assessing the impact of changes to such parameters. Until recently, ergonomic physical exposure evaluation models consisted almost entirely of mono-task models such as Rapid Upper Limb Assessment (RULA) (Mcatamney and Corlett 1993), the Strain Index (SI) (Moore and Garg 1995), and American Conference of Governmental Industrial Hygienists (ACGIH) Hand Activity Level (HAL) Threshold Limit Value (TLV). These tools were created with the intention to primarily analyze single-task jobs and they basically focus analysis on a "snapshot in time." Single-task models assume a constant exposure level for force, posture, and frequency/duration. Figure 4.1 Singe-task Exposure depicts such a constant exposure level, where level and duration of Exposures 1–4 are invariant. A problem from an applications perspective is that very few jobs are actually single task. Most jobs

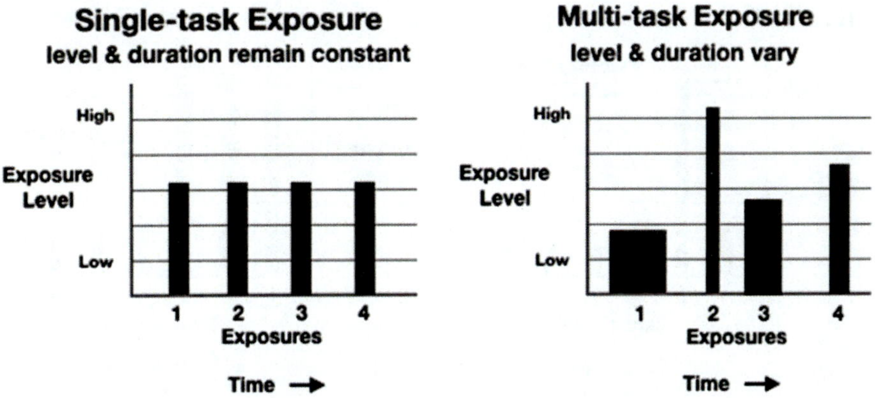

Fig. 4.1 Single-task and Multi-task Exposure. Single-task exposure graphic (upper-left) depicts an assumed constant exposure level, where level and duration of Exposures 1–4 are invariant. The multi-task exposure level graphic (upper-right) depicts a more common exposure scenario, where level and duration of Exposures 1–4 vary

evaluated by ergonomists are complex and are comprised of varying levels of force, posture, and repetition/duration. Figure 4.1 Multi-task Exposure depicts a more realistic scenario where level and duration of Exposures 1–4 vary in both intensity and duration.

Traditional single-task evaluation models fail to address variability in exposure levels. Single-task evaluation models applied to variable task (multi-task) jobs produce errors in the estimation of the risk of these jobs. These errors either overstate or understate the risk of injury. If an average load is used for a variable task, the risk is often understated. Figure 4.2 Average Exposure Level depicts such an "average" exposure level. Using the average risk level often understates the risk because higher force exertion levels have been demonstrated to have a non-linear, disproportionate influence upon acceptable levels of discomfort and muscle fatigue (Gibson and Potvin 2016), and to have greater influence upon tissue damage and subsequent injury risk levels (Gallagher and Schall 2017). In short, higher exposure levels carry disproportionally more weight. Therefore, if the average level/load is used to evaluate a variable task, the injury risk will likely be understated. Inspecting Fig. 4.2 Average Exposure Level, the dotted line for average (the exposure level used in our hypothetical single-task model) is well beneath the level of Exposures 1–3; therefore, our resultant metric will understate the injury risk.

On the other hand, using the peak or worst-case load is likely to overstate the injury risk, particularly when that peak force is relatively infrequent. Figure 4.2 Worst Case Exposure Level depicts the "worst-case" level, Exposure 2 in this example. Using Exposure 2 as the model input would overstate the risk in our hypothetical single-task model because the other exposures, Exposure 1 and Exposures 3–6, are significantly lower than the worst case; but our calculations would treat them as equivalent. The more complex the multi-task job, the greater the potential error in the calculated

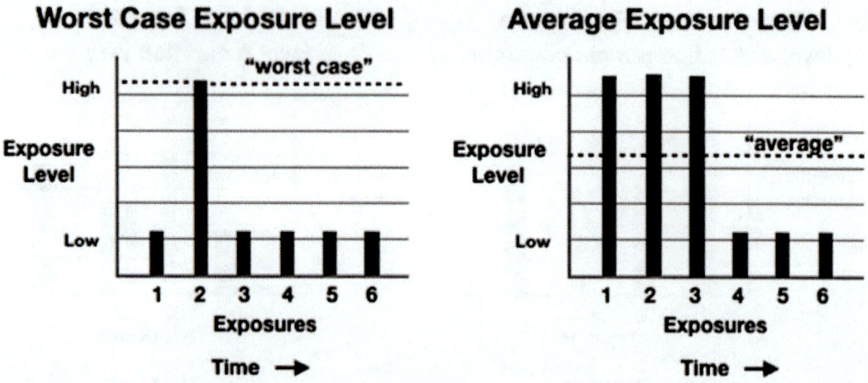

Fig. 4.2 Worst case and average exposure level. Worst case exposure level (upper-left) depicts using the highest exposure level, Exposure 2 (the "worst case") as the model input value. Average Exposure Level (upper-right) depicts using the "average" exposure level (i.e., statistical average exposure level of Exposures 1–6) as the model input value. Neither approach is representative of the true variation in exposure level, nor the resultant risk

evaluation or injury risk metric. Evaluating changes to various elements of a multi-task job using a single-task evaluation model may cause little or no change to the resultant metrics or cause drastic change; but rarely will the model results reflect the true change to the risk.

This error in risk calculations using single-task models can be quite nuanced and often unpredictable, leading to confusion when reevaluating jobs after changes to sub-tasks of the jobs. Whether the changes are abatements intended to reduce risk or increases in exposure (such as item weight increases or required force level increases), single-task evaluation tools can provide risk assessment metrics that lead to wrong conclusions about changes to risk. For example, in Fig. 4.2 Worst Case Exposure Level, increasing the level of Exposures 1 and/or Exposures 3–6 would not change the resultant injury risk metric, as long as the increase did not result in the level of these exposures being greater than that of Exposure 2. This is because we used the worst-case exposure, Exposure 2 (the "worst-case"), in our calculations. These changes would necessarily increase the risk of the job, but the risk metric would not change at all! On the other hand, in Fig. 4.2 Average Exposure Level, something unexpected occurs if we eliminate Exposures 5–6. Removing these exposures would reduce the injury risk because we removed physical work content, but the opposite would likely be predicted by the single-task evaluation tool! The resultant risk metric would likely increase, because eliminating Exposures 5–6 increases the "average" exposure, thereby increasing the resultant risk metric.

Assessment obsolescence is another problem with single-task ergonomics evaluation and risk assessment tools. It can be time-consuming to update pen and paper evaluations and complete the subsequent metrics' calculations. Updating digitized versions of traditional single-task models is not much of an improvement, for the

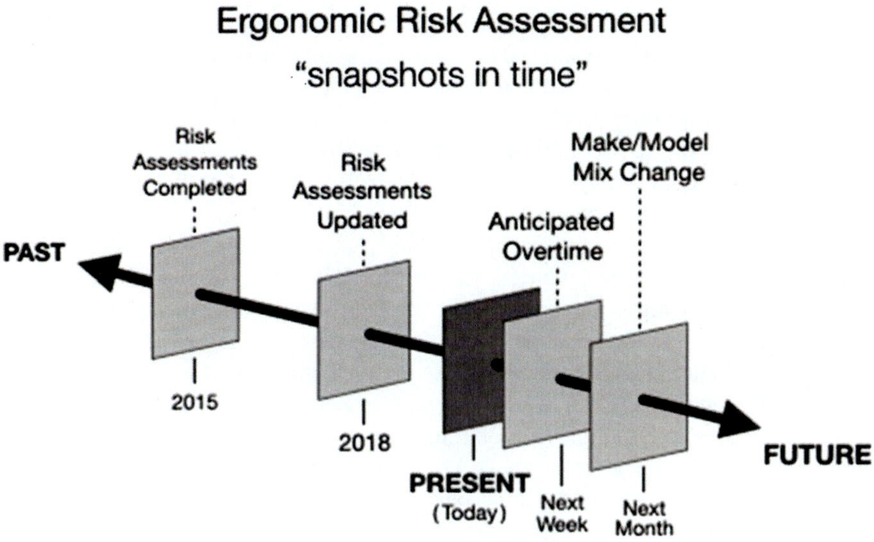

Fig. 4.3 Ergonomic risk assessment ("snapshots in time"). This graphic depicts how traditional ergonomic risk assessment is a static snapshot in time, with risk assessments completed or updated every 2–3 years. These risk assessments can quickly become obsolete and can't be applied in a forward-looking manner to assess impact resulting from anticipated near-term/future changes to operational parameters such as overtime (anticipated to occur one week into the future in the graphic) or changes to make/model mix (anticipated to occur one month into the future in the graphic)

reasons stated above. Assessing the impact of a change to an operational parameter—such as reducing cycle time from 60 s to 55 s—may require updating multiple individual job/station evaluations. This is time-consuming and expensive. Due to these inefficiencies, it is common practice for organizations to only update their ergonomic evaluations or risk assessments every 2–3 years. Refer to Fig. 4.3, Ergonomic Risk Assessment ("snapshots in time"). Due to the difficulty in keeping evaluation and risk assessment metrics current, the tendency is for the metrics to become inaccurate or even obsolete over time. Even if metrics are updated as changes occur (e.g., introduction of a heavier tool, rebalancing of a line, etc.), the metrics nevertheless only represent a snapshot in time. At best, the metrics remain current, with little capability to address proposed future changes.

Assessment obsolescence is another problem with single-task ergonomics evaluation and risk assessment tools. It can be time-consuming to update pen and paper evaluations and complete the subsequent metrics' calculations. Updating digitized versions of traditional single-task models is not much of an improvement, for the reasons stated above. Assessing the impact of a change to an operational parameter—such as reducing cycle time from 60 s to 55 s—may require updating multiple individual job/station evaluations. This is time-consuming and expensive. Due to these inefficiencies, it is common practice for organizations to only update their ergonomic evaluations or risk assessments every 2–3 years. Refer to Fig. 4.3, Ergonomic Risk Assessment ("snapshots in time"). Due to the difficulty in keeping evaluation and risk assessment metrics current, the tendency is for the metrics to become inaccurate or even obsolete over time. Even if metrics are updated as changes occur (e.g., introduction of a heavier tool, rebalancing of a line, etc.), the metrics nevertheless only represent a snapshot in time. At best, the metrics remain current, with little capability to address proposed future changes.

With older single-task evaluation models and risk assessment tools, there is often little incentive to update the metrics. Risk assessment tools commonly have stair-stepped risk levels: high, moderate, and low. These stair-stepped risk approaches often mask the changes to actual risk for both abatements and increases in exposure. In Lean Manufacturing environments, it is common to implement small, incremental improvements. But these small improvements may not be reflected in the risk assessment metric/rating (e.g., a high-risk job remains high-risk after the solution has been implemented).

Even relatively large risk reductions may not appear beneficial if the risk level does not actually drop from one stair-step risk level to the next lower level (i.e., drop from high risk to moderate risk, or moderate risk to low risk). And on occasion a solution resulting in only a very small reduction appears quite beneficial if it lowers the risk level from the current level to the next lower level. Conversely, proposed changes that would increase exposures significantly may not raise any flags of concern if the risk level is not increased to the next higher level (e.g., from moderate to high-risk). And single-task model inputs are often either the average or worst-case exposure level; thus, updates may reflect little, if any, changes to metric, or, as previously discussed, reflect erroneous changes. And due to the inefficiencies of single-task models, it is often not practical to update risk and evaluation metrics in anticipation of future changes (i.e., changes next week, next day, next shift, or next hour). A truly predictive analytics approach requires the ability to look forward, predicting the impact of proposed changes before they are made. In a climate where ergonomics practitioners struggle to keep risk assessment and evaluation metrics current, this type of predictive analytics approach is difficult to attain.

4.3 Multi-task Model Advantages and Limitations

Only recently have multi-task ergonomics evaluation models emerged in the ergonomics profession. Multi-task models such as RCRA and Fatigue Failure enable the ergonomics practitioner to model the exposure levels observed in actual work patterns, rather than "idealized" single-task approximations of tasks. With these new multi-task evaluation models, it is no longer necessary to take the weighted average or use the worst-case exposure level as is necessary when applying single-task models. Multi-task models enable the practitioner to simply evaluate the actual levels of physical work observed. A benefit of these multi-task models is that even small changes in exposure levels are reflected in their metrics.

However, the multi-task evaluation models have their limitations. With a multitask model, the analyst is limited to the observation window in regard to physical work (and operational parameters) in effect at the time the analysis was performed. It is quite common for operational parameters to change over the course of day-to-day operations, particularly in industries such as manufacturing, assembly, and construction. With respect to the prevalence of multi-task work and the changing nature of operational parameters, ergonomics practitioners struggle to efficiently assess the impact of changing operational parameters using either traditional, single-task models or the newer, multi-task models. Multi-task evaluation models can also suffer from the problems associated with using a stair-stepped approach to risk assessment if discrete risk cut-off levels are applied.

4.4 Ergonomic Control Panel

Multi-task models address many of the application limitations of single-task models, but limitations remain. Any variation of the job makes it increasingly difficult to evaluate physical work with any model type. Single-task and multi-task models lack the capability to assess the impact of real-time and future changes to operational parameters. This limitation warrants the need for ergonomic analysis tools that are not only reactive (assess the risk of existing conditions) but also proactive (assess the risk of job changes yet to be implemented) (Drinkaus et al. 2005).

Evaluating changes is a cumbersome task and normally not performed holistically at the time the changes are made. Additionally, changes are often evaluated in isolation without seeing the impact of the job as a whole. In the current paradigm, assessment of changes requires a high level of expertise (i.e., the need to bring in the "ergo expert"). The authors pondered the problems faced by ergonomics practitioners and collaborated to devise a solution, the Ergonomic Control Panel (ECP), that can be applied by practitioners with minimal training. The ECP allows users to monitor operations and evaluate the impact that proposed changes in operational parameters have on resultant ergonomic exposure metrics. The proposed ECP is a tool that can be visualized as a spreadsheet/simulation program that incorporates one or more multi-task ergonomics evaluation models such as Fatigue Failure and RCRA. The ECP updates the exposure metrics (immediately) based upon user inputs. The ECP enables the ergonomics practitioner to apply ergonomic exposure metrics dynamically, in a forward-looking manner.

The ECP's ability to instantly update exposure metrics makes it possible to readily assess the impact resulting from changes to operational parameters such as shift length (i.e., overtime), cycle time, and make/model mix. The ECP's ability to instantly update exposure metrics enables these metrics to be leveraged for more complex applications such as determining optimal job rotation plans and developing effective work-rest schedules.

Several studies have suggested the benefits of job rotation such as an increase in problem-solving skills (Ebeling and Lee 1994; Allwood and Lee 2004) and increased job satisfaction (Dawal et al. 2009; Jeon and Jeong 2016). Looze et al. (2010) concluded that job rotation increases flexibility within a process, especially to accommodate peak demand periods, and that there was no difference in the output per operator. Moreover, in studies where lactic acid was used as an indicator of muscular fatigue, employees involved in job rotation tended to have less lactic acid build-up (Filus and Okimorto 2012) and experienced less physical load on their neck and shoulders (Looze et al. 2010).

Although other studies have shown that job rotation is not a good method for reducing risk (Padula et al. 2017; Mehdizadeh et al. 2020) as it exposes additional workers to the risky elements of the shared job, effectively resulting in high-risk jobs for all of those exposed. These conclusions are normally made based on "stair-stepped" risk assessment models and do not address the reality that work will be performed by one or more employees, working in a job rotation sequence. The reality of the workplace is that many employees will be exposed to some form of job rotation in the execution of normal operations. Ergonomic risk assessment and evaluation tools need to address this job rotation reality. Ergonomic exposure levels are often defined more by rotation paths of workers than by the exposure metrics of individual jobs or stations.

From a top-down approach, the ECP enables the user to adjust multiple high-level operational parameters simultaneously and to evaluate ergonomic physical exposure metrics at multiple levels including departments, lines, stations, and tasks. For example, it would be possible to assess the impact of higher-level decision to work overtime on the exposure metrics for an individual job or station. Conversely, the user can utilize a bottom-up approach, changing the smallest variable at the task level and instantly seeing the impact to ergonomic exposure metrics at the higher job or line levels. The ECP enables managers to evaluate the impact of making these seemingly small changes in the broader context of overall job evaluation. An ECP user would be able to evaluate the interaction of multiple variables and observe how each of them impacts the overall risk of injury. Therefore, the ECP is ideal for lean engineering environments where improvements, by design, are often the result of many small changes over time (Fig. 4.4).

Manipulation of various job and operational parameters can be done both programmatically and manually to assess the best opportunities for improvement. ECP enables the non-ergonomist to see the real-world ergonomic impact of actual and proposed changes to the job on operator risk. Currently, changes that affect risk outcomes are often only discovered after the injuries resulting from those changes manifest themselves. The control panel concept allows injuries to be predicted and, therefore, prevented rather than being discovered after changes are implemented. In situations where operations parameters change, it is beneficial to predict the impact of these changes on resultant ergonomic exposure metrics. These situations include:

- *Cycle time:* In the relentless quest to reduce waste and increase production, it is common in a rapidly changing Lean Manufacturing environments for the work cycle time to be reduced. Changing the operational parameter of cycle time even slightly could impact resultant risk metrics as the overall time taken to perform

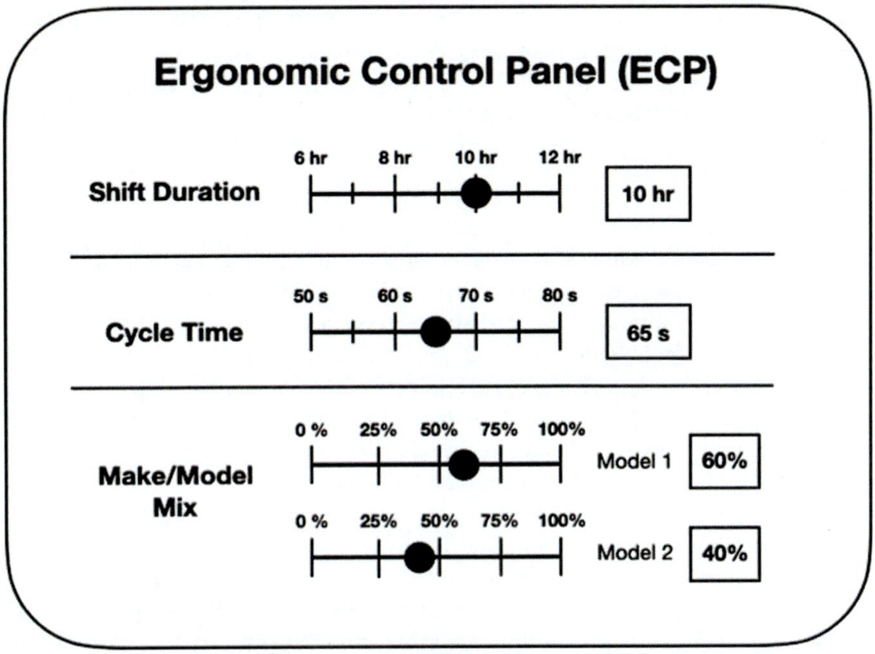

Fig. 4.4 Ergonomic Control Panel (ECP) depicts how the user would use "slider" controls to readily manipulate operational metrics such as shift duration (e.g., changing shift length or working overtime), cycle time, and make/model mix

a task decreases, the task steps do not necessarily decrease resulting in a faster pace and higher repetitions per unit time.

- *Make/model mix:* It is common practice for a manufacturing facility to produce multiple variety of products at the same facility; an example in the automotive setting would be the production of sedans and SUVs on the same assembly line. This is a function of the customer demand where the demand drives the selection of the product to be produced. Different makes/models can result in significantly different ergonomic risk factor exposure levels which makes it difficult to exactly pinpoint a high-risk task. Currently, few ergonomic tools have the capability and the adaptability required to fully analyze such a scenario. The ECP allows for make/model mix changes, both actual and proposed, to be analyzed in their effects on workers. The ECP allows for this adaptability by using cross-training principles to recommend appropriate job rotation schedules (Mccreery and Krajewski 1999).
- *Shift length changes*: Increases in shift length or the addition of weekend shifts can result in workers experiencing a greater number/volume of physical exposures as well as reduced recovery time. Multi-task evaluation models that are additive in nature such as LiFFT (Gallagher and Schall 2017) and DUET (Gallagher et al. 2018) incorporated into the ECP would illustrate these changes to risk as the shift length parameters are adjusted. Another aspect of shift length changes

is working overtime. Overtime work results in workers experiencing a greater number/volume of physical exposures. In cases where overtime work is expected, a traditional single-task analysis tool may or may not account for the increased exposure time. ECP has the flexibility to address either scenario by accumulating the additional risk.

Additionally, it is proposed that ECP could account for actual workforce population demographics, providing a more accurate assessment of risk to the company's actual employee population (e.g., accounting for trends in age, gender, etc.). Such an assessment could also be personalized to the individual level. Recent research has demonstrated that individual characteristics such as age (Houtman et al. 1994) gender (Maj et al. 1999), stature (Smedley et al. 1995), and BMI (Body Mass Index) (Tsai et al. 1992) are critical to accurately determine an injury risk posed to an individual employee (Cox 1989; Badawy et al. 2018; Barim and Sesek 2019). As employee demographics and personal characteristics change, the resultant risk exposures can also change. In many industries, an emerging reality is an aging workforce. Reductions in strength that accompany aging require workers to utilize a higher percentage of their maximum strength performing force exertions, as compared to employee populations comprised of younger workers. This results in increased risk of injury and increased levels of localized muscle fatigue. Moreover, effect of psychosocial risk factors and risks associated with environmental stressors such as exposure to extreme temperatures, vibration, and noise levels cannot be ignored. Additional risk factors can be added to the ECP as new risk assessment tools are developed. With the ECP, the idea is to provide the practitioner a holistic view of the ergonomic-related risks workers face. This readily accessible and conveniently updated risk information will not only enable ergonomic practitioners to develop effective solutions at the station or task level, but also make improved decisions regarding the impact that higher-level operational parameters (i.e., cycle time, shift duration, make/model mix, etc.) have upon risk/exposure metrics.

4.5 Conclusion

In addition to identifying actions to reduce risk to an individual, the Ergonomics Control Panel (ECP) concept facilitates increased safety and productivity in the workplace by providing visualization of risk accumulation. This visualization provides an easy-to-understand methodology for assessing the current state of ergonomic risk, monitoring real-time risk as changes to production are made and allow for evaluation of future risks by allowing a simulation of proposed changes. The ECP makes it clear which aspects of the jobs performed by employees are the riskiest, allowing management to identify actions that could be taken to abate these risks. In addition to identifying actions to reduce risk to an individual, the ECP can also identify the highest risk jobs and even the high-risk tasks within jobs so that effective abatements can be implemented. The ECP has the potential to influence decisions related to whether specific individuals can safely work rotation plans/paths or participate in overtime. The ECP can utilize cross-training principles to enable real-time job and shift changes to facilitate flexibility (Cox 1989) meeting customer-driven demand and tie in seamlessly with a Lean Manufacturing system. By simulating potential work scenarios before making changes to jobs or to scheduling workers to specific rotation paths, problems can be avoided proactively rather than addressing reactively. Tasks that individually have been assessed as acceptable risk may actually be too burdensome when subdivided and combined with other tasks. This assessment is not commonly performed, due to the time and effort required to assess individual rotation paths. The ECP makes it easy for management and employees to see that the risk to employees is result of the actual rotation path that workers perform.

4.6 Future Scope

Moving forward, the ECP solution would be further enhanced by including the capability of assigning monetary values to each unit increase in risk which can enable rapid prioritization of high-risk job abatements with a substantial business justification across different risk factors (Rosecrance 2004) and utilize other proven simulation algorithms (Asensio-Cuesta et al. 2012) to optimize work-rest cycles and reduce localized muscle fatigue. As artificial intelligence (AI) systems and machine learning are continuously improved, this concept could be expanded into an AI expert system. This expert system could both monitor the workplace for changes that affect worker risk and even make recommendations for employee scheduling and job rotation. These recommendations could consider any number of factors including psychosocial, physiological, and economic.

References

Allwood JM, Lee WL (2004) The impact of job rotation on problem solving skills. Int J Prod Res 42:865–881. https://doi.org/10.1080/00207540310001631566

Andersen SN, Broberg O (2015) Participatory ergonomics simulation of hospital work systems: the influence of simulation media on simulation outcome. Appl Ergon 51:331–342. https://doi.org/10.1016/j.apergo.2015.06.003

Asensio-Cuesta S, Diego-Mas JA, Canós-Darós L, Andrés-Romano C (2012) A genetic algorithm for the design of job rotation schedules considering ergonomic and competence criteria. Int J Adv Manuf Technol 60:1161–1174. https://doi.org/10.1007/s00170-011-3672-0

Badawy M, Schall MC, Sesek RF et al (2018) One-handed carrying among elderly and obese individuals: a systematic review to identify research gaps. Ergonomics 61:1345–1354. https://doi.org/10.1080/00140139.2018.1470680

Badler NI, Palmer MS, Bindiganavale R (1999) Animation control for real-time virtual humans. Commun ACM 42(8):64–73

Baines T, Mason S, Siebers PO, Ladbrook J (2004) Humans: the missing link in manufacturing simulation? In: Simulation modelling practice and theory, pp 515–526

Baines TS, Kay JM (2002) Human performance modelling as an aid in the process of manufacturing system design: a pilot study. Int J Prod Res 40:2321–2334. https://doi.org/10.1080/00207540210128198

Barim, MS., Sesek RF (2019) Improving the risk assessment capability of the revised NIOSH lifting equation by incorporating personal characteristics. Appl Ergon 74:67–73

Broberg O, Andersen V, Seim R (2011) Participatory ergonomics in design processes: the role of boundary objects. Appl Ergon 42:464–472. https://doi.org/10.1016/j.apergo.2010.09.006

Case K, Piamonte P (2007) Ergonomics analysis in a virtual environment

Chaffin DB (2005) Improving digital human modelling for proactive ergonomics in design. Ergonomics 48:478–491. https://doi.org/10.1080/00140130400029191

Chang SW, Wang MJJ (2007) Digital human modeling and workplace evaluation: using an automobile assembly task as an example. Human Factors Ergon Manuf 17:445–455. https://doi.org/10.1002/hfm.20085

Cox T Jr (1989) Toward the measurement of manufacturing flexibility. Prod Inventory Manage J 30:68–72

Dawal SZ, Taha Z, Ismail Z (2009) Effect of job organization on job satisfaction among shop floor employees in automotive industries in Malaysia. Int J Ind Ergon 39:1–6. https://doi.org/10.1016/j.ergon.2008.06.005

Drinkaus P, Sesek R, Bloswick DS et al (2005) Job level risk assessment using task level ACGIH hand activity level tlv scores: a pilot study. Int J Occup Saf Ergon 11:263–281. https://doi.org/10.1080/10803548.2005.11076648

Ebeling and Lee (1994) Cross-training effectiveness and profitability. Int J Prod Res 32:2843–2859

Eldar R, Fisher-Gewirtzman D (2019) Ergonomic design visualization mapping- developing an assistive model for design activities. International Journal of Industrial Ergonomics 74. https://doi.org/10.1016/j.ergon.2019.102859

Filus R, Okimorto ML (2012) The effect of job rotation intervals on muscle fatigue - Lactic acid. In: Work, pp 1572–1581

Gallagher S, Schall MC (2017) Musculoskeletal disorders as a fatigue failure process: evidence, implications and research needs. Ergonomics 60:255–269. https://doi.org/10.1080/00140139.2016.1208848

Gallagher S, Schall MC, Sesek RF, Huangfu R (2018) An upper extremity risk assessment tool based on material fatigue failure theory: the Distal Upper Extremity Tool (DUET). Hum Factors 60:1146–1162. https://doi.org/10.1177/0018720818789319

Gibson M, Potvin JR (2016) Association of Canadian ergonomists conference

Hendrick HW (2008) Applying ergonomics to systems: Some documented "lessons learned". Appl Ergon 39:418–426. https://doi.org/10.1016/j.apergo.2008.02.006

Houtman IL, Bongers PM, Kompier MA (1994) Psychosocial stressors at work and musculoskeletal problems. Scand J Work Environ Health 20:139–145

Jeon IS, Jeong BY (2016) Effect of job rotation types on productivity, accident rate, and satisfaction in the automotive assembly line workers. Human Factors Ergon Manuf 26:455–462. https://doi.org/10.1002/hfm.20667

Jung ES, Kee D, Chung MK (1995) Upper body reach posture prediction for ergonomic evaluation models

Karmakar S, Pal MS, Majumdar D, Majumdar D (2012) Application of digital human modeling and simulation for vision analysis of pilots in a jet aircraft: A case study. In: Work, pp 3412–3418

Kim JH, Zhao X, Du W (2018) Assessing the performance of visual identification tasks using time window-based eye inter-fixation duration. Int J Ind Ergon 64:15–22. https://doi.org/10.1016/j.ergon.2017.09.002

Kroemer KE, Kroemer HB (1994) Ergonomics: how to design for ease and efficiency. Prentice-Hall, Englewood Cliffs, NJ

Kumar BU, Bora A, Sanjog J, Karmakar S (2013) Proactive ergonomics through digital human modeling and simulation for product design innovation: a case study. In: 2013 international computer science and engineering conference, ICSEC 2013, pp 319–323

de Looze MP, Bosch T, van Rhijn JW (2010) Increasing short-term output in assembly work. Human Factors Ergon Manuf Serv Ind 20:470–477. https://doi.org/10.1002/hfm.20199
Ma L, Zhang W, Wu S, Zhang Z (2015) A new simple local muscle recovery model and its theoretical and experimental validation. Int J Occup Saf Ergon 21:86–93. https://doi.org/10.1080/10803548.2015.1017961
Maj G,, Berkowitz SM, Usa M, et al (1999) Occupational back disability in U.S. Army personnel
Mayer MP, Odenthal B, Faber M, Schlick CM (2012) Cognitively automated assembly processes: A simulation based evaluation of performance. In: Work, pp 3449–3454
Mccreery JK, Krajewski LJ (1999) Improving performance using workforce flexibility in an assembly environment with learning and forgetting effects. Int J Prod Res 37:2031–2058. https://doi.org/10.1080/002075499190897
Mcatamney L, Corlett EN (1993) RULA: a survey method for the investigation of work-related upper limb disorders
Mehdizadeh A, Vinel A, Hu Q et al (2020) Job rotation and work-related musculoskeletal disorders: a fatigue-failure perspective. Ergonomics 63:461–476. https://doi.org/10.1080/00140139.2020.1717644
Moore JS, Garg A (1995) The strain index: a proposed method to analyze jobs for risk of distal upper extremity disorders. Am Ind Hyg Assoc J 56:443–458. https://doi.org/10.1080/15428119591016863
Padula RS, Comper MLC, Sparer EH, Dennerlein JT (2017) Job rotation designed to prevent musculoskeletal disorders and control risk in manufacturing industries: a systematic review. Appl Ergon 58:386–397
Paquet V, Lin L (2003) An integrated methodology for manufacturing systems design using manual and computer simulation. Human Factors Ergon Manuf 13:19–40. https://doi.org/10.1002/hfm.10026
Örtengren, R, Lämkull, D, Hanson, L (2009) Digital human modeling simulation results and their outcomes in reality: a comparative study within manual assembly of automobiles. AE Int J Passeng Cars Mech Syst 2(1):1600–1613, 2009:14
Rosecrance J (2004) The economics and cost justification of ergonomics
Rychtyckyj N, Stephens A (2009) Assembly ergonomics filters for prevention of injury risk operations at Ford Motor Company. 2009 IEEE Workshop Comput Intell Veh Veh Syst 10:58–65
Schaub KG, Mühlstedt J, Illmann B, et al (2012) Ergonomic assessment of automotive assembly tasks with digital human modelling and the "ergonomics assessment worksheet" (EAWS)
Smedley J, Egger P, Cooper C, Coggon D (1995) Manual handling activities and risk of low back pain in nurses. Occup Environ Med 52:160–163. https://doi.org/10.1136/oem.52.3.160
Song J, Qu X, Chen CH (2016) Simulation of lifting motions using a novel multi-objective optimization approach. Int J Ind Ergon 53:37–47. https://doi.org/10.1016/j.ergon.2015.10.002
Tsai SP, Gilstrap EL, Cowles SR, Waddell Jr LC, Ross CE (1992) Personal and job characteristics of musculoskeletal injuries in an industrial population. J Occup Med 34(6):606–612

Part II
Statistical Analysis of Ergonomic Factors

Chapter 5
Fatigue Study Among 12-Hour and 7-Hour Night Shift Workers in the Manufacturing Industry

Patricia Eugenia Sortillón-González

Abstract In recent years physical and mental fatigue has received much attention due to its impact on work performance and its prevalence in night shift works. The aim of this study is to describe the self-reported fatigue of workers from 7-h and 12-h night shifts to investigate whether fatigue is related to night work schedules. The presence of fatigue was assessed with a multidimensional self-report instrument which includes the sociodemographic information of individuals, the Yoshitake's fatigue symptom checklist and the Corlett-Bishop's musculoskeletal disorder map. We found that fatigue was significantly more prevalent among workers of 12-h shift compared to fatigue in 7-h shift workers. Substantial differences in self-reported fatigue were found through the days and weeks among the workers of the two shifts, an exception was found for fatigue between lapses of 7-h shifts where fatigue tended to the same pattern for all lapses. Our results are consistent with other studies that establish strong links between perceived fatigue and work schedule. Upper body musculoskeletal disorders increased through the weeks as well as perceived fatigue. Further studies are needed for exploring the influence of other factors associated with fatigue on night shift workers, which include sleep deprivation and previous work.

Keywords Fatigue · Shift work · Study · Work schedules

5.1 Introduction

Since the beginning of twenty century the manufacturing industrial sector has been increasingly involved in the operation of long shifts, that is, longer than 8 h a day (Rosekind et al. 1996). These sectors are not only those of the industry but also those dedicated to health care, in the functions of safety, fire, transport, and the military industry; with the particular characteristic that this system of working time expands in the same way as markets and businesses do.

P. E. Sortillón-González (✉)
Department of Manufacturing Industrial Engineering, Universidad Estatal de Sonora, Av. Ley Federal del Trabajo. Col. Apolo, Hermosillo, Sonora, México
e-mail: patricia.sortillon@unison.mx

A. Realyvásquez Vargas et al. (eds.), *New Perspectives on Applied Industrial Ergonomics*,
https://doi.org/10.1007/978-3-030-73468-8_5

According to the Bureau of Labor Statistics (OTA 1991), in the United States, one in five workers has a long, compressed, or irregular working day, and most of them work in manufacturing, transport, or service companies.

There are some disadvantages to the execution of long hours shifts (Rosa et al. 1998). Alert levels can be decreased and with this increase the risks of accidents. On the other hand, reductions in the work rhythms can be generated and longer rest periods may be necessary. Long periods of work, according to Paley et al. (1998), are a risk factor for the appearance of fatigue.

Fatigue is a physical condition often associated with long-term work, usually in the field of parts manufacturing which, by its nature requires hours of night or day work but long hours. The consequences of this condition cover social, family, and even employee performance aspects and therefore have an impact on the overall performance of companies, especially in aspects related to productivity and safety at work. This study aims to compare the physical and mental fatigue experienced by manufacturing industrial night shift workers for both the 7 and 12 h of night shift schedules.

5.2 Literature Review and Hypotheses Statement

5.2.1 Shift Work and Fatigue

According to Hedges and Sekscencki (1979), shift work is a job schedule in which workers are required to work more than 8 to 16 h.

Fatigue is a complex problem that does not respond to the degree of health. Fatigue can be related to cognitive and emotional components, whose symptoms include tiredness, drowsiness, and inability for concentrating, it also can be considered as a multidimensional phenomenon with specifics relationships with shift work.

Shift work can have adverse effects on social relationships and activities due to its effects on physical and mental health. If the shift work is frequent, this could lead to serious dissociation of the circadian pacemaker and ambient synchronizers, such as the light and dark cycle. This can lead to disruptions in the normal sleep and wake rhythm of shift workers, causing fatigue problems (Costa 2003).

Shift work could potentially cause fatigue, thereby increasing the risk of accidents and in some cases falling asleep on the job (Akerstedt et al. 2002).

Previous studies (Dorrian 2011) have shown that fatigue increases through the hours of work; this could suggest that workers may experience different levels of fatigue during work.

According to Folkard and Tucker (2003), efficiency and safety on shift work systems are a major concern, due that shift work is frequently introduced as a strategy to maximize the use of costly equipment, which leads us to economic reasons.

According to Hossain et al. (2003), fatigue could be related to insufficient restorative sleep due to the cumulative effects of chronic sleep deprivation and the mismatch between sleep and wake schedule, shift workers are susceptible to work-related fatigue.

Pigeon et al. (2003) consider that fatigue is related to the following symptoms: weariness, weakness, and depleted energy. Fatigue could be a state in which the efficiency of working is reduced including the capacity to do physical or mental effort (Hossain et al. 2003).

Fatigue, according to Layzer (1990), can be described in four instances: The first one is called objective fatigue which refers to the inability to maintain a specified level of effort during physical activity; the second one is named subjective fatigue which is the discomfort that suppresses the desire to pursue exercise; the third one is the fatigue which relates to prolonged physical effort; and the last one is known as asthenia, which refers to weakness, tiredness, or exhaustion.

Aaronson et al. (1999) suggest that fatigue can be defined as a state in which an individual experiences an overwhelming and permanent feeling of exhaustion and decreased capacity for physical and mental work that cannot be reduced by rest.

Shift workers with a high exposition to work-related fatigue have a risk of decreased alertness and performance that reduces work well-being. Fatigue reduces perceptual skills, reasoning abilities, judgment, and decision-making capabilities (Lal and Craig 2001). Such impairments manifest themselves through general slower brain functioning causing delayed reaction times, memory deficits, and a decrease in cognitive abilities such as logical reasoning and concentration.

Shift workers with work-related fatigue have a greater risk of occupational accidents (Swaen et al. 2003; Berger and Hobbs 2006).

5.2.2 Fatigue Measurement

Fatigue is a complex phenomenon due to subjective nature; however, several studies can lead us to understand its behavior.

Hossain et al. (2003) developed a cross-sectional study to find out if fatigue severity could be used as a predictive tool to become aware of underly sleep problems, which is a risk factor associated with fatigue in shift workers. It was administered the Fatigue Severity Scale (FSS) and two groups of workers were defined based on polysomnographic data, which were treated with Fisher's test, the difference between the two groups was highly significant ($P < 0.0001$). Also, the author did not found a strong correlation between subjective fatigue and sleepiness, suggesting that these two signs can be independent phenomena. The author concluded that fatigue can be a manifestation of sleep feelings, in this way warrants an objective evaluation.

In an attempt to know whether shift work differentially affects fatigue and sleepiness, Shen et al. (2006) used the FSS survey on shift workers from Ontario (United States) and found that the frequency of shift work has a significant effect on subjective fatigue.

In agreement with Hossain et al. (2003) study, a low correlation was found between worker's subjective fatigue and sleepiness scores, which can lead to thinking that fatigue and sleepiness are not correlated phenomena.

Samaha et al. (2007) studied fatigue on shift work nurses and used the Checklist Individual Scale (CIS) (Beurskens et al. 2000) to measure chronic fatigue. The CIS consisted of 20 questions rated on a seven-point Likert scale. The CIS evaluates the degree of fatigue experienced during two weeks. The complete score was derived by way of adding individual items after negatively weighting fatigue-absent items. The level of score reveals greater levels of subjective fatigue including reduced motivation, more difficulty concentrating. According to their findings, multiple regressions showed that mood disturbance, locus of control, and trait anxiety are predictors of fatigue. The author found out that the low quality of sleep is a factor that most strongly contributed to fatigue. Other factors are considered as predictors of fatigue, such as workload worker perception and sedentarism. Jansen et al. (2003) measured fatigue through the checklist individual strength (CIS), which is a 20 item questionnaire developed to measure several aspects of fatigue. The questionnaire is a self-report instrument of four elements which measure twelve subjective factors of fatigue, five factors related to concentration, four factors for motivation, and three items for activity level. All items are scored on seven-point Likert scales, higher scores indicate a higher level of fatigue, concentration problems, reduced motivation, or less activity. For this instrument, the range of results are from 20 to 140. Results showed that the CIS scores for all types of shift workers increased, but it was not significantly higher compared to day workers.

To contribute to the results of the above studies, the hypothesis of this research is as follows:

H1: *Fatigue differs between 12-hour shift workers and those for 7-hour shift workers.*

H2: *Fatigue differs between the weeks in both of two-night shifts.*

H3: *Fatigue differs between the days in both of two-night shifts.*

H4: *Fatigue differs between lapses of the day in both of two-night shifts.*

It was conducted a fatigue study within two groups, each of thirty-nine female workers from 7 and 12 h night shifts, to investigate the association between self-reported fatigue experienced during the night shift through five consecutive weeks, which could let us understand how fatigue evolves through the weeks, the days in a week, and between three lapses during the day for both of two work schedules. Fatigue was measured using the Yoshitake's Fatigue Symptom Checklist (Yoshitake 1978) and it was applied the Corlett and Bishop Musculoskeletal Disorder Map (Corlett and Bishop 1976) to know the associate disorders linked to self-reported fatigue.

5.3 Methodology

The following sections describe the methodology followed to validate the research hypotheses stated in the above paragraphs.

5.3.1 Study Design and Participants

A prospective and observational study was conducted within five weeks in a Mexican manufacturing Industry. Two groups each of 39 female shift workers from two-night shifts were followed using self-administered questionnaires at three-daily intervals (lapses) within five consecutive days in a five continuous week period. Once at the beginning of the study, the workers answered the questionnaire with items on work and non-work-related factors and information on age, educational level, and employment history and the presence of a musculoskeletal disorder through self-report. Fatigue Symptom Checklist was answered at each lapse of the day within the five days of the week during the five-week period. At the end of each week, employees answered the Corlett and Bishop Musculoskeletal Disorder map.

Participants were excluded if they were pregnant or were currently undergoing medical treatment. Temporary employees were excluded since they change from one production process to another, frequently. Also were excluded employees with multiple jobs since we had no information on the working content of the other job. Employees reporting to have switched from another work schedule as from day to night were excluded.

Approval of the study was obtained from the ethics committee at the Mexican company.

Participants received information on the study goals, workers who agreed to participate, signed a consent term. Participants were free to withdraw from the study at any time and were provided with their study results.

5.3.2 Survey Administration

Following informed consent to gather information about short- and long-term fatigue symptoms, participants were provided with a composed fatigue Survey (CFS) to assess the presence of fatigue during the three lapses within the days, the days of the week, and through the weeks. The CFS is a multidimensional self-report instrument comprising three parts. The first part of the instrument is to gather sociodemographic information and it was applied only the first day of assessments, the second part is the Yoshitake's fatigue symptom checklist (YFSC), developed in Japan to measure fatigue in three categories: General symptoms of fatigue, intellectual fatigue, physical fatigue; and the third part of the instrument is the Corlett-Bishop's musculoskeletal

Table 5.1 Survey administration schedule

Shift group	Lapse 1	Lapse 2	Lapse 3
6:00 p.m.–2:00 a.m	8:00 p.m	11:00 p.m	1:00 a.m
7:00 p.m.–7:00 a.m	9:00 p.m	2:00 p.m	5:00 a.m

disorder map (CBDM). The composed survey was translated into Spanish. The reliability and validity of modification were tested. The CFS was applied from Monday to Friday during five continuous weeks in three lapses of time for both of the two groups, as illustrated in Table 5.1.

A total of 2925 surveys were answered for each of 7 h and 12 h shifts during the five-week period which 585 comes from five consecutive days of the week and 117 from each day of the week.

5.3.3 Data Capture and Screening

Once the surveys were completed, a data management and statistical analysis were carried out using the software suite IBM SPSS ® (v24, SPSS Inc., Chicago, USA), where the columns represented the fatigue index for both of the two shifts, as well the independents variables such week, day and lapses and each row represented a shift worker. To screen the data, we first estimated the mean of variables of each case. No missing values (unanswered questions) were found.

5.3.4 Data Analysis

Statistical data processing and analysis regarding to sociodemographic information of participants, YFSC and CBDM, is briefly described below:

- *Sociodemographic information of participants*: Average and standard deviation of workers age was calculated. Percentage of educational levels and employment history were also calculated.
- *Yoshitake's Fatigue Symptom Checklist*: A Fatigue Index (FI) was calculated for the totality of gathered data, summarizing in each CFS the total amount of questions answered with yes and diving it for 30. Data of FI was split into three categories: week FI, day FI, and lapse FI.
- *Corlett and Bishop's Musculoskeletal Disorder Map*: Data from CBDM was screened and percentages of musculoskeletal disorders were calculated.
- *Statistical analysis*: To summarize gathered data, the mean and standard deviation of participant's age was calculated. Mean and standard deviation of FI was calculated for weeks, days, and lapses in the total period of analysis consistent of five weeks. After verifying normally distributed data of FI, the influence of days,

weeks, and lapses on the overall FI mean was analyzed using an one-way analysis of variance (ANOVA) test with a significance level of 0.05 and a Duncan's multiple range test was applied to define fatigue index mean's subsets with no significant difference from one another in the days, weeks, and lapses for both of two shifts.

To compare fatigue index for both groups, 7 h shift and 12 h shift, it was used the one-way Analysis of Covariance (ANCOVA) which can be thought of as an extension of the one-way ANOVA that incorporate a covariate.

A two-way ANOVA was applied to examine the influence of the two different categorical independent variables on one continuous dependent variable with the purpose to assess the main effect of each independent variable but also if there was any interaction between them. Two-way ANOVA was used to examine the influence of week and days and between them, as well as for day and lapses.

Like the one-way ANOVA, the one-way ANCOVA was used to determine whether are differences between two or more unrelated groups on a dependent variable. The ANCOVA test looks for differences in adjusted means. A comparison between one-way ANOVA and the one-way ANCOVA lead us to understand that ANCOVA allows us to statistically control for a third variable (known as a confounding variable), which we believe affect our results. This third variable that could be confounding our results is called the covariate and we included it in our one-way ANCOVA analysis. ANCOVA is usually used when there are differences between the baseline groups (Senn 1994; Overall 1993).

When we choose to analyze data using a one-way ANCOVA, part of the process involves checking to make sure that the data can be analyzed using a one-way ANCOVA. We need to do this because it is only appropriate to use a one-way ANCOVA if our data passes nine assumptions that are required to get a valid result.

All of the following assumptions were verified or tested for the data gathered:

- The dependent variable and covariate variable on a continuous scale.
- The independent variable has two or more categorical, independent groups.
- Independence of observations. This means that there is no relationship between the observations in each group or between the groups themselves.
- No significant outliers, we are talking that all points within our data followed the usual pattern.
- Residuals were approximately normally distributed for each category of the independent variable.
- Homogeneity of variances. It was used Levene's test for homogeneity of variances.
- The linearity of the covariate of the dependent variable at each level of the independent variable.
- Homoscedasticity.
- Homogeneity of regression slopes, which means that there was no interaction between the covariate and the independent variable.

Table 5.2 Descriptive characteristics of the study population

Characteristic	7 h Shift	12 h Shift
N	39	39
Age mean (SD)	28.834 (3.258)	42.892 (5.689)
Highest educational level %		
Higher vocational	97.4	92.3
University	2.6	7.7
Employment history		
≤5 y	23.076	12.820
6–10 y	30.769	25.641
11–15 y	23.076	20.512
>15 y	23.076	41.025

5.4 Results

The survey was administered from January to February 2020 in one manufacturing industry in northern Mexico. A total of 2925 questionnaires for each of both shifts were collected. For one single day of the five-day week, the survey was applied during the three lapses, 117 answered surveys were collected every day, then 585 for every week. None of surveys were removed from the analysis due to missing values. Hence, 2925 surveys for each shift were considered valid and thus were statistically treated. The following subsections discuss the descriptive analysis of the collected data.

5.4.1 Study Population Characteristics

According to Table 5.2, we present descriptive characteristics of the study population; they were 39 female workers for both of two shifts. 7-h shift workers were younger age compared to 12-h shift workers and had received a lower education. Data on employment history revealing that 12-h shift workers had been involved in shift work for more than 15 years, almost twice than 7-h shift workers did.

5.4.2 Statistical Analysis

5.4.2.1 Yoshitake's Fatigue Symptom Checklist

The three categories of the YFSC showed good internal consistency with Cronbach's alpha values of 0.881 (general symptoms of fatigue), 0.834 (intellectual fatigue), and 0.915 (physical fatigue).

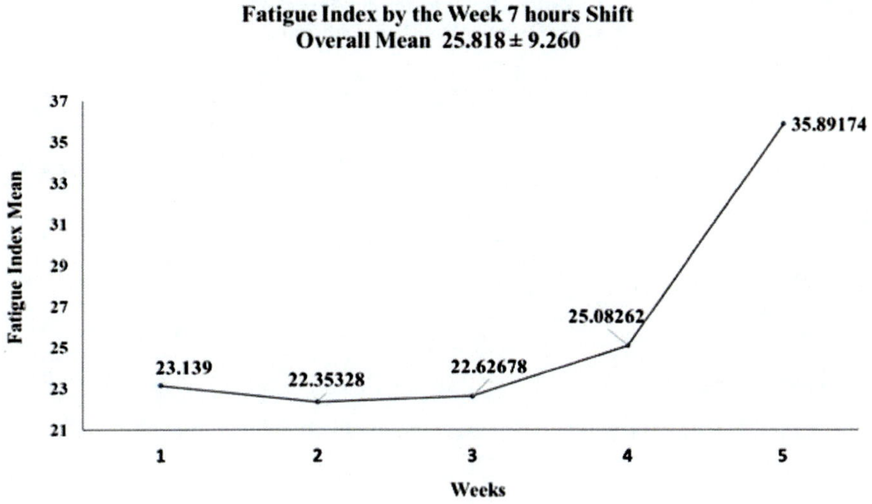

Fig. 5.1 Fatigue index mean by week 7 h shift

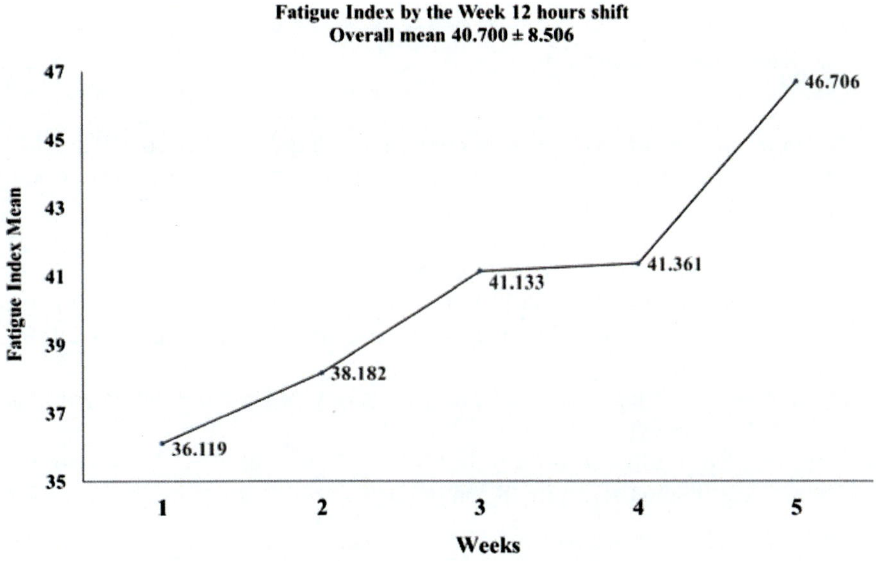

Fig. 5.2 Fatigue index by the week 12 h shift

The corresponding FI mean for weeks in 7 h shift (Fig. 5.1) were 23.139 ± 6.081 ($week_1$), 22.353 ± 5.595 ($week_2$), 22.626 ± 7.209 ($week_3$), 25.082 ± 9.357 ($week_4$), and 35.891 ± 9.484 ($week_5$).

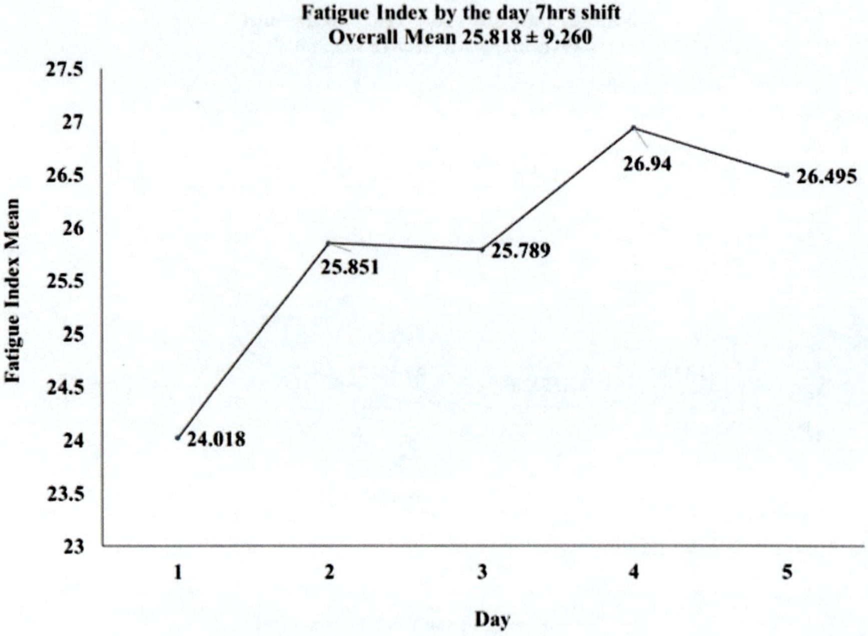

Fig. 5.3 Fatigue index mean by the day 7 h shift

The corresponding FI mean for weeks in 12 h shift (Fig. 5.2) were 36.119 ± 7.532 ($week_1$), 38.182 ± 6.936 ($week_2$), 41.133 ± 7.954 ($week_3$), 41.361 ± 7.533 ($week_4$), and 46.706 ± 8.561 ($week_5$).

FI mean values for days (Fig. 5.3) were 24.017 ± 8.974 (day_1), 25.851 ± 9.155 (day_2), 25.789 ± 8.675 (day_3), 26.940 ± 9.578 (day_4), and 26.495 ± 9.646 (day_5).

FI mean values for days 12 h shift (Fig. 5.4) were 37.658 ± 7.957 (day_1), 41.612 ± 8.636 (day_2), 40.227 ± 8.082 (day_3), 40.336 ± 7.581 (day_4), and 43.669 ± 9.073 (day_5).

FI mean values for lapses for 7 h shift (Fig. 5.5) were 26.160 ± 9.995 ($lapse_1$), 25.357 ± 8.798 ($lapse_2$), 25.938 ± 8.932 ($lapse_3$).

FI mean values for lapses for 12 h shift (Fig. 5.6) were 39.613 ± 8.593 ($lapse_1$), 40.540 ± 8.435 ($lapse_2$), 41.948 ± 8.334 ($lapse_3$).

5.4.2.2 One-Way ANOVA Test Results

One-way ANOVA tests with a significance level of $a = 0.05$ were performed to Fatigue index mean for the following factors: weeks, days, and lapses for both of two shifts: 7 h and 12 h. Table 5.3 resumes results of the one-way ANOVA test for the week factor for the two shifts.

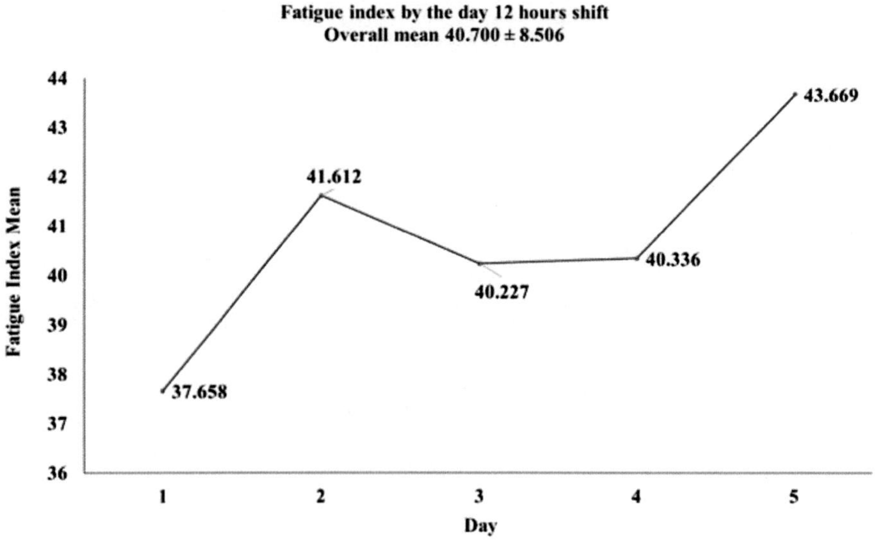

Fig. 5.4 Fatigue index by the day 12 h Shift

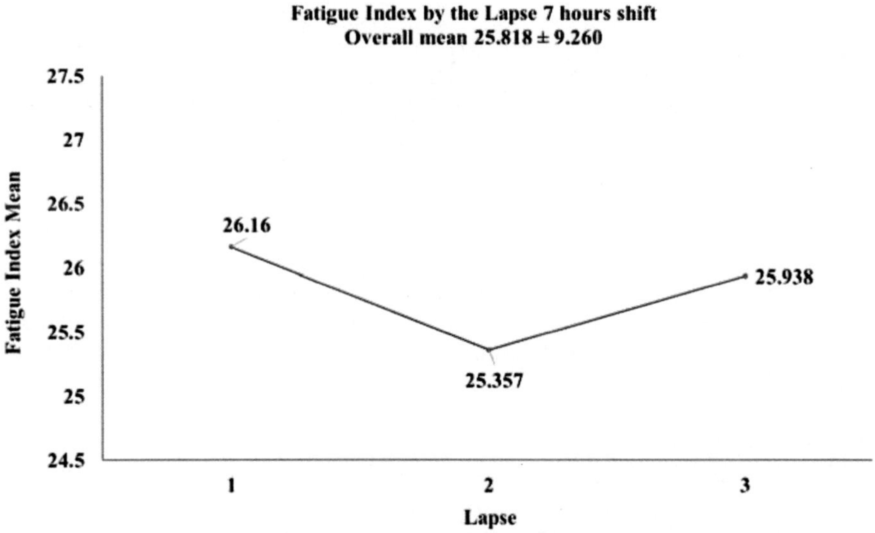

Fig. 5.5 Fatigue index mean by the lapse 7 h shift

For both of the two tests, significance $p = 0.000 < \alpha = 0.05$, tells us the mean of fatigue index mean is different between weeks for both of two shifts. Based on the above results, hypothesis H2 has been proven.

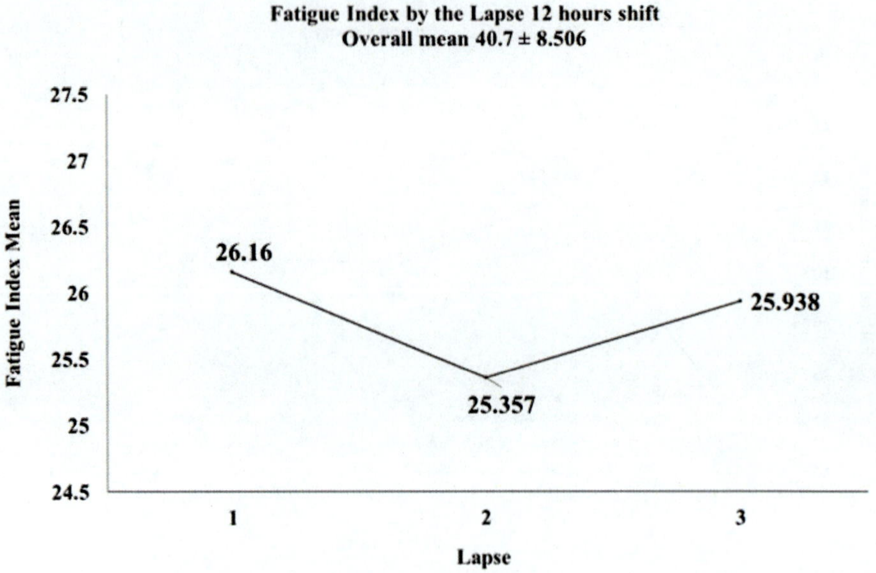

Fig. 5.6 Fatigue index mean by the day 12 h shift

Table 5.3 One-way ANOVA test results for mean of fatigue index 7–12 h shift, factor weeks

Shift	Factor	Significance
7 h	Between Weeks	0.000
12 h	Between Weeks	0.000

Duncan test (Table 5.4) was performed for fatigue index mean on 7 h shift a total of $N = 585$ data were analyzed which comes from three lapses for each individual day in a five-day week. FI is equal for weeks 1, 2, 3, and different for weeks 4 and 5.

Following Duncan test results (Table 5.5) for fatigue index on 12 h shift, FI mean is equal for weeks 3 and 4 and different for the rest of them.

Table 5.4 Post hoc Duncan test fatigue index 7 h shift, factor: weeks

Week	*N*	Subset for alpha = 0.05		
		1	2	3
2	585	22.353		
3	585	22.626		
1	585	23.139		
4	585		25.082	
5	585			35.891

Table 5.5 Post hoc Duncan test fatigue index 12 h shift, factor: weeks

Week	*N*	Subset for alpha = 0.05			
		1	2	3	4
1	585	36.119			
2	585		38.182		
3	585			41.133	
4	585			41.361	
5	585				46.706

Table 5.6 One-way ANOVA test results for fi mean 7–12 h shift, factor: weeks

Shift	Factor	Significance
7 h	Between days	0.000
12 h	Between days	0.000

Results of the one-way ANOVA test are shown in Table 5.6 with a significance level of $\alpha = 0.05$. The independent variable was fatigue index mean and the factor days for both two-night shifts.

For this test, significance = 0.000 < 0.05, tells us means of fatigue index for 7 h and 12 h shifts are different between days for both two shifts. According to the above results, the research hypothesis H3 has been proved.

Duncan test was performed (Table 5.7) and FI mean between days of 7 shift, is equal for weeks 2, 3, 5, and 4 and different for the rest of them.

Duncan test was performed (Table 5.8) and FI mean between days of 12 h shift. Results have shown that FI is equal for weeks 3, 4, and different for the rest of them.

Results of the one-way ANOVA test for fatigue index in 7 h and 12 h shift where the factor is lapses are shown in Table 5.9 with a significance level of $\alpha = 0.05$.

For this test, significance = 0.141 > 0.05, tells us mean of fatigue a 7 h shift is equal between lapses. Therefore, for 12 h shifts fatigue index mean $p = 0.000 < 0.05$, means are not equal for lapses too. The above results compromise the research hypothesis H4 in the case of fatigue of 7-h shift.

Table 5.7 Post hoc Duncan test fatigue index mean 7 h shift factor: days

Week	*N*	Subset for alpha = 0.05		
		1	2	3
1	585	24.017		
3	585		25.789	
2	585		25.851	
5	585		26.495	
4	585			26.940

Table 5.8 Post hoc Duncan test fatigue index mean 12 h shift, factor: days

Days	*N*	Subset for alpha = 0.05			
		1	2	3	4
1	585	37.658			
3	585		40.227		
4	585		40.336		
2	585			41.612	
5	585				43.669

Table 5.9 One-way ANOVA test results for fi mean 7–12 h shift, factor: weeks

Shift	Factor	Significance
7 h	Between lapses	0.141
12 h	Between lapses	0.000

Table 5.10 Post hoc Duncan test fatigue index mean lapses 7 h shift

Week	*N*	Subset for alpha = 0.05
		1
2	975	25.357
3	975	25.938
1	975	26.160

Table 5.11 Post hoc Duncan test fatigue index mean lapses 12 h shift

Lapse	*N*	Subset for alpha = 0.05		
		1	2	3
1	975	39.613		
2	975		40.540	
3	975			41.948

Duncan test was performed for fatigue index mean in 7 h shifts, as follows in Table 5.10, which confirm the above results.

According to Table 5.11, fatigue index means is different for all lapses in the 12 h shift:

5.4.2.3 Two-Way Analysis of Variance Test Results

A two-way ANOVA test was conducted that examined the effect of weeks and days in fatigue mean index and the effect of the days and lapses for both of two shifts: 7 hrs and 12 hrs, results are shown in Table 5.12.

Table 5.12 Two-way ANOVA test results for mean of fatigue index for 7- and 12-h shifts

Shift	Interactions	Significance
7 h	Week*Day	0.000
7 h	Day*Lapse	0.648
12 h	Week*Day	0.000
12 h	Day*Lapse	0.000

Table 5.13 One-way ANCOVA test results, $a = 0.05$

Source	Significance
Week	0.000
Day	0.000
Lapse	0.014

There was found a statistically significant interaction between the effects week-day for both 7 and 12 h shifts and also for day-lapse on 12 h shift, since significance = 0.000 < 0.05. For the interaction day-lapse in 7 h significance = 0.648 > 0.05, thus there was no statistical signification of interaction.

5.4.2.4 One-Way Analysis of Covariance Test Results

Table 5.13 shows results for the one-way ANCOVA test for fatigue index being week, day, lapse the dependent variables and fatigue index mean for 7 h shift the covariate.

Since all p-values from one-Way ANCOVA test from Table 5.13 for week, day and lapse were less than $\alpha = 0.05$ there were found statistically significant differences in fatigue index for 12 h shift (dependent variable), between the weeks, days, and lapses when adjusted for fatigue index mean for 7 h shift (the covariate), the above results agree to the research hypotheses H1.

5.4.2.5 Corlett and bishop's Musculoskeletal Disorder Map

Most relevant musculoskeletal disorders (Fig. 5.7) were found in the neck, shoulders, and low back. There could be a possible relationship between discomfort and standing still during the entire shift. Percentages of musculoskeletal disorders increased through the weeks, for neck, low back, and shoulders. Upper member musculoskeletal disorders could be related to physical fatigue, however, there are three rest periods, which two are fifteen minutes and the other is thirty minutes, which could help to recover from physical fatigue and then relieve some musculoskeletal impairments.

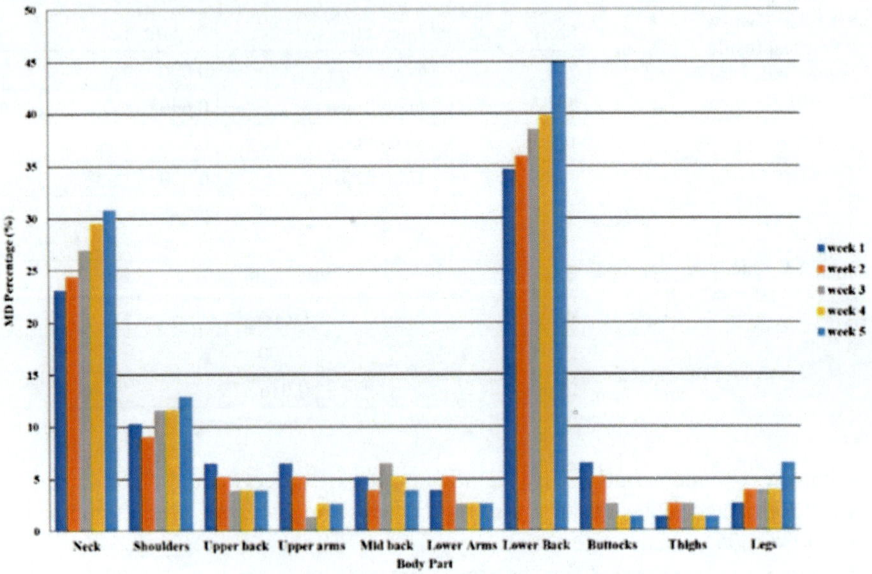

Fig. 5.7 Musculoskeletal disorder percentage by body part through five weeks

5.5 Discussion

Prior work has documented the evolution of physical and mental fatigue among shift workers, Barger et al. (2018) report higher levels of fatigue on medical services workers.

Dorrian et al. (2003) report fatigue in a subjective form in rail workers. Rosa et al. (1998) found that muscular fatigue increased more quickly across the nightshifts.

Schroeder et al. (1998) in a study about the effect of eight versus 10-h work schedule found a progressive increase of fatigue among a five-day week of eight hours and a four-day week of 10-h shifts in air traffic control workers. However, these studies have either been long-term studies or not focused on comparing fatigue evolution through the lapses of a day or the days in a week. In this study, we evaluate the fatigue through a fatigue index, for five continuous weeks, and during three lapses of time during the day, and through the days of a week. We found that fatigue was significantly more prevalent among workers of 12 h shift compared to fatigue in 7 h shift workers. Substantial differences in self-reported fatigue were found through the days and weeks among the workers of the two-night shifts, these results confirm the research hypothesis H_2 and H_3. An exception was found for fatigue between lapses of 7-h shifts where fatigue tended to have the same pattern for all lapses, which compromised the hypothesis H4.

Shen et al. (2006) determined whether the frequency of shifts labored by subjects had an influence on the severity of subjective fatigue they experienced, as measured by using the fatigue severity survey. An ANCOVA test was conducted, where frequency of shift work was used as an independent variable, FSS score was used as the dependent variable and gender was used as the covariant. It was found a significant group effect (frequency) indicating that the frequency of shift work had an effect on the severity of subjective fatigue experienced. The results of pairwise comparisons revealed a significant difference in the FSS scores between groups. A tendency of increasing the mean subjective fatigue scores of the groups as the frequency of shift work increased. In our study three ANCOVA tests were conducted where the fatigue index for 12 h shift was used as an independent variable, the week was used as a dependent variable and the fatigue index for 7 h shift was used as the covariate. Based on our results, the research hypothesis H1 has been proved.

The same ANCOVA test was also be done for the other two factors: day and lapse. Results showed there were found statistically significant differences between the different interventions (week, day, lapse) once their means had been adjusted for fatigue index 7 h. Our study had some limitations, linked to the time the workers had to answer the survey, during the lapses, due to the schedule to apply the survey interfered with their work. In this context we faced difficulties in gathering at specific timestamps during the night activity, as initially planned. Fortunately, collected data are equally distributed throughout the weeks, there was no data missing thus, we worked with balanced data.

Fatigue has been measured by uni-dimensional scales (Åshberg 2000) as they used in this study, in this context, this type of rating gave no information of qualitative differences between individuals which could limit the results.

Results from CBDM revealed that percentages of musculoskeletal disorders increased through the weeks, for neck, low back, and shoulders. Upper member musculoskeletal disorders could be related to physical fatigue, however, there are three rest periods, which two are fifteen minutes and the other is thirty minutes, which could help to recover from physical fatigue and then relieve some musculoskeletal impairments.

Further studies are needed to explore the influence of other factors associated with fatigue feelings of night shift workers, which include sleep deprivation and previous work-related activity, and probably larger samples are necessary to evaluate the effect of individual characteristics such as age and labor experience.

5.6 Conclusions

With most research to date focused on physical and mental fatigue among shift workers from different professional areas, which establish strong links between perceived fatigue and work schedule, specifically in long shifts, our results showed that fatigue among night shift workers had a similar trend, however, 12 h night shift workers could be somewhat more fatigued as a result of higher working hours per week as compared to 7 h night shift workers. Accumulated sleep debt could also be an explanation of 12 h night shift workers presented higher fatigue levels among employees.

An association was not found between fatigue and the lapses of the day, the mean of fatigue index followed the same pattern between the different lapses, which could lead us to think those rest periods are enough to recuperate from fatigue through the lapses of the day. The findings with regard to fatigue through the days of the week, revealed, that week fatigue index increased during days of the week for both of two-night shifts, this result has a possible association that the fact fatigue is an accumulative physical phenomenon. This investigation is an approach that may be useful to analyze rest breaks schedules for shift workers as a strategy to reduce fatigue feelings and improve productivity. However, some other aspects could be considered such those related to family conflicts, previous home activity and age, which could impact the relationship between fatigue and night work schedules.

References

Aaronson LS, Teel CS, Cassmeyer V, Neuberger GB, Pallikkathayil L, Pierce J, Press AN, Williams PD, Wingate A (1999) Defining and measuring fatigue. Image J Nurs Sch 31:45–50. https://doi.org/10.1111/j.1547-5069.1999.tb00420.x

Åshberg E (2000) Dimensions of fatigue in different working populations. Scandinavian Journal of Psychology 41: 231–241. https://doi.org/10.1111/1467-9450.00192

Akerstedt T, Fredlund P, Gillberg M, Jansson B (2002) Workload and work hours in relation to disturbed sleep and fatigue in a large representative sample. J Psychosom Res 53:585–588. https://doi.org/10.1016/S0022-3999(02)00447-6

Barger LK, Runyon MS, Renn ML, Moore GG, Weiss PM, Condle JP, Flickinger KL, Divecha AA, Coppler PJ, Sequeira DJ, Lang ES, Higgins JS, Patterson PD (2018) Effect of fatigue training on safety, fatigue, and sleep in emergency medical services personnel and other shift workers: a systematic review and metanalysis. Prehosp Emerg Care 22(Suppl. 1):58–68. https://doi.org/10.1080/10903127.2017.1362087

Berger AM, Hobbs BB (2006) Impact of shift work on the health and safety of nurses and patients. Clin J Oncol Nurs 10:465–471

Beurskens AJ, Bultmann U, Kant I, Vercoulen JH, Bleijenberg G, Swaen GM (2000) Fatigue among working people: validity of a questionnaire measure. Occup Environ Med 57:353–357. https://doi.org/10.1136/oem.57.5.353

Corlett EN, Bishop RP (1976) A technique for measuring postural discomfort. Ergonomics 9:175–182. https://doi.org/10.1080/00140137608931530

Costa G (2003) Shift-work and occupational medicine: an overview. Occup Med 53:83–88. https://doi.org/10.1093/occmed/kqg045

Dorrian J, Lamond N, Holmes AL (2003) The ability to self-monitor performance during a week of simulated night shift. Sleep 26(7):871–877. https://doi.org/10.1093/sleep/26.7.871

Dorrian J, Baulk SD, Dawson, D (2011) Work hours, workload, sleep and fatigue in Australian Rail Industry employees. Applied Ergonomics, 42(2): 202–209. https://doi.org/10.1016/j.apergo.2010.06.009

Folkard S, Tucker P (2003) Shift work, safety and productivity, Occupational Medicine, Volume 53: 95–101. https://doi.org/10.1093/occmed/kqg047

Hedges JN, Sekscenski ES (1979) Workers on late shifts in a changing economy. Mo Labor Rev 10:431–436

Hossain JL, Reinish LW, Kayumov L, Bhuiya P, Shapiro CM (2003) Underlying sleep pathology may cause chronic high fatigue in shift-workers. J Sleep Res 12:223–230. https://doi.org/10.1046/j.1365-2869.2003.00354.x

Jansen NWH, Van Amelswoort LGPM, Kristensen TS, Van den Brandt PA, Kant IJ (2003) Work schedules and fatigue: a prospective cohort study. Occup Environ Med 60(Suppl I):i47–i53. https://doi.org/10.1136/oem.60.suppl_1.i47

Lal SK, Craig A (2001) A critical review of the psychophysiology of driver fatigue. Biol Psychol 55:173–194. https://doi.org/10.1016/S0301-0511(00)00085-5

Layzer RB (1990) Muscle metabolism during fatigue and work. Baillieres Clin Endocrinol Metab 4:441–459

Office of Technology Assessment (1991) Biological Rhytms: Implications for the worker

Overall J (1993) Letter to the editor: the use of inadequate correlations for baseline imbalance remains a serious problem. J Biopharm Stat 3:271. https://doi.org/10.1080/10543409308835066

Paley MJ, Price JM, Tepas DI (1998) The impact of change in rotating shift schedules, a comparison of the effects of 8,10 and 14 hrs work shift schedules, International Journal of Industrial Ergonomics: 293–305. https://doi.org/10.1016/S0169-8141(97)00048-6

Pigeon WR, Sateia MJ, Ferguson RJ (2003) Distinguishing between excessive daytime sleepiness and fatigue: toward improved detection and treatment. J Psychosom Res 54:61–69. https://doi.org/10.1016/S0022-3999(02)00542-1

Rosa RR, Bonnet MH, Cole LL (1998) Work Schedule and task factors in upper-extremity fatigue. Hum Factors 40:150–158. https://doi.org/10.1518/001872098779480523

Rosekind MR, Gander PH, Gregory K (1996) Managing fatigue in operational settings: Physiological considerations and countermeasures. Behavioral Medicine 21: 256–264. https://doi.org/10.1080/08964289.1996.9933753

Samaha E, LaL S, Samaha N, Wyndham J (2007) Psychological, lifestyle and coping contributors to chronic fatigue in shift-worker nurses. J Adv Nurs 59(3):221–232. https://doi.org/10.1111/j.1365-2648.2007.04338.x.

Schroeder D, Rosa RR, Witt A (1998) Some effects of 8-versus 10-hour work schedules on the test performance/alertness of air traffic control specialists. Int J Ind Ergon 21:307–321. https://doi.org/10.1016/S0169-8141(97)00044-9

Senn S (1994) Testing for baseline balance in clinical trials. Stat Med 13(17):1715–1726. https://doi.org/10.1002/sim.4780131703

Shen J, Botly LCP, Chung SA, Gibbs AL, Sabanadzovic S, Shapiro CM (2006) Fatigue and Shift Work. J Sleep Res 15:1–5. https://doi.org/10.1111/j.1365-2869.2006.00493.x

Swaen GMH, Van Amelsvoort LGPM, Bultmann U, Kant IJ (2003) Fatigue as a risk factor for being injured in an occupational accident: results from the Maastricht Cohort Study. Occup Environ Med 60(Suppl. 1):i88–92. https://doi.org/10.1136/oem.60.suppl_1.i88

Yoshitake H (1978) Three characteristic patterns of subjective fatigue symptoms. Ergonomics 21(3):231–233. https://doi.org/10.1080/00140137808931718

Chapter 6
Different Conceptions of Burnout and Its Relationships with Job Strain and Emotional Intelligence

Miguel A. Serrano, Yasmina El Arbi, and Raquel Costa

Abstract Burnout is an emotional syndrome that involves a prolonged response to stress. Its study in the work environment is of interest, since it manifests itself through negative attitudes toward work and affects the emotional, attitudinal, and physical level of the person who suffers it. The most relevant instruments for measuring burnout are the Maslach Burnout Inventory (MBI), most commonly used, and the Shirom-Melamed Burnout Measure (SMBM). Likewise, taking into account the emotional nature of the Burnout, it is related to Emotional Intelligence (EI), playing a very important role in the adaptive capacity of individuals in stressful situations. The aim of the present study is (1) to analyze the differences and similarities of the questionnaires that evaluate Burnout, (2) as well as to study how these scales relate to Job Strain and EI. 60 workers (37 women) answered 4 scales measuring the main variables. Results show that both the MBI and the SMBM are significantly related. Secondly, Job Strain predicts the subscales of both questionnaires, unlike EI, which does not correlate significantly with these Burnout indicators. We conclude that both questionnaires are valid instruments for measuring burnout syndrome and they have a significant relationship with work stress but no with EI.

Keywords Burnout · Shirom-Melamed Burnout Measure · Maslach Burnout Inventory · Emotional Intelligence · Job Strain

6.1 Introduction

Burnout is a well-studied psychosocial risk recently highlighted in scientific literature (Caballero et al. 2010), which consequences are not only evident in the worker, affecting people at the cognitive, emotional, behavioral, and physical levels (Beltrán et al. 2004), but also presents negative consequences for the organization, manifesting itself in the form of progressive deterioration of communication, productivity, and quality of work. Despite being a frequent phenomenon, its origins as a concept are not

M. A. Serrano (✉) · Y. El Arbi · R. Costa
Department of Psychobiology, University of Valencia, 46010 Valencia, Spain
e-mail: m.angel.serrano@uv.es

A. Realyvásquez Vargas et al. (eds.), *New Perspectives on Applied Industrial Ergonomics*,
https://doi.org/10.1007/978-3-030-73468-8_6

new, since the term was used for the first time by Freudenberger (Martínez 2010), who in 1974 used this concept to refer to a behavioral pattern characterized by a progressive loss of energy, demotivation, and lack of interest in work (Freudenberger 1974). However, Burnout is not only a term that has been used at a clinical or organizational level, but has been extrapolated to other areas (Brill 1984), which, along with a lack of conceptual limitation with respect to occupational stress, has been highly criticized and questioned. Therefore, one of the most notable differences regarding occupational stress is that the Burnout is a chronic state difficult to overcome, where the high expectations of the worker that have not been carried out, lead to a great disappointment that makes people feel "burned out," little involved in their work and exhausted, hence the concept of "burnout syndrome" (Martínez 2010).

There are different conceptualizations for measuring the Burnout. First is carried out by Maslach, who defined this term as a behavioral manifestation of occupational stress (Maslach 1976). This conception understands burnout syndrome as a three-dimensional phenomenon, which is characterized by emotional fatigue—exhaustion—depersonalization in dealing with clients and a difficulty in personal realization—professional effectiveness—and from these components Maslach et al. elaborated the "Maslach Burnout Inventory" (MBI). This scale has become the most accepted due to its wide use in different populations and in different types of studies, hence its multiple revisions, in the last of which the term depersonalization has been replaced by "cynicism." On the other hand, a different conceptualization is the one offered by Shirom. Shirom (2003) defended that the Burnout comes to be developed after a prolonged time of exposure to a work stress and, for that reason Shirom considers burnout as a consequence or symptom of stress. Shirom's burnout conception is related exclusively to the energetic resources and includes physical energy, and emotional and cognitive dimensions. These dimensions are part of the conceptualization of the Burnout, since he distinguishes in his theory three determining facets for its development. The first of the three facets refer to physical fatigue—sensation of tiredness and low energy-; the second refers to emotional exhaustion -lack of energy to show empathy to others-; and, finally, cognitive fatigue, understood as the sensation of reduced mental agility (Shirom 2009).

In spite of the conceptual differences from both conceptions and its development, there are not many studies that have been in charge of comparing both forms of measuring the Burnout. Based on the previous definitions offered by both authors, it is possible to understand the Burnout as the consequence of a high level of stress, as we can see reflected in studies that relate work stress to Burnout. Karasek, defined Job Strain from an organizational point of view as a high demand for job—greater pressure and act quickly to tasks—and less control—less ability to decide to face the situation—all presenting a higher cardiovascular risk for the person (Karasek 1979). Therefore, a conceptualization of Burnout focused on the organizational environment is understood as a prolonged response to occupational stress through negative behaviors or attitudes toward work (Gil-Monte 2002). In addition, it influences the person on an emotional, attitudinal, and physical levels (Marrau 2004). Specifically, it has been found that the Job Strain—concept that describes working conditions that combine high demands and low opportunities of control—is highly related to Burnout indicators (Ortiz et al. 2014).

Finally, considering that Burnout is an emotional syndrome, it is important to study how it is related to emotions in work environments. Thus, a fundamental aspect of emotions is Emotional Intelligence (EI), which has been related to healthy working environments (Garrido and Pacheco 2012). This term was widely conceptualized by Goleman (1995), who considers EI as a combination of a series of attributes related to personality and which is closely linked to the competencies linked to professional achievement (Pérez Pérez 2006). Literature that links Burnout with EI indicates that EI acts as a protective factor of this concept (Mikolajczak et al. 2007) where, in addition, related studies indicate that this relationship of EI on Burnout directly affects the dimensions that characterize this syndrome (Extremera et al. 2003; Görgens-Ekermans and Brand 2012; Álvarez-Ramírez et al. 2017).

Taking into account all the above, we aimed, firstly, to analyze the relationships between two Burnout scales (Maslach and Shirom questionnaires), hypothesizing that both scales would be related taking into account that both conceptions consider burnout as a consequence of work stress. The second aim of this study was to study the relationship between Burnout (measured with both scales) and Job Strain and Emotional Intelligence. It is hypothesized that Job Strain and Emotional Intelligence would explain Burnout scores.

6.2 Method

6.2.1 Sample

Sample was composed of 60 workers (37 women), aged between 20 and 65 (M = 39.28; SD = 11.7) who voluntarily participated in this study. After explaining the main aims of the study, participants read and signed the informed consent approved by the Ethics Committee of Universidad Miguel Hernández de Elche (Spain).

6.2.2 Procedure

To carry out this study, a set of questionnaires was administered to employees of different companies of different sectors (education and convenience stores). Questionnaires were answered individually, voluntarily, and anonymously.

6.2.3 Measures

First, Burnout was measured using Maslach Burnout Inventory-General Survey (Schaufeli et al. 1996), adapted to Spanish by Salanova et al. (2000). This questionnaire consists of 16 items, where a total Burnout score can be obtained (Total Burnout = Exhaustion + Cynicism - Professional Efficacy), as well as a score for each of the three subscales: Exhaustion, Cynicism, and Professional Efficacy. Salanova et al. (2000) found alphas coefficients of 0.85 (emotional burnout), 0.78 (cynicism), and 0.73 (professional efficacy).

Secondly, Shirom-Melamed Burnout Measure questionnaire was also used (Melamed et al. 1999), which consists of 12 items. Items show different sensations/symptoms that can be experienced at work, thus evaluating three dimensions of Burnout: physical fatigue, emotional exhaustion, and cognitive weakness. In the sample a Cronbach's alpha of 0.88 was obtained in the physical fatigue factor, 0.81 in emotional exhaustion, and 0.89 in the cognitive weakness factor.

Job Strain was evaluated using the Work Stress Questionnaire by Karasek and Theorell (1990) in the version adapted by Steptoe et al. (1999). The questionnaire consists of 15 items where the dimensions of demands at work, latitude of decision or control and social support are evaluated. A total index of Job Strain was obtained: demands / (control + use of skills). Cronbach's alpha obtained by Steptoe et al. (1999) were 0.72 (demands at work), 0.64 (latitude of decision or control), and 0.76 (social support).

TMMS-24 was used to measure Emotional Intelligence. This version translated into Spanish by Fernández-Berrocal and Ramos (1999) is based on Trait Meta-Mood Scale (TMMS) from Salovey and Mayer's research group (Salovey et al. 1995). The original scale evaluates the meta-knowledge of emotional states—skills with which we can be aware of our own emotions as well as our capacity to regulate them—through 24 items. The TMMS-24 scale contains three dimensions: Emotional Attention, Clarity of Feelings, and Emotional Repair. The reliability obtained for each component is an alpha of 0.90 for Emotional Attention, 0.90 for Clarity of feeling, and 0.86 for Emotional Repair.

6.2.4 Statistical Analyses

Pearson's correlations were used to analyze the relationship between the MBI and SMBM subscales. When correlations resulted significant, simple linear regressions were performed, being Job Strain and EI variables the predictors of the total burnout in the case of the MBI and of the 3 scales in the case of the SMBM questionnaire. Statistical analysis was carried out with the SPSS 20.0 package with a level and significance of 0.05.

6.3 Results

6.3.1 Relationship Between Maslach and Shirom Burnout Subscales

The physical fatigue of the SMBM questionnaire correlates significantly with exhaustion ($r = -0.744$, $p < 0.001$), cynicism ($r = -0.433$, $p < 0.001$), and the Burnout Total variable ($r = -0.587, p < 0.001$) of the MBI questionnaire. On the other hand, emotional fatigue correlates with exhaustion ($r = -0.781, p < 0.001$), cynicism ($r = -0.430, p < 0.001$), and the variable Burnout Total ($r = -0.596, p < 0.001$). Finally, cognitive weakness correlates with exhaustion ($r = -0.722, p < 0.001$), cynicism ($r = -0.551, p < 0.001$), and Total Burnout variable ($r = -0.679, p < 0.001$).

6.3.2 Job Strain as a Predictor of the SMBM and MBI

First, Pearson's correlations were performed between Job Strain and the subscales of both burnout tests (MBI and SMBM). Job strain was related to Total Burnout from MBI score were positively ($r = 0.335; p = 0.001$). In the case of SMBM, Job Strain correlated significantly with Physical Fatigue ($r = -0.431$; $p = 0.001$), Emotional Fatigue ($r = -0.448; p = 0.001$), and Cognitive Weakness ($r = -0.316; p = 0.001$).

Secondly, four different linear regressions were carried out to measure whether the work stress influenced the subscales of both Burnout tests equally. Total Burnout, Physical Fatigue, Emotional Fatigue, and Cognitive Weakness were used as dependent variables and Workplace Stress as independent. In all four cases, the results were significant.

As for the results obtained, they show how Job Strain significantly predicts both the MBI and SMBM subscales (see Table 6.1). Specifically, it significantly explains 11.2% of the variance in Total Burnout ($R^2 = 0.112, p < 0.001$), 18.6% for Physical Fatigue ($R^2 = 0.186$, $p < 0.001$), 20.1% of the variance in Emotional Fatigue (R^2

Table 6.1 Job Strain predicting Total Burnout (Maslach) and scales of Burnout (Shirom-Melamed). DV = dependent variable; IV = independent variable

Variables	R^2	*Beta*	*t*
DV: Burnout Total	0.112***		
IV: Job Strain		0.335	2.709***
DV: Physical Fatigue	0.186***		
IV: Job Strain		−0.431	−3.639***
DV: Emotional exhaustion	0.201***		
IV: Job Strain		−0.448	−3.817***
DV: Cognitive fatigue	0.100***		
IV: Job Strain		0.316	−2.539**

$= 0.201, p < 0.001$) and, finally, 10% of the variance in Cognitive Weakness ($R^2 = 0.100, p < 0.001$).

6.3.3 Emotional Intelligence as a Predictor of MBI and SMBM

No significant relationships were found between EI and both Burnout scales.

The scores obtained for the EI subscale, Emotional Attention were, in relation to the SMBM subscales; physical fatigue ($r = -0.191, p > 0.1$), emotional exhaustion ($r = -0.168, p > 0.1$), cognitive weakness ($r = -0.182, p > 0.1$) and for the MBI subscales; exhaustion ($r = 0.141, p > 0.1$), cynicism ($r = 0.010, p > 0.1$), professional efficiency ($r = 0.152, p > 0.1$).

In relation to the EI subscale, Emotional Clarity was, for the MBI subscales; physical fatigue ($r = 0.207, p > 0.1$), emotional exhaustion ($r = 0.165, p > 0.1$), cognitive weakness ($r = 0.167, p > 0.1$) and in relation to the MBI subscales; exhaustion ($r = -0.227, p > 0.1$), cynicism ($r = -0.092, p > 0.1$), professional efficacy ($r = 0.134, p > 0.1$).

Finally, the EI subscale, Emotional Repair as for the SMBM subscales were physical fatigue ($r = 0.142, p > 0.1$), emotional exhaustion ($r = 0.000, p > 0.1$), cognitive weakness ($r = 0.132, p > 0.1$), and for the MBI subscales; exhaustion ($r = -0.126, p > 0.1$), cynicism ($r = 0.002, p > 0.1$), and professional efficacy ($r = -0.048, p > 0.1$).

6.4 Discussion

In view of the increase in studies related to Burnout (Leiter 2017; Pines 2017; Shanafelt et al. 2017), it has been considered especially relevant to analyze the characteristics of the two most studied instruments in a theoretical way (Bianchi et al. 2018), the MBI and SMBM, the former being the most considered in its practical utility in work contexts and scientific environment (Bianchi et al. 2015). From this controversy arises the first hypothesis of the present study.

Results showed how this controversy is reduced to terms of greater scientific diffusion (Faúndez 2017), since both instruments are equivalent in both validity and reliability and, in their factor-to-factor comparison, positive and significant data result. Although it should be mentioned that there is limited scientific literature that confronts both questionnaires in their practical use, the study carried out by Shirom and Melamed (Shirom and Melamed 2006) shows how the SMBM is characterized by a better adjustment between factors, considered other aspects irrelevant between the distinction of both tests. This more detailed analysis on purely statistical issues confirms the relevance and results of the present study.

One of the possible factors related to the use of MBI versus SMBM is considered to be the global score index that arises from the three levels of analysis used by the MBI and that would not be possible to carry out in SMBM (Moreno-jimenez 2001). Despite this, results obtained in the research emphasize that the factorization process in both instruments is significant.

On the other hand, there is a fact that should not generate dissonance in the reader, since it is the object of study of a growing number of scientific articles due to its great impact and interest in the labor environment, that is the study of Job Strain as a predictive factor of Burnout in environments with high levels of stress (Wong and Spence Laschinger 2015; Jiang et al. 2017). It is worth mentioning, therefore, the relevance of the present research, being considered as fundamental implication of both measures of burnout (MBI and SMBM) as predicted factors. Results show that Job Strain is a significant predictor of both the MBI and SMBM. This fact is not surprising, since some studies indicate that a possible trigger of high demands and little control in the work environment (Karasek 1998) lead to suffer Burnout as time goes by and the accumulation of such tension (Chirico 2016). For that reason, the acquisition of new skills and changes in coping styles is recommended for people whose levels of Job Strain are high and interfere their daily activities as well as their normative development at work (Vander Elst et al. 2016).

Finally, in relation to EI, all the correlational analyses showed no relationship between EI and BO scales. These results emphasize the importance of continuing to study EI, since there are studies which show results opposite to those obtained in the present research (Extremera et al. 2003, 2007). Although these studies do not emphasize the predictive power of EI on Burnout, others consider personality factors as a modulating factor between both do show significant results (González et al. 2014). Thus, personality traits such as Neuroticism and Extraversion (Jiménez Morales and López Zafra 2008) acquire special relevance on how EI could predict i.e. emotional exhaustion—a characteristic feature of Burnout. Therefore, not contemplating such personal characteristics could affect the results obtained in this research.

Some limitations should be remarked from this study. First, the reduced number of participants. In order to increase the validity of these results, bigger samples are needed. In this line, bigger samples would contribute to the development of normative data that could help to compare burnout levels transnationally. Another limitation is that in our study only two different types of jobs were evaluated, so different types of jobs would increase the extension of these measures and results to different types of workers. In this sense, the majority of studies on Burnout are performed in professions as teachers, nurses, or physician so more studies in industrial settings are necessary in order to describe burnout and its relationships in jobs with different contextual characteristics.

With reference to possible future research, it would be interesting to review the MBI and SMBM instruments in greater depth, in order to determine if they are totally valid measures. In addition, it is necessary to contribute to the analysis of new factors, especially emotional ones, which can predict Burnout, since it is a syndrome that is increasingly found in work environment. It is also necessary to emphasize not only on Job Strain as a predictive factor of Burnout and EI in future research, but also to focus studies on a more positive view on how to face such stress and, in case of suffering it, how to reduce the impact on other aspects of the individual's life.

In short, in spite of the number of participants' limitation, results should serve to promote new studies focusing on new ways of approaching and predicting Burnout since the ultimate goal is to entail the improvement of quality of life and workers' health. Furthermore, results also show the need to study more deeply the effects of EI and Job Strain on Burnout. Therefore, promoting an adequate work climate and providing new skills (associated with reduction of work stress) to workers (Ortega and Ortega 2017), and emotional strategies should be key to reduction of Burnout probability risk (Back et al. 2016).

References

Back AL, Steinhauser KE, Kamal AH, Jackson VA (2016) Building Resilience for Palliative Care Clinicians: An Approach to Burnout Prevention Based on Individual Skills and Workplace Factors. J Pain Symptom Manage 52:284–291. https://doi.org/10.1016/J.JPAINSYMMAN.2016.02.002

Beltrán C, Pando M, Pérez M (2004) Social support and burnout: a revision [In Spanish]. Psicol Salud 14:79–87

Bianchi R, Schonfeld IS, Laurent E (2018) Burnout Syndrome and Depression. In: Kim Y-K (ed) Understanding Depression, vol 2. Clinical Manifestations. Diagnosis and Treatment. Springer Singapore, Singapore, pp 187–202

Bianchi R, Schonfeld IS, Laurent E (2015) Burnout–depression overlap: A review. Clin Psychol Rev 36:28–41. https://doi.org/10.1016/J.CPR.2015.01.004

Brill PL (1984) The need for an operational definition of burnout. Fam. Community Heal. J. Heal. Promot. Maint. 6:12–24

Caballero C, Hederich C, Palacio J (2010) Academic Burnout: syndrome delimitation and associated factors [In Spanish]. Rev Lat Psicol 42:131–146

Chirico F (2016) Job stress models for predicting burnout syndrome: a review. Ann Ist Super Sanita 52:443–456. https://doi.org/10.4415/ANN_16_03_17

de Álvarez-Ramírez M, los R, Pena Garrido M, Losada Vicente L, (2017) Mission possible: improve the welfare of counselors through their emotional intelligence [In Spanish]. REOP - Rev Española Orientaci6n Y Psicopedag 28:19–32. https://doi.org/10.5944/reop.vol.28.num.1.2017.19356

Extremera N, Durán A, Rey L (2007) Emotional intelligence and its relationship with the levels of burnout, engagement and stress in university students [In Spanish]. Rev Educ 342:239–256

Extremera N, Fernández-Berrocal P, Durán A (2003) Emotional intelligence and burnout in teachers [In Spanish]. Enc Psicol. Soc. 1:260–265

Faúndez V (2017) Cristina Maslach, understanding burnout [In Spanish]. Ciencia & Trabajo 2017 [acceso 20 de setiembre de 2018]; 19(58):59–63. https://doi.org/10.4067/S0718-24492017000100059

Fernández-Berrocal P, Alcaide R, and Ramos N (1999) The influence of emotional intelligence on the emotional adjustment in highschool students. Bull Kharkov State Univ 439(1–2):119–123

Freudenberger HJ (1974) Staff Burn-Out. J Soc Issues 30:159–165. https://doi.org/10.1111/j.1540-4560.1974.tb00706.x

Garrido MP, Pacheco NE (2012) IPerceived emotional intelligence in primary and secondary schools and its relationship with burnout levels and work illusion [In Spanish]. Rev Educ 359:604–627. https://doi.org/10.4438/1988-592X-RE-2011-359-109

Gil-Monte PR (2002) Factorial validity of the Spanish adaptation of Maslach Burnout Inventory-General Survey [In Spanish]. Sal Publ Mex 44:33–40

Goleman DP (1995) Emotional intelligence: why it can matter more than IQ for character, health and lifelong achievement. Bantam, New York
González J, Ros AB, Jiménez MI, De Los G, Fayos E (2014) Analysis of burnout levels in athletes according to the level of perceived emotional intelligence: The moderating role of personality [In Spanish]. CDP 14:39–48. https://doi.org/10.4321/S1578-84232014000300005
Görgens-Ekermans G, Brand T (2012) Emotional intelligence as a moderator in the stress-burnout relationship: A questionnaire study on nurses. J Clin Nurs 21:2275–2285. https://doi.org/10.1111/j.1365-2702.2012.04171.x
Jiang L, Tripp TM, Hong PY (2017) College instruction is not so stress free after all: A qualitative and quantitative study of academic entitlement, uncivil behaviors, and instructor strain and burnout. Stress Heal 33:578–589. https://doi.org/10.1002/smi.2742
Jiménez Morales M, López Zafra E (2008) Emotional self-concept as an emotional risk factor in college students: gender and age differences [In Spanish]. Bol Psicol 21–39
Karasek RA (1979) Job demands, job decision latitude, and mental strain: implications for Job Redesign. Adm Sci Q 24:285–308. https://doi.org/10.2307/2392498
Karasek RA (1998) Demand/Control model: a social-emotional, and psycho-logical approach to stress risk and active behavior development. ILO, In ILO encyclo-pedia of occupational health and safety
Karasek R, Theorell T (1990) Healthy work: stress, productivity, and the reconstruction of working life. Basic Books, New York
Leiter MP (2017) Burnout as a developmental process: consideration of models. In Professional burnout. Routledge, pp 237–250
Marrau M (2004) Burnout Syndrome and its possible consequences on teachers [In Spanish]. Fundam Hum 10:53–68
Martínez A (2010) Burnout Syndrome. Conceptual evolution and current state of the issue [In Spanish]. Rev Comun Vivat Acad 112:42–80. https://dx.doi.org/10.15178/va.2010.112.42-80
Maslach C (1976) Burned-out. Hum Behav 5:16–22
Melamed S, Ugarten U, Shirom A et al (1999) Chronic burnout, somatic arousal and elevated salivary cortisol levels. J Psychosom Res 46:591–598. https://doi.org/10.1016/s0022-3999(99)00007-0
Mikolajczak M, Menil C, Luminet O (2007) Explaining the protective effect of trait emotional intelligence regarding occupational stress: exploration of emo-tional labour processes. J Res Pers 41(5):107–117
Moreno-jimenez B (2001) Professional burnout evaluation. MBI-GS factorialization. A preliminary analysis [In Spanish]. Ansiedad y Estrés 69–78
Ortega MJ, Ortega OJ (2017) Work climate: effect of professional burnout on quality of life at work [In Spanish]. Revista Empresarial 11(1):6–13
Ortiz VG, Toro LEP, Rodríguez AMH (2014) Moderation of the relationship between labor tension and discomfort of university professors: role of conflict and facilitation between work and family [In Spanish]. Rev Colomb Psicol 24:185–201. https://doi.org/10.15446/rcp.v24n1.42081
Pérez Pérez N (2006) Relationships between emotional intelligence and IQ with academic performance in college students [In Spanish]. Reme 9:6
Pines AM (2017) Burnout: an existential perspective. In Professional burnout. Routledge, pp 33–51
Schaufeli WB, Leiter MP, Maslach Ch, Jackson SE (1996) Maslach Burnout Inventory – General Survey. In: Maslach Ch, Jackson SE, & Leiter MP (eds), The Maslach Burnout Inventory – Test Manual, 3rd end, Palo Alto: Consulting Psychologists Press
Salanova M, Schaufeli W, Llorens S, Peiró JM, Grau R (2000) From "Burnout" to "Engagement". A new perspective [In Spanish]. Revista de Psicología del Trabajo y de las Organizaciones 16(2):117–34
Salovey P, Mayer JD, Goldman SL, Turvey C, Palfai TP (1995) Emotional attention, clarity, and repair: exploring emotional intelligence using the Trait Meta-Mood Scale. In: Pennebaker JW (ed) Emotion, disclosure, & health, pp 125–154
Shanafelt TD, Dyrbye LN, West CP (2017) Addressing physician burnout: The Way Forward. JAMA 317:901–902. https://doi.org/10.1001/jama.2017.0076

Shirom A (2003) Job-related Burnout: A review. In: Quick JC & Tetrick LE (eds), Handbook of occupational health psychology. American Psychological Association, pp 245–264
Shirom A (2009) About the validity of the construct, predictors and consequences of Burnout in the workplace [In Spanish]. Ciencia & Trabajo 32(11):44–54
Shirom A, Melamed S (2006) A comparison of the construct validity of two burnout measures in two groups of professionals. Int J Stress Manag 13:176–200. https://doi.org/10.1037/1072-5245.13.2.176
Steptoe A, Cropley M, Joekes K (1999) Job strain, blood pressure and re-sponse to uncontrollable stress. J Hypertens 17(2):193–200
Vander Elst T, De Cuyper N, Baillien E et al. (2016) Perceived control and psychological contract breach as explanations of the relationships between job insecurity, job strain and coping reactions: towards a theoretical integration. Stress Health 32:100–116. https://doi.org/10.1002/smi.2584
Wong CA, Spence Laschinger HK (2015) The influence of frontline manager job strain on burnout, commitment and turnover intention: a cross-sectional study. Int J Nurs Stud 52:1824–1833. https://doi.org/10.1016/j.ijnurstu.2015.09.006

Chapter 7
Ergonomics Implementation in Manufacturing Industries: Management Commitment for Financial Benefits

Aide Aracely Maldonado-Macías, Cesar Roberto Alferez-Padrón, Manuel Alejandro Barajas-Bustillos, Oziely Daniela Armenta-Hernández, Arturo Realyvásquez Vargas, and Cesar Omar Balderrama-Armendáriz

Abstract Ergonomics is a multidisciplinary science that seeks to adapt systems and products to human capacities and limitations. Therefore, Ergonomics leads to benefits in terms of health, quality of life, and safety, as well as to financial benefits. However, its implementation process is slow and complex as it takes effective analysis to determine critical factors for a successful implementation and the attainment of such benefits. This chapter offers an analysis by structural equations modeling to identify those critical success factors as well as the way they relate to financial benefits. Four stages in the implementation process were considered and a 114-item questionnaire was developed and validated for data collection within manufacturing companies. The questionnaire was administered to a sample of 139 individuals in middle and upper management positions, showing direct, indirect and total effects of Ergonomics implementation and the role of management commitment standing out as a critical factor to obtain financial benefits.

A. A. Maldonado-Macías (✉)
Industrial Engineering and Manufacturing Department, Universidad Autónoma de Ciudad Juárez, Ciudad Juárez, Chihuahua, Mexico
e-mail: amaldona@uacj.mx

C. R. Alferez-Padrón
Industrial Engineering and Manufacturing Department, Autonomous University of Ciudad Juárez, Ciudad Juárez, Chihuahua, Mexico

M. A. Barajas-Bustillos · O. D. Armenta-Hernández
Electrical and Computing Department, Autonomous University of Ciudad Juárez, Ciudad Juárez, Chihuahua, Mexico

A. Realyvásquez Vargas
Department of Industrial Engineering, Tecnológico Nacional de México/I.T. Tijuana, Tijuana, Baja California, Mexico
e-mail: arturo.realyvazquez@tectijuana.edu.mx

C. O. Balderrama-Armendáriz
Industrial Design Department, Autonomous University of Ciudad Juárez, Ciudad Juárez, Chihuahua, Mexico

A. Realyvásquez Vargas et al. (eds.), *New Perspectives on Applied Industrial Ergonomics*,
https://doi.org/10.1007/978-3-030-73468-8_7

Keywords Critical Success Factors · Structural Modeling

7.1 Introduction

Ergonomics is the science responsible for adjusting work to the worker (Laos et al. 2007). This science seeks to ensure spaces, products, and systems based on users' capabilities and limitations (Harne and Deshmukh 2016). With the purpose of successfully implement Ergonomics, the manufacturing industry has begun to execute Ergonomics programs through the involvement of a multidisciplinary team (Broberg et al. 2011). The correct implementation of an Ergonomics programs begins with careful planning, followed by a rational use of material, human and financial resources, yet even after that, implementing the program can still be a slow, painstaking process (Liker and Chaffin 1987).

Several authors have proposed different models for the implementation of Ergonomics. For example, Fig. 7.1 shows one diagram from the perspective of a small company (Stuart-Buttle 2006). Figure 7.2 shows the model proposed by Kilbom and Petersson (2006), which outlines six steps in the successful implementation of Ergonomics. On the other hand, Liker and Chaffin (1987) set forth a model consisting of four stages: planning, starting the process, work-improvement cycle,

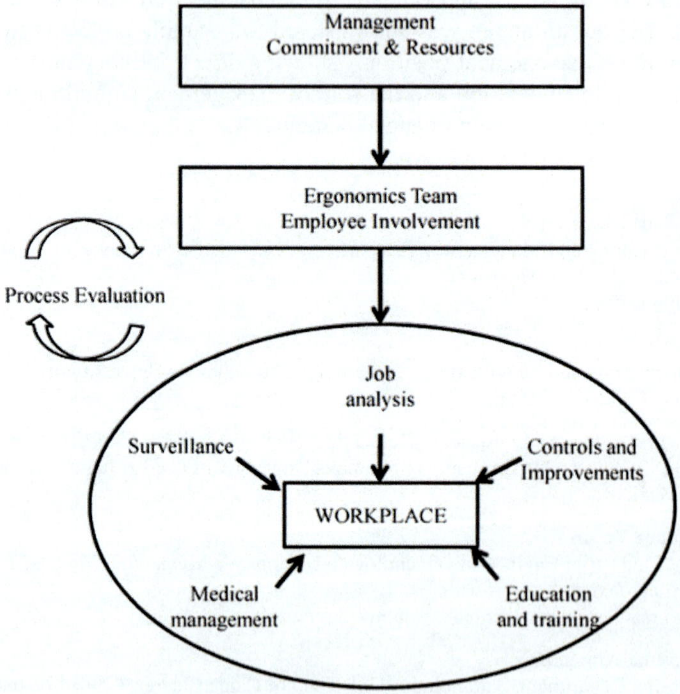

Fig. 7.1 Elements of an Ergonomics process (*Source* Stuart-Buttle [2006])

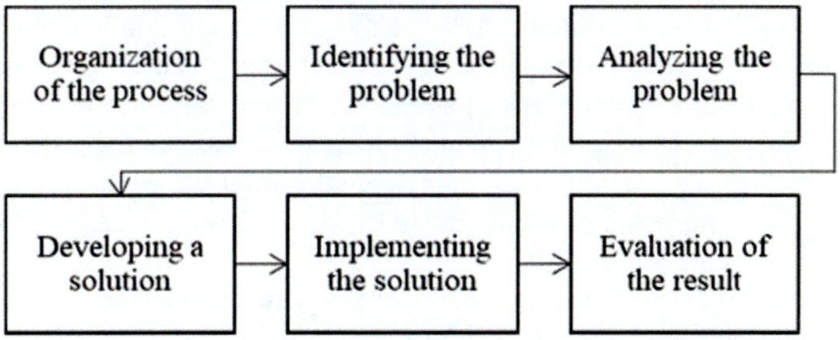

Fig. 7.2 Steps in the conduction of an Ergonomics process according to Kilbom and Petersson (2006)

and long-term development. The large number of models and critical factors scattered throughout the literature makes it difficult for a company to clearly identify a model that can describe the main critical success factors as well as the benefits associated with them.

7.2 Problem Statement

Although companies are making increasing efforts to maintain their competitiveness in a globalized world through the implementation of Ergonomics, there is evidence of the difficulties they are still encountering in the process. This section will outline both the main factors hindering the implementation of Ergonomics and the direct and indirect costs resulting from not implementing it. Furthermore, the section explains a few of the problems that companies have faced even after implementing Ergonomics programs. These evidences offer the opportunity to research and find out the critical success factors involved in the implementation of Ergonomics.

7.2.1 Difficulties and Costs Associated with the Implementation of Ergonomics

In industry, the obstacles for the implementation of Ergonomics impede the improvement of multiple aspects and requirements in the work area, for example, minimizing stress and workload, facilitating task execution, ensuring occupational safety and health, and achieving the ease of use of the elements in a work area are al challenges of implementation (Marmaras and Nathanael 2012). Authors have informed that difficulties in implementing Ergonomics programs entail several elements. Among them are, the underprovided performance of the management (Ip and Rostykus 2014). Additionally, the scarce communication among staff difficults proper feedback for ergonomic problems (Skepper et al. 2000) and, the awareness that the program can be expensive and time-consuming usually delays an adequate implementation (Dul and Neumann 2009). Other issues can be the lack of management commitment, problems in the organization and the lack of attention of the members of the Ergonomics committee, the lack of consultant support, the cost of interventions, the lack of a specific methodologies and policies to implement the program. All these factors have thus far affected the proper implementation of Ergonomics programs (Haslegrave and Holmes 1994; Green 2002).

Complications mentioned can cause companies to respond leisurely to health and safety challenges that may increase their costs accordingly. When work accidents, or illnesses occur, companies can incur in two types of costs: direct and indirect costs. Direct costs are related with the medical care, the quality of insurance coverage, and the recovery cost of the injured personnel. Examples of these are the treatment costs (from primary care, specialized care, and therapies), insurance compensation, and court costs (Douphrate and Rosecrance 2004; Tappin et al. 2016). On the other hand, there are also indirect costs, which are difficult to pinpoint. Though intangible, however, these costs affect the company the most. Some examples of indirect costs are (1) the day lost at the time the injury occurred, which affects employees and productivity; (2) the time lost by supervisors' investigating the accident; (3) the time lost from training a person to replace the injured worker; (4) the loss in production time when a machine is on hold or waiting either for the injured employee to be replaced or for an employee who is just learning; (5) low morale among employees; (6) increased absenteeism, (7) presentism: When an injured employee returns to work but is less productive than he was in a healthy state (Oregon Occupational Safety and Health, n/a; Schulze 2010; d'Errico et al. 2013).

Due to the above reasons, there is a need for companies to have clear and simple information in order to plan successfully and manage resources appropriately to correctly implement the program and enjoy its benefits.

7.2.2 Problems and Costs Associated with the Implementation of Ergonomics

This next section lists some of the difficulties that companies face when implementing an Ergonomics program.

A first main problem, as Hendrick (2003) points out, is that companies are not clear about the prioritization of activities in order to implement the program. Additionally, some ergonomists lack the necessary knowledge in administration or resource management to implement a program. Haines and Wilson (1998) describe that, often, the expert in charge of the program has worked isolated on the project, thus making its implementation unsuccessful. Other problems that have been identified are the lack of a specific methodology to follow as well as the lack of program implementation policies (Haslegrave and Holmes 1994; Green 2002). Regarding management, the committee or the person in charge of carrying out the program often fails to identify the savings that implementing the program creates for the company (Beevis 2003).

7.3 Objectives

The general objective of this work was to identify the critical success factors that directly influence the financial benefits of implementing Ergonomics programs during the stages of planning, program launching, work-improvement cycle, and long-term development, with a focus on the exporting manufacturing industry in Ciudad Juárez. Following are the specific objectives of this research:

- To develop a methodology to identify critical success factors and financial benefits.
- To develop structural models to relate CSF with the financial benefits.
- To identify the critical success factors and their impact on the financial benefits of implementing Ergonomics.

7.4 Methodology

This section describes both the materials and the method that were used in this research. Regarding materials, a questionnaire developed by Maldonado-Macías Alferez-Padron, García-Alcaraz, and Avelar-Sosa (2017) was used to identify the critical success factors and the benefits of implementing Ergonomics in the manufacturing industry in Ciudad Juárez. The method, on the other hand, proposed following seven stages based on Lévy and Varela (2003). These stages are: (1) data preparation, (2) model specification, (3) model identification, (4) parameter estimation, (5) model adjustment, (6) model interpretation, and (7) model re-specification.

7.4.1 Materials

The questionnaire used to identify the critical success factors and the benefits of implementing Ergonomics in the manufacturing industry in Ciudad Juárez measured the frequency with which activities are carried out to implement Ergonomics programs (Maldonado-Macías et al. 2017). The frequency was measured on a five-point Likert scale, where 1 was never and 5, always. The instrument encompassed 17 areas and was made up of a total of 114 questions. Table 7.1 shows the content of the questionnaire.

The questionnaire was administered to staff in middle and upper management positions in the manufacturing industry in Ciudad Juárez. Such staff were selected as they hold important positions in the company. They are key people in the development and performance of the organization, for they have knowledge about the company's

Table 7.1 Content of the critical success factors and benefits of implementing Ergonomics in the manufacturing industry questionnaire

Stage	Dimension	Number of items
1. Planning Stage	Planning costs	5
	Prevention activities	5
	Work environment	6
	Management's commitment	6
2. Process Launch	Organization	5
	Training	5
	Project-launching costs	5
3. Work-Improvement Cycle	Identification of priority jobs	5
	Investigation of Risks	5
	Solution development	6
	Implementation costs	5
4. Long-term Development	Project's cost–benefit analysis	5
	Health and safety follow-up	5
	Feedback	5
5. Benefits of Ergonomics	Health and safety benefits	11
	Benefits on the quality of life at work	15
	Financial benefits	15
Total		114

financial aspects and are in charge of personnel. Middle managers are people in positions such as supervisors, technicians, interns, group leaders, and engineers.

7.4.2 Method

7.4.2.1 Step 1: Data Acquisition and Preparation

For data acquisition, the questionnaire previously described was used. The instrument was properly validated for gathering data. This instrument includes 73 items designed and validated properly to gather data related to 4 stages concerning the Ergonomics program implementation (Maldonado-Macías et al.2017).

The data preparation process occurred in two phases, which are described below.

- Once the data was entered, the first phase consisted of correcting errors, dealing with lost values, and validating the questionnaire. The lost values of each question were replaced by the median, since the questionnaire used a scale to obtain measures (Garcia-Alcaraz et al. 2016). Finally, taking the corrections made in case of errors while entering the data and the lost values, the questionnaire was

validated calculating a Cronbach's Alpha index higher than 0.7 and lower than 0.9. In addition, a descriptive, statistical frequency analysis was carried out to highlight the activities that had a consensus.
- In the second phase, a factor analysis was done, which aimed to reduce the size of the data set.

7.4.2.2 Step 2: Specification of the Model

To formulate a theoretical model, the researcher must conduct a literature review to identify latent variables (constructs) that can be measured directly and must be able to specify the relationships between such variables (Lévy Mangin and Varela Mallou 2003; Cupani 2012). One way of specifying a model is through graphic representations (Kline 2015).

7.4.2.3 Step 3: Model Identification

In order to identify the model, the researcher must verify that the theoretical model is correct (Lévy Mangin and Varela Mallou 2003). One way to do so is by comparing the model's degrees of freedom, which must be greater or equal to zero; this means that a model is either identified or over-identified:

- An identified model has exactly zero degrees of freedom ($gl = 0$) although this reflects a model's perfect fit.
- An over-identified model is the goal of all structural equation models. It has more information in the data matrix than the number of parameters to estimate; this means that it has a positive number of degrees of freedom ($gl > 0$).
- An underestimated model has negative degrees of freedom ($gl < 0$), which reflects an attempt to estimate more parameters than the available information allows. In this case, the model cannot be estimated (Lévy Mangin and Varela Mallou 2003).

Consequently, Cupani (2012) explains that the greater the degrees of freedom, the more parsimonious the model, and the more it can serve as a support for the researcher in the demonstration of associations between variables. In certain WarpPLS 4.0 versions, it can be directly obtained from the model fit and quality indices section, located in Tenenhaus goodness of fit (GoF) (Kock 2012).

7.4.2.4 Step 4: Parameter Estimation

To perform the parameter estimation, it is necessary to establish values for the unknown parameters as well as their respective measurement error (Cupani 2012). A widely used technique is that of maximum likelihood estimation, which refers, is efficient and is unbiased whenever the assumptions of multivariate normality are met (Lévy Mangin and Varela Mallou 2003; Cupani 2012).

7.4.2.5 Step 5: Model Fit

A way to determine a relationship between recognizing or rejecting a relationship between two or more variables is represented by an error known as *P* value. There are two hypotheses available to accept or reject a relationship between two dimensions:

H0: There is evidence of a relationship between two dimensions.
H1: There is no evidence of a relationship between two dimensions.

According to Kock (2015), to determine if a hypothesis is recognized or rejected, the *P* value can be compared to 0.05% to establish a reliability of 95% or non-rejection of the null hypothesis. Regarding the quality of fit (Hair et al. 2014) it can be of three types:

1. Absolute fit measures, a parameter that evaluates the global model fit.
2. Incremental fit measures, which compare the proposed model to other specified models.
3. Measures of parsimony fit, which are measures that offer adjustment by comparing models to different coefficient values with the purpose of determining the adjustment obtained for each estimated coefficient.
4. Decomposition of direct effects, which is represented by the influence that one variable has over another (Manzano et al. 2010). This section analyzes the values that make up the square R of each dimension in order to identify the percentage of each variable as explained to another.
5. The sum of indirect effects that is of the effects obtained through other dimensions or latent variables with trajectories of two or more segments. The sum of these effects results in the total effects.
6. Sum of total effects: This sum allows the quantification of the change observed in the variable in which the induced effect occurs (Manzano et al. 2010). The analysis of the sum of the total effects as well as the analysis of direct effects and the sum of the effects.

The adjustment indices are shown in Table 7.2. The comparison and a conclusion are also presented.

7.4.2.6 Step 6: Model Re-specification

It is infrequent when the proposed model provides the best fit (Cupani 2012). In this way, the researcher must look for methods and/or theory that might improve it (Cupani 2012). Re-specification entails adding or eliminating estimated parameters to or from the original model (Cupani 2012). Nonetheless, it is important to note that any changes should be made with caution as well as taking into considerations theoretical rather than empirical justifications (Cupani 2012).

Table 7.2 Adjustment indices through WarpPLS

Adjustment index	Comparison	Conclusion
Average path coefficient (APC)	$p < 0.05$	Indicates an efficient and predictive model
R^2 average (ARS)	$p < 0.05$	Indicates an efficient and predictive model
R^2 adjusted average (AARS)	$p < 0.05$	Indicates that in average, all parameters measuring the relationships between the latent variables
Average variance inflation factor (AVIF)	Acceptable if < = 5, ideal <= 3.3	Indicates there needs to be a rejection of collinearity problems
Total collinearity average (AFVIF)	Acceptable if < = 5, ideal < = 3.3	Indicates there needs to be a rejection of collinearity problems
Tenenhaus goodness of fit (GoF)	Small > = 0.1, medium > = 0.25, large >= 0.36	Indicates the explicative reach of the model
Simpson's Paradox (SPR)	Acceptable if > = 0.7, ideal = 1	Indicates that the model's paths are paradox-free
Contribución de la relación R cuadrada (RSCR)	Acceptable if > = 0.9, ideal = 1	Indicates that the model is free from negative contributions
Relación de supresión estadística (SSR)	Acceptable si > = 0.7	Indicates that the model is free from statistical suppression

7.4.2.7 Step 7: Model Interpretation

Based on the results obtained from the model, relationships found were sustained on literature review on the critical factors and the stages for the implementation of Ergonomics.

7.5 Results

Based on the methodology and the questionnaire to determine the critical success factors and financial benefits of the application of Ergonomics in the maquiladora industry in Ciudad Juárez, the questionnaire was administered in the maquiladora industry, and a structural equations model encompassing success factors for the implementation of Ergonomics in the manufacturing industry of exportation in Ciudad Juárez was generated.

7.5.1 Step 1: Data Preparation

This section shows the results of the administration of the questionnaire once the instrument was validated and confirmed. First, it gives a general descriptive report of the questionnaire results. Then a descriptive analysis of the stages that make up the Ergonomics program is offered.

7.5.1.1 Descriptive Study of the Planning Stage

The descriptive analyses of the collected data help to better understand the behavior of the participants' responses and to better interpret the results obtained. To do so, this paper describes the questionnaire items' quartiles, interquartile ranges (RI) and median values to determine both the critical success factors and the financial benefits of implementing Ergonomics in the manufacturing industry in Ciudad Juárez.

The planning stage consists of determining planning costs, carrying out prevention activities, improving work environment, and engaging management commitment. The main actions described by the respondents were four: (1) writing up reports on costs associated with employees' disabilities, accidents, injuries, or illnesses at work, (2) creating adequate and comfortable environmental conditions and workplaces for the employee, (3) identifying ergonomic risk factors, and 4. complying with the activities established by the Mexican laws and regulations. Table 7.3 shows the percentiles and the interquartile ranges; in addition, all the activities with a low consensus were marked with an asterisk (*), while activities with a high consensus were doubly marked (**).

7.5.1.2 Descriptive Study of the Process-Launching Stage

The main measures described to carry out an Ergonomics program were: (1) to document the activities carried out, (2) to train in real study cases, and (3) one of the frequent costs is the time to conduct meetings. The results of the quartiles and interquartile range of the responses are shown in Table 7.4.

7.5.1.3 Descriptive Study of the Work-Improvement Cycle Stage

According to the results obtained from the questionnaires, the predominant activities in the implementation of Ergonomics were, (1) giving priority to the departments or stations that exhibit repetitiveness, energy efforts, stressful postures, mechanical stress, extreme temperatures, and vibration, (2) identifying risks and work priorities by means of tours, (3) developing solutions by the individual or group in charge of Ergonomics group and taking into consideration the employee's opinion, and (4)

Table 7.3 Percentiles and interquartile ranges in the planning stage

Dimension	Abbreviation	Percentiles			RI
Planning Costs	COSPLA	25	50	75	
Records are kept regarding costs associated with					
1. Work-related disabilities, accidents, injuries, or illnesses in employees	P1COSPLA1	3.93	4.59 **	4.95	1.02
2. Employees absenteeism due to work-related physical disabilities	P2COSPLA2	3.31	4.37	4.93	1.62
3. Medical attention (internal and external, including specialized attention) to work-related injuries or illnesses	P3COSPLA3	3.59	4.48	4.92	1.33
4. Fines by Social Security (IMSS), the Ministry of Labor and Social Prevention, or other health agencies	P4COSPLA4	2.34	3.9 *	4.75	2.41
5. Training of the new employees replacing those injured and/or disabled due to work-related accidents or illnesses	P5COSPLA5	3.04	4.09	4.83	1.79
Prevention Activities	ACTPRE				
6. Creation of adequate and comfortable environmental conditions for the employee	P6ACTPRE1	3.38	4.18**	4.81	1.43
7. Placing employees in a workstation or site according to their capacities and limitations	P7ACTPRE2	3.33	4.17**	4.83	1.5
8. Arranging for the purchase of adequate and safe machines and tools, according to employees' capacities and limitations	P8ACTPRE3	3.22	4.02	4.7	1.48
9. Employees' periodical medical check-ups	P9ACTPRE4	3.01	4.02	4.78	1.77
10. Scheduling resting time according to the physical or mental wear caused by employees' activities	P10ACTPRE 5	2.34	3.48*	4.52	2.18
Work Environment	ENTLABO				
11. Decrease in complaints regarding work conditions	P11ENTLABO1	2.97	3.72	4.48	1.51
12. Improvements in the plant's layout	P12ENTLABO2	3.24	4.06	4.73	1.49
13. Identification of Ergonomics risk factors in the work environment	P13ENTLABO3	3.32	4.12**	4.77	1.45
14. Considering implementing automatization in operations which are risky for humans	P14ENTLABO4	3.2	4.03	4.75	1.55

(continued)

Table 7.3 (continued)

Dimension	Abbreviation	Percentiles			RI
Planning Costs	COSPLA	25	50	75	
15. Decrease in human error in tasks	P15ENTLABO5	3.11	3.8	4.53	1.42
16. Identification and reduction of tasks information load	P16ENTLABO6	2.9	3.66*	4.44	1.54
Management Commitment	COMPRO				
17. Management commitment	P17COMPRO1	3.33	4.13	4.78	1.45
18. Foster relationships between employees and management	P18COMPRO2	2.63	3.64*	4.53	1.9
19. Appoint resources to avoid work-caused disabilities, accidents, or injuries	P19COMPRO3	3.09	3.99	4.72	1.63
20. Comply with Mexican laws and regulations	P20COMPRO4	3.85	4.51**	4.96	1.07
21. Strengthen market competitiveness	P21COMPRO5	3.39	4.23	4.86	1.47
22. Comply with new Ergonomics-related guidelines (proactive approach)	P22COMPRO6	3.14	4.05	4.77	1.63

The question with the lowest median value was marked with an asterisk (*), while the question with the highest median value was doubly marked (**)
Source Prepared by the author

recording and taking into account the cost of comparing a device or machine. Table 7.5 reports on the results of the questionnaires of the work-improvement cycle stage.

7.5.1.4 Descriptive Study of the Long-Term Development Stage

The quartiles and the interquartile range make up the descriptive analysis. The objective of this section is to identify those activities that are carried out more frequently, to define the factors that lead to a successful implementation of Ergonomics. Three activities were identified as the most recurrent: (1) keeping records considering the costs associated with the completion of the entire project, (2) promoting the prevention of occupational accidents, injuries, or diseases, and (3) providing feedback of the solution to those in charge of solving the problem. The descriptive analysis of questionnaire responses is shown in Table 7.6.

7.5.1.5 Descriptive Study of the Financial Benefits

According to the questionnaire results, the participants indicated that the main financial benefits of implementing an Ergonomics program are the improvement of product

Table 7.4 Percentiles and interquartile range during the process launch stage

Dimension	Abbreviation	Percentiles			RI
Organization	ORGAN	25	50	75	
23. To develop a work plan for the person or group (committee) in charge of Ergonomics	P23ORGAN1	2.57	3.62	4.51	1.94
24. To document in Minutes the activities carried out (in the case of a committee)	P24ORGAN2	2.84	3.78**	4.59	1.75
25. To implement a work plan for the individual or group (committee) in charge of Ergonomics	P25ORGAN3	2.77	3.72	4.56	1.79
26. To ensure communication between the individual or group and the management	P26ORGAN4	2.60	3.67	4.51	1.91
27. To ensure communication between the individual or group and employees	P27ORGAN5	2.62	3.60*	4.46	1.84
Training	CAPAC				
To carry out training in					
28. Topics only, (without analyzing real cases)	P28CAPAC1	2.48	3.47	4.38	1.90
29. Real study cases with the individual or group in charge of Ergonomics	P29CAPAC2	2.59	3.55**	4.42	1.83
30. Developing manuals or guides in Ergonomics	P30CAPAC3	2.20	3.10	3.99	1.79
31. The use of Ergonomics-specialized computer software	P31CAPAC4	1.80	2.89 *	4.02	2.22
32. The conduction of post-training evaluations	P32CAPAC5	2.13	3.09	4.09	1.96
Project-launching costs	COSARRA				
Records are kept and costs are considered in relation to					
33. The promotion of the Ergonomics program in the company	P33COSARRA1	2.47	3.54	4.45	1.98
34. The time spent in meetings	P34COSARRA2	2.50	3.56**	4.49	1.99
35. The purchase of Ergonomics-specialized software	P35COSARRA3	1.63	2.73	3.78	2.15
36. Training in the use of specialized software	P36COSARRA4	1.56	2.62*	3.69	2.13

(continued)

Table 7.4 (continued)

Dimension	Abbreviation	Percentiles			RI
Organization	ORGAN	25	50	75	
37. Training and teaching in Ergonomics topics (congresses, specialization, master's degrees, doctorate programs, etc.)	P37COSARRA5	1.90	2.88	3.82	1.92

The question with the lowest median value was marked with an asterisk (*), while the question with the highest median value was doubly marked (**)

quality, the improvement of the company's image or reputation, and an increase in productivity. The results can be seen in Table 7.7.

7.5.2 Step 2: Model Specification

To specify the model, a review of the available literature was conducted in search of the main success factors. Based on the information, hypotheses were defined regarding critical success factors (CSF) and financial benefits. Thirdly, the component load and the inflation index of the variance were reviewed. Both points were key to the definition of the final model.

To identify the critical success factors, a literature review was conducted, from which the authors' most widely described factors were selected. Only the factors with the greatest consensus were marked and included in the study. Success factors involving employees and ensuring communication were also included throughout the study.

Next, based on the literature available, several hypotheses regarding critical success factors were proposed.

In order to carry out an Ergonomics program, it is necessary to form groups that include workers and directors with diverse abilities and capacities (Weick 1987; Eerd et al. 2010).

H1: Management commitment will have a direct and positive effect on the organization (organ).

The first step toward the implementation of Ergonomics is that managers agree to implement the program. Likewise, the management's commitment must be clear and firm, resulting in a decisive and valuable aspect for the implementation of Ergonomics (Institute for Work & Health, n/a; Blanco et al. 2014). Commitment is one of the main steps to implement Ergonomics, and one of the ways to identify situations of risk is through training (Yazdani et al. 2015). Thus,

H2: The management's commitment (compro) will have a direct and positive effect on training.

Table 7.5 Percentiles and interquartile range of the work-improvement cycle stage

Dimension	Abbreviation	Percentiles			RI
		25	50	75	
Identification of priority Jobs	IDENTIF				
Priority is given to those departments or stations where					
38. Repetitiveness, energy efforts, stressful postures, mechanical stress, extreme temperatures, and vibration were found	P38IDENTIF1	3.00	3.90**	4.71	1.71
39. There was a higher rate of work-related accidents, injuries, and/or illnesses	P39IDENTIF2	2.70	3.77	4.66	1.96
40. Recommendations from the medical staff had been made	P40IDENTIF3	2.81	3.70*	4.58	1.77
41. New processes, materials, equipment, and work tools had been introduced	P41IDENTIF4	3.02	3.83	4.62	1.60
42. There was employee participation and involvement	P42IDENTIF5	2.99	3.89	4.68	1.69
Risk Investigation	INVEST				
Risks were identified through					
43. Tours	P43INVEST1	3.27	4.17**	4.85	1.58
44. The use of questionnaires	P44INVEST2	2.23	3.21	4.22	1.99
45. The conduction of interviews	P45INVEST3	2.31	3.27	4.25	1.94
46. An analysis using an Ergonomics method (REBA, RULA, OWAS, or others)	P46INVEST4	2.08	3.09	4.18	2.10
47. The use of Ergonomics-specialized software such as Intelex Ergonomics Analysis, Jack and Process Simulate Human, among others	P47INVEST5	1.51	2.68*	3.81	2.30
Solution development	SOLUC				
Solutions are developed by					
48. Both the individual or group in charge of Ergonomics' as well as the employee's opinion	P48SOLUC1	2.59	3.65**	4.54	1.95
49. Both the individual or group in charge of Ergonomics as well as the area manager	P49SOLUC2	2.46	3.54	4.47	2.01
50. The use of Ergonomics-specialized software	P50SOLUC3	1.62	2.69*	3.80	2.18
51. Research (journal articles, Internet files, etc.)	P51SOLUC4	1.99	2.97	3.95	1.96

(continued)

Table 7.5 (continued)

Dimension	Abbreviation	Percentiles			RI
		25	50	75	
52. The person responsible and experts (advisors, ergonomists, or university experts)	P52SOLUC5	1.98	3.00	4.12	2.14
53. The collection of ideas and information by comparing aspects within and outside the company (Benchmarking)	P53SOLUC6	2.11	3.03	3.97	1.86
Implementation Costs	COSIMPL				
Records are kept, and the following costs are considered					
54. The purchase of tools	P54COSIMPL1	3.10	3.97	4.71	1.61
55. The purchase of equipment or machines (whether automatic or semi-automatic)	P55COSIMPL2	3.15	4.05**	4.76	1.61
56. Hiring an expert (advisor, ergonomist, university, etc.)	P56COSIMPL3	2.18	3.15	4.19	2.01
57. Stopping the line when an issue is found	P57COSIMPL4	2.84	3.89	4.68	1.84
58. The purchase of a software to identify solutions	P58COSIMPL5	1.76	2.83*	3.89	2.13

The question with the lowest median value was marked with an asterisk (*), while the question with the highest median value was doubly marked (**)

Some mechanisms for the implementation of an Ergonomics program are informative and training meetings for workers and supervisors (Blanco et al. 2014). A few examples of training topics in Ergonomics are the definition and solution of problems that originate from lack of Ergonomics, the formation of work teams (Ergo Team), and promotion of Ergonomics (Eerd et al. 2010). Therefore, the following statement is established:

H3: The organization will have a direct and positive effect on education and training.

H4: Organization will have a direct and positive effect on the identification of risks.

H5: Teaching and training will have a direct and positive effect on the identification of risks.

Committees are working groups made up by individuals who work either permanently or temporarily. The aim of these working groups is to increase productivity, effectiveness, and quality (Jiang et al. 2009; Sajjadi et al. 2011). The committee has the mandate to review employees' opinion and to inform executives of the problems (Jiang et al. 2009; Sajjadi et al. 2011).

Table 7.6 Percentiles and interquartile range in the long-term development stage

Dimension	Abbreviation	Percentiles			RI
		25	50	75	
Project's cost–benefit analysis	CB				
59. Feedback is offered to the management to obtain further investment in the Ergonomics program	P59CB1	2.42	3.46	4.35	1.93
60. A comparison is made between the total investment and the benefits obtained from the program	P60CB2	2.38	3.32*	4.24	1.86
61. Records are kept, and the costs associated with the entire project are taken into consideration	P61CB3	2.71	3.71**	4.55	1.84
62. Records are kept, and costs in the production system before and after the project are considered	P62CB4	2.67	3.65	4.50	1.83
63. Records are kept, and the costs associated with internal medical attention after the program has been implemented are analyzed	P63CB5	2.61	3.64	4.52	1.91
Health and safety follow-up	SEG				
Activities are carried out to					
64. Follow up on the development of a culture of prevention	P64SEG1	3.08	4.01	4.74	1.66
65. Follow up on Ergonomics risks	P65SEG2	2.91	3.81	4.65	1.74
66. Follow up on Ergonomics risks whose solutions have not been finalized	P66SEG3	2.89	3.77*	4.61	1.72
67. Foster the prevention of work-related accidents, injuries, or illnesses	P67SEG4	3.34	4.20**	4.87	1.53
68. Make a comparison between the rate of accidents, injuries, or illnesses before and after implementing the program	P68SEG5	3.00	3.89	4.67	1.67
Feedback	RETRO				
Activities are carried out to provide feedback on the solution of ergonomic risks to:					
69. The entire corporate	P69RETRO1	2.76	3.85*	4.68	1.92
70. Employees	P70RETRO2	3.00	3.93	4.70	1.70
71. The individual(s) in charge of supporting the solution to the program	P71RETRO3	2.84	3.99**	4.77	1.93
72. The manager of the department affected by the ergonomic risks	P72RETRO4	2.92	3.95	4.74	1.82
73. The directors, regarding the effectiveness of the Ergonomics program	P73RETRO5	2.71	3.87	4.69	1.98

The question with the lowest median value was marked with an asterisk (*), while the question with the highest median value was doubly marked (**)

Table 7.7 Percentiles and interquartile range of the financial benefits

Dimension	Abbreviation	Percentiles			RI
		25	50	75	
Financial Benefits	EB				
1. Improvements in the quality of the product	B1EB1	3.14	3.89**	4.60	1.46
2. Decrease in employees' work-related absenteeism, accidents, injuries or illnesses	B2EB2	2.77	3.68	4.51	1.74
3. Decrease in mistakes in the product's manufacturing	B3EB3	2.94	3.75	4.54	1.60
4. Decrease in lost days due to work-related accident, injuries, or illnesses	B4EB4	2.82	3.67	4.49	1.67
5. Increase in productivity	B5EB5	3.08	3.82**	4.58	1.50
6. Efficient use of time at work	B6EB6	3.06	3.77	4.51	1.45
7. Decrease in scrap	B7EB7	2.71	3.61	4.45	1.74
8. Reduction in costs of providing internal medical attention to employees	B8EB8	2.80	3.65	4.47	1.67
9. Reduction in material handling	B9EB9	2.77	3.62	4.43	1.66
10. Decrease in monotony	B10EB10	2.43	3.43*	4.27	1.84
11. Prevention of harm risks at companies' facilities	B11EB11	3.03	3.81	4.57	1.54
12. Increase in the quality of equipment maintenance	B12EB12	2.78	3.62	4.42	1.64
13. Improvement on the company's image/reputation	B13EB13	2.97	3.87**	4.63	1.66
14. Prevention/reduction in warnings /fines from the authorities	B14EB14	2.79	3.73	4.56	1.77
15. Decrease in delays	B15EB15	2.83	3.75	4.57	1.74

The question with the lowest median value was marked with an asterisk (*), while the question with the highest median value was doubly marked (**)
Source Prepared by the author

H6: Teaching and training will have a direct and positive effect on the investigation of risks.

Blanco et al. (2014) point out that the knowledge offered by Ergonomics is beneficial. In the development of an Ergonomics program, training and training are very important points to identify risk conditions, lead to an evaluation of the work site, and investigate solutions to the problem (Occupational Health & Safety Agency, n/a; Hägg 2003; Stuart-Buttle 2006). Based on the literature, the following points are established:

H7: Teaching and training will have a direct and positive effect on the development of solutions.

The Canadian Center for Occupational Health and Safety (2020) describes that some factors leading to musculoskeletal disorders can be the frequent and deficient manipulation of materials. Furthermore, the exposure to dangerous temperatures (both low and high), exposure to excessive vibration, repetitive movements during the workday, uncomfortable or stationary work postures, the use of excessive pressure or force to perform a task, unnecessary lifting of uncomfortable items, and insufficient breaks. That is why the following is considered:

H8: Prioritizing jobs will have a direct and positive effect on risk investigation.

Workers know the problems in each area; therefore, they are the best agents in offering solutions (Blanco et al. 2014, Occupational Health & Safety Agency, n/a). Thus, the following is considered:

H9: The identification of priority jobs will have a direct and positive effect on the development of solutions.

One of the main objectives of the Ergonomics program is the investigation of musculoskeletal disorders in order to avoid them or reduce the risks of acquiring them (Bridger 2017). Thus, the following hypothesis is proposed:

H10: Risk research will have a direct and positive effect on the development of solutions.

Once the solution has been developed, the recommendation is to offer feedback on the activities carried out (Stuart-Buttle 2006).

H11: The development of solutions will have a direct and positive effect on the company's financial benefits.

To guarantee the success of the Ergonomics program, it is necessary to establish a mechanism to develop solutions (Blanco et al. 2014). Thus:

H12: The development of solutions will have a direct and positive effect on the company's financial benefits.

At the end of the program, the committee or the person in charge of the program will have to provide feedback to the managers regarding the benefits of implementing Ergonomics in the company (Yazdani et al. 2015). The following is established:

H13: Feedback will have a direct and positive effect on the company's financial benefits.

Table 7.8 shows abbreviations of the dimensions in order to simplify the use of text in the model.

A factor analysis was conducted, in order to reduce dimensions and questions describing each dimension. Table 7.9 shows the results of the factor load, the variance inflation index (VIF), and the effect size (ES). For example, regarding the factor load, the minimum number of questions that make up the organization dimension

Table 7.8 Dimensions and abbreviations

Dimension	Abbreviation
Management commitment	Compro
Organization	Organi
Training	Capaci
Prioritizing jobs	Identi
Risk investigation	Invest
Development of solutions	Soluc
Feedback	Retro
Financial Benefit	EB

are questions 23, 24, and 27. On the other hand, the variance inflation index describes the degree of correlation among predictors. Those questions with VIF values greater than 4 were eliminated (Pan and Jackson 2008).

Based on the hypotheses found in the literature, the factor analysis, and the variance inflation indices, a hypothetical model is proposed, which is shown in Fig. 7.3.

7.5.3 Step 3: Model Identification

In this step, the type of relationship and the degrees of freedom of the model are identified. In the case of the type of relationship, it can be observed in Fig. 7.3 that the relationships are not correlated but rather, unidirectional, from which it can be implied that there is a unique value for each of the parameters; thus, the model is recursive. Regarding the degrees of freedom (GoF), it is 0.643; thus, the model is over-identified.

7.5.4 Step 4: Parameters Estimation

Once the model is specified and identified, the structural equations model (SEM) analysis is performed. The case of the structural equations models as analyzed by WarpPLS is shown in Fig. 7.4. To indicate the significant relationships among the dimensions, for a confidence level of 95% it indicates to obtain a P value lower than 0.05.

Table 7.9 Results of the factor loads and VIF

Items	Factor loads								
	Compro	Organi	Capaci	identi	invest	soluc	retro	EB	VIF
P17compro1	0.201	0	0	0	0	0	0	0	2.194
P18compro2	0.196	0	0	0	0	0	0	0	2.192
P19compro3	0.222	0	0	0	0	0	0	0	3.322
P20compro4	0.203	0	0	0	0	0	0	0	2.179
P21compro5	0.207	0	0	0	0	0	0	0	2.459
P22compro6	0.194	0	0	0	0	0	0	0	2.049
P23Organi1	0	0.364	0	0	0	0	0	0	3.636
P24Organi2	0	0.365	0	0	0	0	0	0	3.748
P27Organi5	0	0.353	0	0	0	0	0	0	2.672
P29Capaci2	0	0	0.361	0	0	0	0	0	2.181
P30Capaci3	0	0	0.379	0	0	0	0	0	2.982
P32Capaci5	0	0	0.371	0	0	0	0	0	2.595
P38Identif1	0	0	0	0.233	0	0	0	0	2.201
P39Identif2	0	0	0	0.217	0	0	0	0	1.888
P40Identif3	0	0	0	0.242	0	0	0	0	2.749
P41Identif4	0	0	0	0.249	0	0	0	0	3.404
P42Identif5	0	0	0	0.244	0	0	0	0	2.819
P43Invest1	0	0	0	0	0.35	0	0	0	1.685
P44Invest2	0	0	0	0	0.385	0	0	0	2.781
P45Invest3	0	0	0	0	0.4	0	0	0	3.287
P51Soluc4	0	0	0	0	0	0.362	0	0	2.307
P52Soluc5	0	0	0	0	0	0.369	0	0	2.628
P53Soluc6	0	0	0	0	0	0.376	0	0	2.948
P59CB1	0	0	0	0	0	0	0.336	0	1.499
P69Retro1	0	0	0	0	0	0	0.399	0	3.793
P70Retro2	0	0	0	0	0	0	0.399	0	3.817
EB2	0	0	0	0	0	0	0	0.226	2.306
EB	0	0	0	0	0	0	0	0.237	2.647
EB	0	0	0	0	0	0	0	0.237	2.566
EB1	0	0	0	0	0	0	0	0.236	2.698
EB13	0	0	0	0	0	0	0	0.236	2.583

Note Variance inflation factor (VIF)

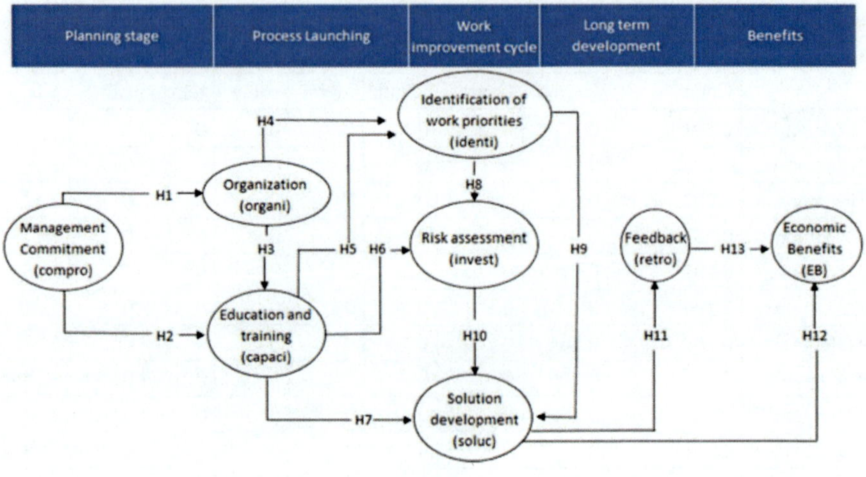

Fig. 7.3 Hypothetical model proposed

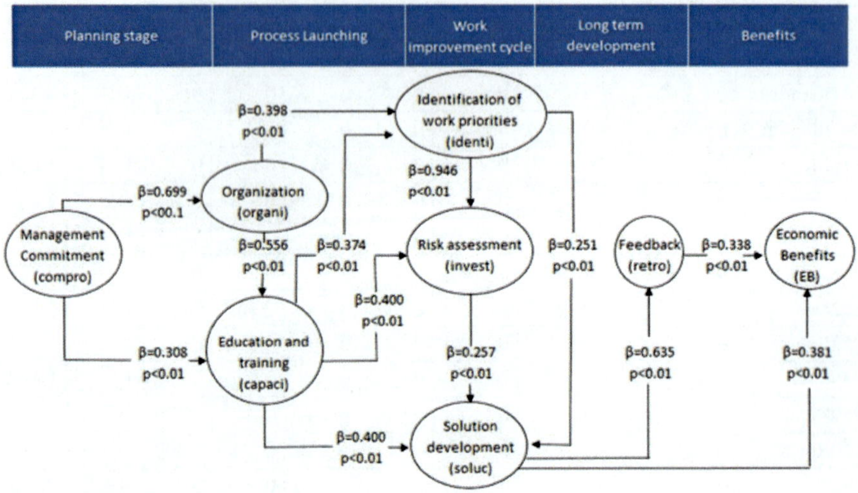

Fig. 7.4 Structural equations model

7.5.5 Step 5: Model Adjustment

The results of the indexes of model adjustments at the planning stage are given below.

1. Average trajectory coefficient (APC) obtained value 0.420, *P* value < 0.001.
2. Average R2 (ARS) obtained value 0.540, *P* value < 0.001.
3. Average adjusted R2 (AARS) obtained value 0.535, *P* value < 0.001.
4. Factor of average blocks of variance inflation (AVIF); obtained value was 2,108.

5. Average of total collinearity (AFVIF); obtained value was 3043.
6. Tenenhaus goodness of fit (GoF); obtained value 0.643.
7. Simpson's paradox (SPR); obtained value of 1000.
8. Contribution of the R2 relationship (RSCR); value obtained was 1000.

The P values of the average trajectory coefficient (APC), average R2 (ARS), average adjusted R2 (AARS) were lower compared to 0.05; therefore, it can be concluded that the model is efficient as it holds a predictive power of 95% confidence. On the other hand, the registered values of the average of variance inflation block factor (AVIF) and the average of total collinearity (AFVIF) were 2108 and 3043, respectively; these values are lower than 3.3; therefore, collinearity problems can be discarded. Regarding the Tenenhaus goodness of fit indicator (GoF), the value obtained was 0.643, which compared to 0.360, means that the model has great explanatory power. Furthermore, the value obtained from the Simpson Paradox (SPR) was of 1. In comparison with 0.7, this value means that at least 70% of the routes of the paths are free of paradox. Finally, the contribution of the R-square relation (RSCR) obtained was 1; its being greater than 0.9 it indicates that the model is exempt from negative contributions. In conclusion, the model is efficient and predictive.

7.5.5.1 Effect Size by Path Coefficients

The effect size are absolute values of the individual contributions of the predictive latent variables corresponding to the *R*-squared coefficients (*it reflects the percentage of the variance explained for each of those variables*) of the criterion's latent variable in each latent variable block (Kock 2015). In other words, it is the contribution of the predictive latent variable to the criterion variable. From the sizes of the effects, the researcher can determine whether the indicated effects are small, medium, or large (Kock 2015). The normally recommended values are 0.02, 0.15, and 0.35; values below 0.02 suggest effects that are too weak to be considered relevant for all practical purposes, even when the corresponding *P* values are statistically significant (Cohen 1988). Table 7.10 shows the size effects for trajectory coefficients and their corresponding *P* values. Note that all size effects are higher than 0.02, and all the *P* values are less than 0.01. The sum of all contributions indicates the value of the R-square (Kock 2015). For example, the seven dimensions contribute to the financial benefits; when adding the sizes of the effects, a value of 0.54 is obtained, which is the same as the R-squared of the benefits. On the other hand, the development of solutions is the element that most contributes to obtaining financial benefits.

7.5.5.2 Direct Effects

Direct effects indicate direct relationships between dimensions. Table 7.11 shows the values of dependence and the *P* value. For example, the relationship between the *Management commitment (compro)* and the *organization (organ)* has a *P* value of

Table 7.10 Size effects for trajectory coefficients

To	From						
	Compro	Organi	Capaci	Identi	Invest	Soluc	Retro
Compro							
Organi	0.489 ($P < 0.01$)						
Capaci	0.217 ($P < 0.01$)	0.431 ($P < 0.01$)					
Identi		0.273 ($P < 0.01$)	0.255 ($P < 0.01$)				
Invest			0.284 ($P < 0.01$)	0.338 ($P < 0.01$)			
Soluc			0.301 ($P < 0.01$)	0.178 ($P < 0.01$)	0.185 ($P < 0.01$)		
Retro						0.403 ($P < 0.01$)	
EB						0.227 ($P < 0.01$)	0.196 ($P < 0.01$)

Table 7.11 Direct effects of the model

To	From						
	Compro	Organ	Capaci	Identi	Invest	Soluc	Retro
Compro							
Organi	$\beta = 0.699$ $P < 0.001$						
Capaci	$\beta = 0.308$ $P < 0.001$	$\beta = 0.556$ $P < 0.001$					
Identi		$\beta = 0.398$ $P < 0.001$	$\beta = 0.374$ $P < 0.001$				
Invest			$\beta = 0.400$ $P < 0.001$	$\beta = 0.462$ $P < 0.001$			
Soluc			$\beta = 0.400$ $P < 0.001$	$\beta = 0.251$ $P < 0.001$	$\beta = 0.257$ $P < 0.001$		
Retro						$\beta = 0.635$ $P < 0.001$	
EB						$\beta = 0.381$ $P < 0.001$	$\beta = 0.338$ $P < 0.001$

less than 0.001, which, with a confidence level of 95%, represents a significant effect. Thus, the dependency value between both relationships is 0.699, which means that a variation of 1.00 standard deviation in management commitment activities would lead to a 69.9% increase in the variation of the organization dimension. In addition, as seen in Table 7.11, *P* values of less than 0.05 mean that these relationships have a high explanatory power with a 95% confidence level.

7.5.5.3 Total Effects

The total effects are the sum of the direct and indirect effects. Table 7.12 shows the dependency value (β, beta value), the *P* value, and the number of roads. For example, the relationship between support and financial economic has a direct and indirect relationship with 20 road types, a significant *P* value and a dependency value of 0.327. Moreover, support and organization are strongly linked, with a value of 0.699. Other relationships with high values of dependence are support and training, organization and identification, training and research, training and the development of solutions, and the development of solutions and the evaluation of progress. Because of the research, the seven dimensions affect the financial benefits, as long as support activities, development of solutions and evaluation of progress are carried out.

7.5.6 Step 6: Model Re-specification

Based on the results above, it can be concluded that the proposed model shows efficiency and quality indexes values that indicate an efficient and predictive model out of the range of paradoxes and negative contributions, consequently it has a good statistical adjustment. In addition, significant *P* values were obtained.

7.5.7 Step 7: Model Interpretation

Next, the results of the β and *P* values for the hypotheses proposed in step 2 are presented.

H1: *Management commitment (Compro)* will have a direct and positive effect on the *organization (organ)*. Result $\beta = 0.699$, *P* value < 0.001.

In conclusion, there is enough statistical evidence to affirm that providing support has a direct and positive effect on the organization, with a 95% reliability since when the *management commitment* increases by one standard deviation, *organization* increases by 0.387.

Table 7.12 Total effects of the model

To	From						
	Compro	Organi	Capaci	Identi	Invest	Soluc	Retro
Compro							
Organi	$\beta = 0.699$ $P < 0.001$ paths (1)						
Capaci	$\beta = 0.697$ $P < 0.001$ paths (2)	$\beta = 0.556$ $P < 0.001$ paths (1)					
Identi	$\beta = 0.539$ $P < 0.001$ paths (3)	$\beta = 0.606$ $P < 0.001$ paths (2)	$\beta = 0.374$ $P < 0.001$ paths (1)				
Invest	$\beta = 0.528$ $P < 0.001$ paths (5)	$\beta = 0.503$ $P < 0.001$ paths (3)	$\beta = 0.573$ $P < 0.001$ paths (2)	$\beta = 0.462$ $P < 0.001$ paths (1)			
Soluc	$\beta = 0.550$ $P < 0.001$ paths (10)	$\beta = 0.504$ $P < 0.001$ paths (6)	$\beta = 0.641$ $P < 0.001$ paths (4)	$\beta = 0.370$ $P < 0.001$ paths (2)	$\beta = 0.257$ $P < 0.001$ paths (1)		
Retro	$\beta = 0.349$ $P < 0.001$ paths (10)	$\beta = 0.320$ $P < 0.001$ paths (6)	$\beta = 0.407$ $P < 0.001$ paths (4)	$\beta = 0.235$ $P < 0.001$ paths (2)	$\beta = 0.163$ $P < 0.001$ paths (1)	$\beta = 0.635$ $P < 0.001$ paths (1)	
EB	$\beta = 0.327$ $P < 0.001$ paths (20)	$\beta = 0.300$ $P < 0.001$ paths (12)	$\beta = 0.382$ $P < 0.001$ paths (8)	$\beta = 0.220$ $P < 0.001$ paths (4)	$\beta = 0.153$ $P < 0.001$ paths (2)	$\beta = 0.595$ $P < 0.001$ paths (2)	$\beta = 0.338$ $P < 0.001$ paths (1)

H2: *Management commitment (Compro)* will have a direct and positive effect on *training (capac)*. Result $\beta = 0.308$, P value < 0.001.

This shows that there is enough statistical evidence to affirm that providing support has a direct and positive effect on training, with a 95% reliability since when *management commitment* increases by one standard deviation, *training* increases by 0.308.

H3: The *organization (organ)* will have a direct and positive effect on *training (capac)*. Result $\beta = 0.556$, P value < 0.001.

It can be stated that there is enough statistical evidence to affirm that *organization* has a direct and positive effect training, with a 95% reliability since when organization increases by one standard deviation, *training* increases by 0.556.

H4: The *organization (organ)* will have a direct and positive effect on the *identification of work priorities (ident)*. Result $\beta = 0.398$, P value < 0.001.

In conclusion, there is enough statistical evidence to affirm that *organization* has a direct and positive effect on identification of work priorities with a reliability of 95% because when *organization* increases by one standard deviation *the identification of work priorities* increases by 0.398.

H5: *Training (capac)* will have a direct and positive effect on *identification of work priorities (ident)*. Result $\beta = 0.374$, $P < 0.001$.

This offers enough statistical evidence to affirm that training has a direct and positive effect on identification of work priorities with a 95% reliability because when training increases by one standard deviation, identification of work priorities increases by 0.374.

H6: *Training (capac)* will have a direct and positive effect on *risk assessment (invest)*. Result $\beta = 0.400$, $P < 0.001$.

Thus, there is enough statistical evidence to affirm that the *training* has a direct and positive effect on risk research, with a 95% reliability since when training increases by one standard deviation, *risk assessment* increases by 0.400.

H7: *Training (capac)* will have a direct and positive effect on *development of solutions (soluc)*. Result $\beta = 0.400$, $P < 0.001$.

It can be concluded that there is enough statistical evidence to affirm that *training* has a direct and positive effect on the *solution development* with a 95% reliability since when training increases by one standard deviation, the development of solutions increases by 0.400.

H8: The *identification of work priorities (ident)* will have a direct and positive effect on *risk assessment*. Result $\beta = 0.462$, P value < 0.001.

It can be concluded, there is enough statistical evidence to affirm that identification of work priorities (ident) or job priorities has a direct and positive effect on risk research with 95% reliability since when identification of work priorities increases by one standard deviation, risk investigation increases by 0.462.

H9: The *identification of work priorities (ident)* will have a direct and positive effect on the *development of solutions (soluc)*. Result $\beta = 0.251$, P value < 0.001.

It can be concluded that there is enough statistical evidence to affirm that the *identification of work priorities* has a direct and positive effect on the *development of solutions*, with a reliability of 95% since when the identification of risks increases by one standard deviation, the development of solutions increases by 0.251.

H10: Support to *risk assessment (invest)* will have a direct and positive effect on the *development of solutions (soluc)* Results $\beta = 0.257$, P value < 0.001.

This provides enough statistical evidence to affirm that *risk assessment* has a direct and positive effect on the development of solutions, with a reliability of 95% since when *risk assessment* increases by one standard deviation, the development of solutions increases by 0.257.

H11: The *development of solutions (soluc)* will have a direct and positive effect on the evaluation of progress by giving proper *Feedback (retro)*. Result $\beta = 0.635$, P value < 0.001.

Thus, there is enough statistical evidence to affirm that the *development of solutions* has a direct and positive effect on the *Feedback (retro)* with a reliability of 95% since when the development of solutions increases by one standard deviation, the evaluation of progress increases by 0.635.

H12: The *development of solutions (soluc)* will have a direct and positive effect on the *financial benefits* of the implementation of Ergonomics. Result $\beta = 0.381$, P value < 0.001.

It can be concluded that there is enough statistical evidence to affirm that the *development of solutions* has a direct and positive effect on *financial benefits*, with a reliability of 95% since when the development of solutions increases by one standard deviation, the economic benefits increase by 0.381.

H13: The evaluation of progress by fiving *Feedback (retro)* will have a direct and positive effect on the *financial benefits* of implementing Ergonomics. Result $\beta = 0.338$, P value < 0.001.

In conclusion, there is enough statistical evidence to affirm that the evaluation of progress and feedback has a direct and positive effect on financial benefits, with a 95% reliability since when the development of solutions increases by one standard deviation, the benefits increase by 0.338.

7.6 Conclusions and Recommendations

In sum, the methodology used in this study including an instrument for gathering data developed by Maldonado-Macías, Alferez-Padron, García-Alcaraz, and Avelar-Sosa (2017) to determine the critical success factors and benefits of implementing Ergonomics in the manufacturing industry in Ciudad Juárez was a reliable instrument to inquire about the activities being carried out to implement Ergonomics programs. On the other hand, both the literature review and the methodology to conduct structural equations models were properly and effectively conducted since from them; it was possible to show the effects between variables and their effect on the financial benefits.

Finally, the conclusions to the stages are described in the following points:

In order to implement an Ergonomics program, it is recommended to start with the management's commitment, which through work and continuity will be reflected in all areas and will affect the financial benefits of the program.

Organization is a key factor of success, significantly affecting training and education as the committee or the person in charge of carrying out the program needs to be trained to carry out the program (Hägg 2003; Occupational Health & Safety Agency, n/a).

Once the Ergonomics program has been established and has the support of the management and of an organized, trained committee, the next step is to identify, investigate, and develop solutions for the ergonomic risks detected.

- Identification means description or classification of priority jobs. This has an effect on research and the development of solutions.
- Risk research is characterized by having an effect on the development of solutions.
- The development of solutions for the findings has an effect on the evaluation of progress and on the financial benefits of the program.

In addition, it is recommended to conduct a progressive and continue evaluation, which significantly affects the financial benefits of implementing an Ergonomics program.

From this research, it can be concluded that the development of the structural equation methodology provided valuable information for effective planning and successful implementation of an Ergonomics program in industries. Accordingly, the procurement of the economical benefits that the program can offer is properly related to their correspondent critical factors to lead to better organizational decisions. It was observed that the model was able to provide valuable information about the four main stages to conduct the implementation of an Ergonomics program in the studied context. In addition, this study extends the knowledge regarding the critical success factors that lead to the benefits of the implementation of Ergonomics programs.

References

Beevis D (2003) Ergonomics—Costs and Benefits Revisited. Applied Ergonomics 34:491–496. https://doi.org/10.1016/S0003-6870(03)00068-1

Blanco G, Castroman R, Chacón L, Hernández P, Ferrer P (2014) Prevention program based on participatory ergonomics to minimize the effects of the physical workload on workers of a company hardware store. Revista Electrónica De Terapia Ocupacional Galicia, TOG 11:1–23

Bridger R (2017) Introduction to Human Factors and Ergonomics, 4th edn. CRC Press, Boca Raton

Broberg O, Andersen V, Seim R (2011) Participatory ergonomics in design processes: The role of boundary objects. Applied Ergonomics 42:464–472

Canadian Centre for Occupational Health and Safety (2020) Work-related Musculoskeletal Disorders (WMSDs) - Risk Factors : OSH Answers. https://www.ccohs.ca/. Accessed 1 Jun 2020

Cohen J (1988) Statistical Power Analysis for the Behavioral Sciences, 2nd edn. Elsevier

Cupani M (2012) Structural Equation Analysis: concepts, development stages and an example of application (In Spanish). Revista Tesis 2:186–199

d'Errico A, Viotti S, Baratti A, Mottura B, Barocelli AP, Tagna M, Sgambelluri B, Battaglino P, Converso D (2013) Low back pain and associated presenteeism among hospital nursing staff. Journal of Occupational Health 55:276–283

Douphrate DI, Rosecrance J (2004) The economics and cost justification of ergonomics. In: Proceedings of the 2nd Annual Regional National Occupational Research Agenda Young Investigators Symposium. University of Utah Press: Salt Lake City, UT, USA, pp 29–40

Dul J, Neumann WP (2009) Ergonomics contributions to company strategies. Applied Ergonomics 40:745–752

van Eerd D, Cole D, Irvin E, Mahood Q, Keown K, Theberge N, Village J, Vincent MS, Cullen K (2010) Process and implementation of participatory ergonomic interventions: a systematic review. Ergonomics 53:1153–1166. https://doi.org/10.1080/00140139.2010.513452

Garcia-Alcaraz JL, Maldonado Macias AA, Prieto Prie DJ, Blanco Fernández J, de López A, JG, Jiménez Macías E (2016) Main benefits obtained from a successful JIT implementation. Int J Adv Manuf Technol 86:2711–2722. https://doi.org/10.1007/s00170-016-8399-5

Green P (2002) Why safety and Human Factors/Ergonomics standards are so difficult to establish. Human Factors in Transportation, Communication, Health and the Workplace, edited by de Waard D, Brookhuis KA, Moraal J and Toffetti A. Maastricht, the Netherlands: Shaker Publishing, http://wwwumichedu/~driving/publications/PGreen-Turin pdf

Hägg GM (2003) Corporate initiatives in ergonomics—an introduction. Applied Ergonomics 34:3–15. https://doi.org/10.1016/S0003-6870(02)00078-9

Haines HM, Wilson JR (1998) Development of a framework for participatory ergonomics. Health and Safety Executive report. HMSO for HSE Books

Hair JF, Black WC, Babin BJ, Anderson RE (2014) Multivariate Data Analysis, 7th edn. Pearson Education Limited, Harlow, Essex

Harne MS, Deshmukh SV (2016) Integration of Ergonomics in Continuous Passive Motion Machine. International Advanced Research Journal in Science, Engineering and Technology 3:

Haslegrave CM, Holmes K (1994) Integrating ergonomics and engineering in the technical design process. Applied Ergonomics 25:211–220

Hendrick HW (2003) Determining the cost–benefits of ergonomics projects and factors that lead to their success. Applied Ergonomics 34:419–427

Institute for Work & Health (n/a) Factors for success in participatory ergonomics. https://www.iwh.on.ca/summaries/sharing-best-evidence/factors-for-success-in-participatory-ergonomics. Accessed 1 Jun 2020

Ip W, Rostykus W (2014) Five Critical Elements for Managing an Ergonomics Program. In: ASSE Professional Development Conference and Exposition. American Society of Safety Engineers

Jiang Hj, Lockee C, Bass K, Fraser I, Norwood EP (2009) Board Oversight of Ouality: Any Differences in Process of Care. Journal of Healthcare Management 54:

Kilbom A, Petersson NF (2006) Elements of the ergonomic process. In: Marras WS, Karwowski W (eds) Interventions, Controls, and Applications in Occupational Ergonomics. CRC Press
Kline RB (2015) Principles and practice of structural equation modeling. Guilford publications
Kock N (2012) WarpPLS 5.0 user manual. Laredo, TX: ScriptWarp Systems
Kock N (2015) One-tailed or two-tailed P values in PLS-SEM? Int J e-Collaboration (IJeC) 11(2):1–7
Laos M, Rengifo L, Rodríguez B, Valencia C, Chabes A, Garcia A (2007) Level of knowledge about ergonomic working conditions in office workers from Ricardo Palma University (In Spanish). Revista de La Facultad de Medicina Humana, 7:
Lévy Mangin JP, Varela Mallou J (2003) Multivariate analysis for the Social Sciences (In Spanish). Pearson Educación
Liker J, Chaffin D (1987) Improvement through Ergonomics: A guide for the development and implementation of an effective Ergonomics program in your plant (In Spanish). Ford Motor Company, Michigan
Maldonado-Macías AA, Alferez-Padron C, García-Alcaraz J, Avelar-Sosa L (2017) Knowledge Management and Ergonomics Implementation in Manufacturing Systems: Development and Validation of a Questionnaire for Critical Success Factors In Knowledge Integration Strategies for Entrepreneurship and Sustainability. IGI Global, Chapter 11:188–213
Manzano A, Zamora S, Salvador R (2010) System of structural equations: a research tool (In Spanish). Centro Nacional de Evaluación para la Educación Superior, México DF
Marmaras N, Nathanael D (2012) Workplace Design. In: Handbook of Human Factors and Ergonomics. John Wiley & Sons, Ltd, pp 597–615
Occupational Health & Safety Agency (n/a) Safety and Health Topics | Ergonomics | Occupational Safety and Health Administration. https://www.osha.gov/SLTC/ergonomics/. Accessed 1 Jun 2020
Oregon Occupational Safety and Health (n/a) The Advantages of Ergonomics. https://osha.oregon.gov/OSHAPubs/ergo/ergoadvantages.pdf. Accessed 1 Jun 2020
Pan Y, Jackson RT (2008) Ethnic difference in the relationship between acute inflammation and serum ferritin in US adult males. Epidemiol Infect 136:421–431. https://doi.org/10.1017/S095026880700831X
Sajjadi HS, Hadi M, Hariri MH, Harirchiyan M, Toghiani A (2011) The impact of productivity committees being established in different hospitals of the Isfahan University of Medical Sciences in 2008. Zeitschrift Für Evidenz, Fortbildung Und Qualität Im Gesundheitswesen 105:708–713
Schulze L (2010) Case Study: Indirect Costs Associated with a Back Injury Incurred in a Manufacturing Facility. American Society of Safety Engineers 3:
Skepper N, Straker L, Pollock C (2000) A case study of the use of ergonomics information in a heavy engineering design process. Int J Ind Ergon 26:425–435. https://doi.org/10.1016/S0169-8141(00)00017-2
Stuart-Buttle C (2006) Ergonomics process in small industry. In: Marras WS, Karwowski W (eds) Interventions, controls, and applications in occupational ergonomics, 2nd edn. Taylor & Francis, Boca Raton, pp 8.1–8.17
Tappin DC, Vitalis A, Bentley TA (2016) The application of an industry level participatory ergonomics approach in developing MSD interventions. Applied Ergonomics 52:151–159. https://doi.org/10.1016/j.apergo.2015.07.007
Weick KE (1987) Organizational Culture as a Source of High Reliability. California Management Review. https://doi.org/10.2307/41165243
Yazdani A, Neumann P, Imbeau D, Bigelow P, Pagell M, Theberge N, Hilbrecht M, Wells R (2015) How compatible are participatory ergonomics programs with occupational health and safety management systems? Scand J Work Environ Health 41:111–123. https://doi.org/10.5271/sjweh.3467

Part III
Applied Physical Ergonomics

Chapter 8
Applying Hierarchical Task Analysis (HTA) to Identify Ergonomic Risks in Exhibition Booths Set-Up Process

Román E. Méndez

Abstract In order to apply the Hierarchical Task Analysis (HTA) to the exhibition booths set-up process inside a convention center in Mexico City, six assemblers were videotaped from the start of the day to the end of set-up. Only activities that required some type of physical effort to perform or in which there was manual material handling by the assemblers were analyzed. From the observations made, the HTA was applied and a total of 14 tasks were identified: material arrival, material carrying and dragging, set-up site preparation, floor installation, panelized walls installation, working at heights, electrical and electronic installation, paint application, graphics set-up, furniture and accessories placement, final details, display products placement, materials removal and set-up end. With this study, it was possible to know the interaction that assemblers have with their work tools, their work environment, their tasks with its descriptions, and the possible ergonomic risks to which they are exposed when set-up exhibition booths and installing panelized walls.

Keywords HTA · Exhibition booths set-up · Ergonomic evaluation · Manual material handling · Panelized walls

8.1 Introduction

8.1.1 Some Definitions and Examples of HTA Application in Different Research Fields

Hierarchical Task Analysis (HTA) is a subjective assessment method used by ergonomists and human factors engineers to identify the tasks that constitute a given work, as well as make visible the levels of importance of these tasks; also, with this analysis it is possible to examine, detect and avoid problems, errors, and failures in the tasks that make up a system or process and thus propose solutions that improve the

R. E. Méndez (✉)
Department of Technology and Production, CyAD (Sciences and Arts for Design), Universidad Autónoma Metropolitana – Xochimilco, Mexico City, Mexico
e-mail: rmendezg@correo.xoc.uam.mx

A. Realyvásquez Vargas et al. (eds.), *New Perspectives on Applied Industrial Ergonomics*,
https://doi.org/10.1007/978-3-030-73468-8_8

performance and implementation of a certain work. These problems, errors and failures can be cognitive or physical (Promann and Zhang 2015). According to Stanton (2006: 56), HTA offers "a means of describing a system in terms of goals and subgoals, with feedback loops in a nested hierarchy." As a result, when applying an HTA, a complete description of the activities within a task is obtained (MacLeod et al. 2005).

Some examples of its application in different professional fields, such as medicine, have been carried out in the UK by Lane et al. (2006), where they use the HTA to understand, find and eliminate or reduce as much as possible medication administration errors. Another application of the analysis was developed by Sarker et al. (2007) with the aim of training surgeons to perform any operation or surgical procedure.

Another professional field where the use of HTA is resorted to is computing, in the field of human–computer interaction; the authors Promann and Zhang (2015) use it as a complement to increase efficiency of usability and user-centered design evaluations, because simplifies workflows in the complex computer systems of Purdue University virtual libraries, in the United States of America (USA). Likewise, in the field of agriculture, Italian researchers Fargnoli et al. (2019) apply HTA in activities related to the use of pesticides in vineyard cultivation to give a proposal that optimizes safety measures and reduces health risks in farmers.

In Mexico, HTA is also applied in the field of medicine and health, an example of this, according to López Hernández et al. (2015), is the recommendation of solutions based on the identification of risk factors and errors in the laboratory area of a private hospital. Another example is the study performed by Moreno García et al. (2018), where they analyze the tasks performed by older adults when using a drug dispenser to identify problems and satisfy needs of nurses who provide them with medications. In the field of product design, HTA is also used to analyze all possible stages and sub-stages in human–device interaction, an example of this is the work carried out by Durán-Coronado et al. (2019), who study the use of wireless headphones so that product designers find guidance to improve the user experience in the use of these devices.

8.1.2 Labor Context in Which This Research Is Developed

The booths set-up is developed in the meetings' industry context, which includes conventions, congresses, fairs and exhibitions, among others (Entorno Turístico 2016) and that currently contributes 1.8% of Gross Domestic Product (GDP) in Mexico (Klever 2020). The booths set-up is the work on site, the action of installing prefabricated and pre-assembled structures of the booth in the assigned space within the venue or convention center, before the start of the exhibition, during the hours indicated in the exhibitor's regulations of the convention centers (Meetings Mexico 2015; Reed Exhibitions 2016; Vanexpo 2016). The exhibition booth is defined as an "installation within a market or fair, for the exhibition or sale of products" (RAE 2014) and is assembled by means of prefabricated wooden or metallic structures, or other materials that are generally of large dimensions and weights. These aforementioned structures are handled by assemblers and they may be exposed to health and ergonomic risks that can result in postural and strength loads, falls, musculoskeletal symptoms or other occupational injuries, consequently, the exhibition booths assembly can be of interest and of great contribution when performing ergonomic evaluations and knowing their dynamics to provide ergonomic and design improvements and solutions in this labor field (Kim et al. 2006a, b). Therefore, the main objective of this work is to apply the hierarchical task analysis (HTA) to the exhibition booths set-up process inside a convention center in Mexico City, and once this is achieved, to describe the tasks that make up this process and identify among them the task and stage with the greatest ergonomic risk for assemblers.

8.2 Materials and Methods

8.2.1 Participants

With the signed authorization of the assemblers, two video recordings of tasks were made to 6 male workers, employees of a private company dedicated to the booths set-up for fairs and exhibitions, located in the west of Mexico City in 2016. The first video recording was made inside a convention center in Mexico City, within the hours (9:00 am–10:00 pm) and assembly period (three business days) before the start of the exhibition. The assemblers performed the set-up of two 24 m^2 (6 m $\times$ 4 m) booths and one 16 m^2 (4 m $\times$ 4 m) booth. In the second video recording, made during a single day, the subjects set up a 24 m^2 (6 m $\times$ 4 m) booth. The final structure of the 4 booths reached a total height of 4 m.

8.2.2 Instruments

To perform this research, the following instruments were used: GoPro HERO4 Digital Video Camera—HD Widescreen with extra battery and power cord, tripod, pen and the HTA format on sheets of paper.

8.2.3 Procedure

An observational exploratory study was carried out in order to identify the tasks and stages in which the process of setting-up booths for exhibitions is developed, the set-up was performed by 6 workers inside a convention center in Mexico City. This information was used to collect data for a previous research project (Méndez 2018) on the work performed by booth assemblers.

Only activities that required some type of physical effort or in which there was manual material handling (MMH) were analyzed (Gómez-Bull et al. 2015) with the aim of finding possible ergonomic risks, therefore food breaks or organization times were ruled out. An adaptation of the guidelines for ergonomic analysis proposed by Gómez-Bull et al. (2015) and Herrera Lugo (2015) was carried out in this study, shown in a sequence of 10 steps. The sequence of the final procedure is listed below:

- Authorization signature by the assemblers to be videotaped.
- Installation of the video camera without invading the workspace of the assemblers or interrupting their activities according to Herrera Lugo (2015).
- Video recording throughout the working day; from the beginning to the end of the set-up, making changes to the camera's battery when necessary (Gómez-Bull et al. 2015; Herrera Lugo 2015).
- Review of videos and observation of their content and duration time (Gómez-Bull et al. 2015), to proceed with the task analysis (HTA).
- HTA application (Annett 2004; Stanton 2006) in two stages: HTA 1 and HTA 2 as follows:
- A. HTA 1: the main tasks were identified.
- B. HTA 2: the stages were identified for each main task, later the possible ergonomic risks were observed.
- Time record in minutes of each task.
- Observation and recording of possible ergonomic risks for each identified task.
- Analysis of task that has taken the longest time to complete.
- Determination of exposure times to postural risk and riskiest stage.
- Recording of the data obtained in tables where the tasks and stages analyzed from the video recordings were compiled.

8.3 Results

The results are shown in five sections, the first section (Table 8.1) corresponds to the first analysis carried out to identify the main tasks, the second section (Table 8.2) corresponds to the second, more detailed analysis, to find the stages and characteristics of the tasks, this section is where the HTA was fully applied. The third section (Table 8.3) shows the general execution times for each task with their percentages according to the total set-up time and the possible ergonomic risks observed for each task. The fourth section (Table 8.4) is the analysis of the task with the longest duration of time to be completed. The fifth section covers the times of exposure to postural risk by assemblers (Table 8.6) and the identification of the riskiest stage (Fig. 8.1).

Table 8.1 HTA 1: first section of the analysis. Frames showing the main tasks in which the total set-up process takes place

#	Task	Code	Frame
1	Material arrival	MA	
2	Material carrying and dragging	MC	
3	Set-up site preparation	SP	

(continued)

8.3.1 *Main Tasks of the Booth Set-Up Process*

From the first video recordings, with a 720 min approximate duration, made to the workers and their activities, 14 different tasks were observed. Tables 8.1 and 8.2 compile the 14 main tasks and their stages or sub-tasks in which the entire set-up

Table 8.1 (continued)

#	Task	Code	Frame
4	Floor installation	FI	
5	Panelized walls installation	PW	
6	Working at heights	WH	

(continued)

process is developed, from the moment the material arrives at the convention center until the workers leave at the end of their activities.

Table 8.1 shows images of each of the 14 main tasks identified in the first part of the HTA from the video recordings made.

Fourteen main tasks were identified with their stages by means of an HTA recorded in Tables 8.1 and 8.2 shown below: (1) material arrival, (2) material carrying and

Table 8.1 (continued)

#	Task	Code	Frame
7	Electrical and electronic installation	EE	
8	Paint application	PA	
9	Graphics set-up	GS	

(continued)

dragging, (3) set-up site preparation, (4) floor installation, (5) panelized walls installation, (6) working at heights, (7) electrical and electronic installation, (8) paint application, (9) graphics set-up, (10) furniture and accessories placement, (11) final details, (12) display products placement, (13) materials removal, and (14) set-up end.

Table 8.1 (continued)

#	Task	Code	Frame
10	Furniture and accessories placement	FP	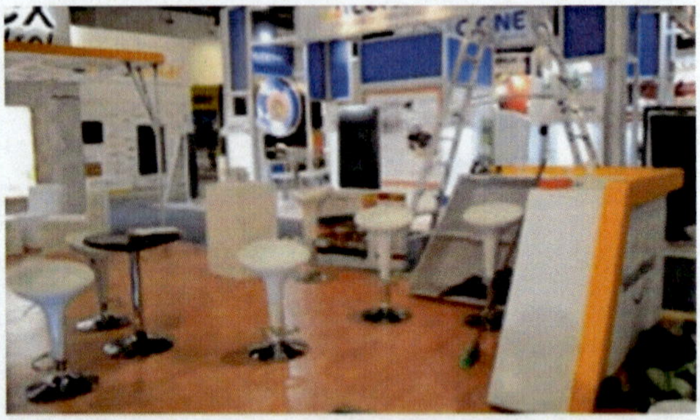
11	Final details	FD	
12	Display products placement	DP	

(continued)

8.3.2 Stages and Characteristics of Each Task in the Set-Up Process

After having identified the 14 main tasks from the first analysis, a second HTA was carried out for each task to break it down into its stages or sub-tasks according to

Table 8.1 (continued)

#	Task	Code	Frame
13	Materials removal	MR	
14	Set-up end	SE	

the guidelines proposed by Stanton (2006) and Annett (2004) to identify the goals and sub-goals of the system to be analyzed. The results and their observations were recorded in Table 8.2.

The tasks shown in Tables 8.1 and 8.2 are presented in a linear order, however in practice there are tasks that can be executed in parallel, such as the material arrival, the display products placement, the final details, the furniture and accessories placement, to name a few.

8.3.3 Time to Perform Each Task and Possible Ergonomic Risks

The times of each task were documented to determine which of these required more time to complete, likewise when observing the videos, it was also possible to detect possible ergonomic risks to which the assemblers were exposed during the performance of their work. The data is shown in Table 8.3.

Table 8.2 HTA 2: second section of the analysis. Stages of each task of the total set-up process

#	Code	Task		Stages	Observations
1	MA	Material arrival			
			1.1	Assemblers prepare to unload materials to set-up	The assemblers go to the convention center loading docks, where the transport arrives with the materials for the booth set-up, coming from the fabrication shop or warehouse
			1.2	Opening the trailer box doors	
			1.3	A group of workers enters the trailer box to unload the packed materials	
			1.4	Some other group of workers receives the material and places it along the floor	
			1.5	Another group of workers carries and drags the material to the set-up site	
2	MC	Material carrying and dragging			The material is carried through the aisles, avoiding materials scattered along the floor; from the loading docks to the set-up site where the booth will be installed
			2.1	Carrying of furniture and small structures: chairs, armchairs, tables, lamps, trash cans, toolboxes and small machines, rolls of light cables, paint cans, brooms	
			2.2	Carrying carpet rolls, pallets, laminate flooring boxes	Carpet rolls are carried between 2 or among 4 assemblers, depending on the length and weight of the rolls

(continued)

Table 8.2 (continued)

#	Code	Task		Stages	Observations
			2.3	Carriage of panelized walls, doors, carpentry structures, large display cabinets, metal scaffolding, stairs	
3	SP	Set-up site preparation			
			3.1	Structures are located around the set-up site	
			3.2	Panelized walls are stacked on top of each other	
			3.3	Set-up site cleared for floor or carpet installation 3.3.1 The floor area is swept and cleaned to work 3.3.2 Masking tape is taped over the set-up site perimeter to mark pallets and/or carpet and/or laminate flooring installation location	
			3.4	Furniture, structures and materials are moved from one side to another to free the aisles or the set-up site interior	
4	FI	Floor installation			
			4.1	The measurements of the set-up site area are rectified for the installation of pallets and/or carpet and/or laminate flooring	
			4.2	Carpet or floor packaging is removed	
			4.3	Pallets are placed and/or laminate flooring is installed and/or carpet is laid	After installation, laminate flooring or carpet is usually covered with a plastic film for protection during wall installation. It doesn't always happen

(continued)

Table 8.2 (continued)

#	Code	Task		Stages	Observations
			4.4	The installation is rectified by cutting excess material, leveling slopes and correcting bulges in the floor or carpet	Hand tools such as pocket knives and cutters and machine tools such as jigsaw are used
5	PW	Panelized walls installation			
			5.1	Carrying of panelized walls from their temporary location to the site where they will be installed 5.1.1. The plastic film that surrounds the panelized walls is removed to begin the installation	
			5.2	Lifting of walls	Two or more assemblers lift the walls
			5.3	Holding of walls	Wooden poles nailed to the floor are used to hold structures up. Afterward, an assembler is usually responsible for holding the wall while it is assembled with another wall
			5.4	Wall-to-wall fixing to form the booth structure	Tools such as an electric screwdriver, a pneumatic gun, a carpenter's hammer, and nails are used
6	WH	Working at heights			
			6.1	Installation of plafonds and/or marquees in elevated parts of the booth	Metal scaffolds, ladders and/or cranes are used, depending on the height and weight of the structures
7	EE	Electrical and electronic installation			The use of stairs may be required
			7.1	Installation of lamps	
			7.2	Installation of light boxes	

(continued)

Table 8.2 (continued)

#	Code	Task		Stages	Observations
			7.3	Installation of TV screens	
8	PA	Paint application			
			8.1	Walls are painted	Brushes and rollers with extensions are used. The use of ladders and/or scaffolding may be required
			8.2	Wall edges are painted	
9	GS	Graphics set-up			
			9.1	Large format graphics are placed. They can be self-adhesive vinyl, sublimated fabric or printed canvas, among others	The use of ladders and/or scaffolding may be required
			9.2	Exhibiting company large logos are placed in designated spots on the booth	
			9.3	Small graphics are placed on furniture or designated spots of the booth	
10	FP	Furniture and accessories placement			
			10.1	The furniture is located in the specified areas, according to the booth design 10.1.1 The heaviest furniture, such as showcases, light boxes, large armchairs and high tables, are placed first 10.1.2 The lighter furniture, such as chairs, low tables and benches, are then placed 10.1.3 Complementary accessories and ornaments such as planters, flower pots, lamps, centerpieces and trash cans are placed last	

(continued)

Table 8.2 (continued)

#	Code	Task		Stages	Observations
11	FD	Final details			
			11.1	General cleaning of the booth 11.1.1 Laminate floor is swept and cleaned 11.1.2 Furniture and glass are cleaned	
			11.2	Rectifications of: 11.2.1 Electrical installation and electronic devices 11.2.2 Paint on walls and furniture	
12	DP	Display products placements			It is usually done by the exhibitors or designated workers of the exhibiting company. Assemblers can help if required
13	MR	Materials removal			
			13.1	Surplus material and used tools are taken to the loading docks 13.1.1 Materials and tools are stored in transport for a return to the fabrication shop or warehouse	
14	SE	Set-up end			The assemblers leave the set-up site and the convention center. Only the exhibitors stay inside the booth

It can be seen in Table 8.3 that the PW, EE and PA tasks are the longest within the booth set-up time, especially the PW task, which takes 181.2 min (3 h) to complete, which represents 25.16% of the entire set-up process. In this task, possible severe postural and strength risks could be observed, which indicates the need for more in-depth studies about the panelized walls installation in future analyzes. Considering these findings important, the analysis continued by making a second video recording focused specifically on the PW task in the next assembly carried out by the assemblers, presented in the next section.

Table 8.3 Code of task, execution time in minutes and percentage, and possible ergonomic risks of each task

#	Task code	Time (min)	%	Ergonomic risks
1	MA	29.4	4.1	MMH Handling of heavy objects Postural load Visual discomfort: very sunny outside the convention center, very dark inside Aspiration of toxic fumes released by trailers and trucks Possible falls and slips Bumping into staff from other set-ups
2	MC	15.1	2.1	MMH Handling of heavy objects Postural load Possible falls and slips Bumping into staff from other set-ups
3	SP	14.8	2.0	MMH
4	FI	20.6	2.9	Suffering cuts and scrapes on the hands, arms and elbows Knees bent for a long time Discomfort in shoulders and spine
5	PW	181.2	25.2	MMH Handling of heavy objects Severe postural load Severe strength load Long-standing time Discomfort in shoulders, arms, hands, neck and back Possible falls and slips
6	WH	19.9	2.8	Risk of falls Long standing time Bent back, twisted neck and shoulder discomfort
7	EE	119.6	16.6	Bent back, twisted neck, discomfort in shoulders and hands Repetitive movements
8	PA	119.2	16.5	Risk of falls Repetitive movements Bent back, twisted neck, discomfort in shoulders and hands
9	GS	29.9	4.1	Risks of falls and slips Postural load Suffering cuts on hands
10	FP	24.8	3.4	MMH
11	FD	89.8	12.5	MMH Postural load Repetitive movements
12	DP	25.6	3.5	No possible risks were observed

(continued)

Table 8.3 (continued)

#	Task code	Time (min)	%	Ergonomic risks
13	MR	19.1	2.7	MMH
14	SE	11.0	1.6	No possible risks were observed
Total		720 min	100%	

Table 8.4 Analysis of the stages of the panelized walls installation task

Task	Stage	Code	Description	Frame
Panelized walls installation		PW	The walls that make up the basic structure that give the main external characteristics of the booth are assembled	
	Carrying	C	The walls are moved from their temporary location (usually in the aisles) and near the set-up site, into the space designated to build the booth	
	Lifting	L	The wall is lifted from the floor to its vertical position (90°) to be assembled with another wall	

(continued)

8.3.4 Set-Up Process Task with Longer Execution Time

The PW task analysis was developed from the observation of the videos made to the assemblers work with a total duration of 150 min and the information obtained

Table 8.4 (continued)

Task	Stage	Code	Description	Frame
	Holding	H	The wall is kept up, avoiding its fall until it is joined with the other wall	
	Fixing	F	Wall to wall is spliced by means of mechanical joints (drywall dowels, wood screws, nails, etc.) and tools (hammer, pneumatic gun, etc.)	
	Work organization	O	At this stage, it is proposed how to continue with the installation of walls. It is part of the set-up process but no physical effort is made	
	Complementary activities	X	They are various activities carried out within the time of the installation of walls. Joints are rectified, damaged parts are searched for repairing or painting, etc	

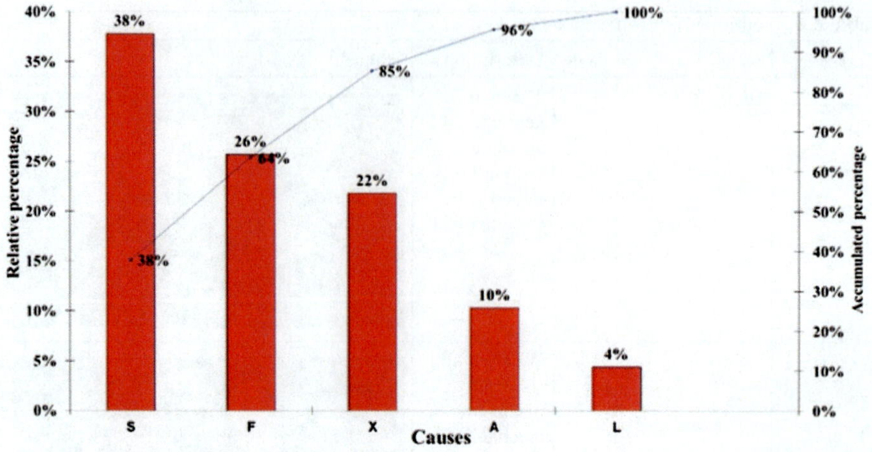

Fig. 8.1 Pareto chart for the causes of risk to the assembler

is presented in Table 8.4. Each stage was assigned a code for its registration and analysis.

The last two recorded stages: work organization (O) and complementary activities (X); although they are presented in a hierarchical way, in practice they can occur at any time during the installation process in no apparent order, depending on the construction progress, the previous organization, the quantity of materials and workers and the remaining set-up time.

8.3.5 The Riskiest Stage in the Task of Panelized Walls Installation

After task analysis, the postural risk exposure percentage of each stage of the task and its time in minutes was determined. Only the stages in which there was physical effort were considered; therefore, the work organization stage (O) was discarded. Then the time analyzed was reduced to 120 min before obtaining the exposure frequencies, obtaining 60 min equivalent to 100% of the time of postural risk exposure. The most representative frame was selected to illustrate each stage of the task. Below is an example (Table 8.5) with the data obtained for participant 1. Subsequently, another table is presented with the data collected from the six participants.

Due to the nature of the work, one assembler was not followed with video at a time, the six subjects interacted and exchanged activities during the entire duration of the task. Some frames showed several workers at the same time doing the same or different stage of the task. The frames were cut, separated and classified by subject and stage of the task. After obtaining the percentages of postural exposure and the

Table 8.5 Exposure to postural risk for each stage of the PW task. Participant 1

PW				
Code	Task	Frame	Postural risk exposure percentage	Postural risk time exposure (minutes)
C	Carrying		9.8	5.9
L	Lifting		4.5	2.7
H	Holding		37.8	22.7

(continued)

time in minutes for each subject and stage of the task, the data was collected in Table 8.6.

PW = panelized walls installation; % = exposure to postural risk percentage; T_{min} = time in minutes; C = carrying; L = lifting; H = holding; F = fixing; X = complementary activities.

When obtaining these results, the mean of each percentage of postural risk exposure was used to classify it in order of influence degree and establish which cause

Table 8.5 (continued)

PW				
Code	Task	Frame	Postural risk exposure percentage	Postural risk time exposure (minutes)
F	Fixing		25.9	15.5
X	Complementary activities		22.0	8.2
Total			100	60

Table 8.6 Percentage and time of postural risk exposure for each stage of the PW task of the six participants

PW										
Participant	Stage									
	C		L		H		F		X	
	%	T_{min}	%	T_{min}	%	T_{min}	%	T_{min}	%	T_{min}
1	9.8	5.9	4.5	2.7	37.8	22.7	25.9	15.5	22.0	8.2
2	9.9	5.9	4.5	2.7	37.7	22.6	25.8	15.5	22.1	13.3
3	11.1	6.7	4.0	2.4	37.6	22.6	25.7	15.4	21.6	12.9
4	11.2	6.7	4.1	2.5	37.6	22.6	25.5	15.3	21.6	12.9
5	8.7	5.2	4.8	2.9	38.2	22.9	25.9	15.5	22.4	13.5
6	11.3	6.8	4.2	2.5	37.6	22.6	25.5	15.3	21.4	12.8
Mean	10.3	6.2	4.4	2.6	37.8	22.7	25.7	15.4	21.8	12.3

PW = Panelized walls installation

Table 8.7 Ordered frequency

Order	Causes	%	% Normalized	% Cumulative
1	H	37.8	38	38
2	F	25.7	26	64
3	X	21.8	22	85
4	C	10.3	10	96
5	L	4.4	4	100

% = cause frequency; H = holding; F = fixing; X = complementary activities; C = carrying; L = lifting

intervenes to a greater extent in the ergonomic postural risk of the subjects and is presented in the following Pareto chart.

Using the Pareto chart (Fig. 8.1), it was determined which stage (cause) of the PW task represented the highest risk to the assembler. This diagram is an element of analysis that is used to determine which factor or factors intervene to a greater extent in the problem studied (Zegarra and Andara 2012). In this case, it was used to establish the riskiest stage in the panelized wall installation task. The Pareto chart is used when there is a need to identify the causes of problems systematically, it is useful in determining the main cause during problem-solving (Sales 2002).

Below is a table of ordered frequencies (Table 8.7) where the influence degree of the causes presented in the Pareto chart is indicated.

It was determined with the Pareto chart that the cause that represents the highest risk to the assemblers when performing the panelized walls installation task (PW) is the holding stage (H), occupying the first place in the diagram with a percentage of the 38%. Therefore, with this information and having identified the riskiest stage, it is possible and necessary to implement a postural analysis of the subjects in greater depth with tools such as RULA or REBA to determine their level of postural risk, and ergonomically intervene this stage within the PW task.

8.4 Discussion

With this study, it was possible to know the interaction that assemblers have with their work tools and their work environment, as reported by Lane et al. (2006). The order and hierarchy of the tasks identified in this study can serve as a guide to train apprentices, as well as update staff and experts in this field of work, as described by Sarker et al. (2007) and Moreno García et al. (2018).

Six stages were found, classified and defined within the PW task by means of an HTA (Annett 2004; Stanton 2006): (1) carrying, (2) lifting, (3) holding, (4) fixing, (5) work organization and (6) complementary activities. The first four stages had already been defined by Jia et al. 2011; Kim et al. 2006a, b, 2010, 2012; Nussbaum et al. 2009, as part of the panelized walls installation process classified as risky.

The last two stages classified and defined in this research: work organization (O) and complementary activities (X) were not considered or categorized by the authors; however, they retain analogy in their operations and definition with the stages of conversation and error repair, respectively, categorized by Lobb and Woods (2012), and these appear throughout the construction process with panelized walls studied by the authors.

To determine the factor that most intervenes in the problem of the task and its influence degree, the Pareto chart focused on ergonomic risk analysis was used according to Zegarra and Andara (2012). The stage identified that represents the greatest risk to the worker was the holding stage (H) with a 38% degree of influence in the panelized walls installation task. This stage had already been classified as risky by Kim et al. 2006a, b, and although its definition is like that provided by this research, its classification as a risky stage is not given by postural exposure but by the danger of a possible fall of the panelized wall and of the worker(s) when holding it due to the weight and dimensions of the wall, in addition to the lack of lifelines.

The information obtained with this analysis may be useful to improve safety measures and reduce risks to the assemblers' health, as detailed by Fargnoli et al. (2019) and López Hernández et al. (2015). Likewise, the results of this study can contribute to simplify set-up workflows and improve the work experience of assemblers according to what was mentioned by Durán-Coronado et al. (2019) and Promann and Zhang (2015).

It is important to note that the results obtained in this study were the synthesis of the work carried out through repeated observations of the video recordings made to the assemblers and the performance of their activities, which means that the total of identified tasks, its stages and sub-stages, as well as its hierarchy, could vary when studying other assemblers from other companies with different set-up processes, carrying out the same type of booths, for which it is necessary to carry out more studies to obtain a statistically significant result.

8.4.1 Conclusions and Recommendations

It is suggested to continue with the research process by carrying out an in-depth postural evaluation to know which body segments of the assemblers are affected and in what way when carrying out the task of installing panelized walls and with these results implement a design proposal that can be tested by the users for whom it is intended and continue to act on the ergonomic risks encountered. It is recommended sensitizing assemblers and their employers about the meaning of postural hygiene and the proper execution of their tasks to avoid injuries.

It is also recommended approaching the subject of study from the perspective of other lines of research in ergonomics to generate other intervention proposals, such as studies on energy expenditure of workers; research in environmental ergonomics to study the work environment, lighting, temperature, ventilation, visual load, noise, vibrations; cognitive ergonomics to find out what kind of mental load workers are under and ergonomics training teaching workers about taking care of their health.

This study also wants to make known the use of HTA as a fundamental piece for the beginning of an ergonomic evaluation to give way to other more in-depth evaluations as well as the use of complementary instruments to enrich the ergonomic analysis.

The use of HTA is also noted as an easy to learn and apply tool that allows the ergonomics practitioner to know how a system or process works in a practical, useful and economical way, and that this method can be applied in works other than the areas of health, informatics and product design.

Acknowledgements Author wants to acknowledge to the workers of the assembly company, to the Universidad of Guadalajara, to M. S. Enrique Herrera Lugo and Ph. D. Elvia L. González M., advisors of the research project. To the CECAD of the Universidad Autónoma Metropolitana - Xochimilco. Moreover, the author wants to acknowledge to Ph. D. Marcelo Soares, to M. S. Guillermo Maldonado Sandoval, to Ph. D. Jesús Rodríguez Diego and Gustavo Maza, meeting industry expert, for reading, comments, and advice on this document. This work was carried out with the support of Conacyt (National Council of Science and Technology [in Spanish] in Mexico).

References

Annett J (2004) Hierarchical task analysis. In: Diaper D, Stanton N (eds) The handbook of task analysis for human–computer interaction. Lawrence Erlbaum Associates, Mahwah, NJ, pp 67–82

Durán-Coronado A, Maldonado-Macías A, Barajas-Bustillos M, Hernández-Arellano J (2019) Cognitive analysis of mental load and identification of human error to improve the user experience [In Spanish]. CienciaUAT 14:71. https://doi.org/10.29059/cienciauat.v14i1.1173

Entorno Turístico (2016) What is the Meetings Industry and how is it classified? [In Spanish] In: Entorno Turístico. https://www.entornoturistico.com/que-es-la-industria-de-reuniones-y-como-se-clasifica/. Accessed 13 Aug 2020

Fargnoli M, Lombardi M, Puri D (2019) Applying hierarchical task analysis to depict human safety errors during pesticide use in vineyard cultivation. Agriculture 9:158. https://doi.org/10.3390/agriculture9070158

Gómez-Bull K, Hernández-Arellano J, Ibarra-Mejía G (2015) A proposed methodology for task analysis in ergonomic evaluations. Proc Manuf 3:4756–4760. https://doi.org/10.1016/j.promfg.2015.07.573

Herrera Lugo E (2015) Video in ergonomic analysis of workstations. In: Prado León L (ed) Methods and techniques for the ergonomic analysis of the workplace [In Spanish], 1st edn. Universidad de Guadalajara, Guadalajara, pp 77–81

Jia B, Kim S, Nussbaum M (2011) An EMG-based model to estimate lumbar muscle forces and spinal loads during complex, high-effort tasks: development and application to residential construction using prefabricated walls. Int J Ind Ergon 41:437–446. https://doi.org/10.1016/j.ergon.2011.03.004

Kim H, Nussbaum M, Seol H et al (2006a) Risk assessment of panelized wall systems in residential construction using critical incident technique. Proc Hum Fact Ergon Soc Ann Meet 50:2222–2224. https://doi.org/10.1177/154193120605001907

Kim S, Hurley M, Nussbaum M et al (2006b) Residential wall panel designers' knowledge and attitudes toward ergonomics. Proc Hum Fact Ergon Soc Ann Meet 50:1892–1895. https://doi.org/10.1177/154193120605001737

Kim S, Nussbaum M, Jia B (2010) Low back injury risks during construction with prefabricated (panelised) walls: effects of task and design factors. Ergonomics 54:60–71. https://doi.org/10.1080/00140139.2010.535024

Kim S, Nussbaum M, Jia B (2012) The benefits of an additional worker are task-dependent: assessing low-back injury risks during prefabricated (panelized) wall construction. Appl Ergon 43:843–849. https://doi.org/10.1016/j.apergo.2011.12.005

Klever (2020) Exhibition industry, key to economic reactivation [In Spanish]. In: Klever Business Media. https://www.klever.com.mx/industria-de-exposiciones-clave-para-reactivacion-economica/. Accessed 13 Aug 2020

Lane R, Stanton N, Harrison D (2006) Applying hierarchical task analysis to medication administration errors. Appl Ergon 37:669–679. https://doi.org/10.1016/j.apergo.2005.08.001

Lobb B, Woods G (2012) In search of a representative sample of residential building work. Appl Ergon 43:868–875. https://doi.org/10.1016/j.apergo.2011.12.008

López Hernández N, Maldonado A, García Alcaraz J et al. (2015) Evaluation of human error in the laboratory area of a private hospital in Cd. Juárez, Chihuahua [In Spanish]. Cult Cien Tecnol 55:192–200. In: https://erevistas.uacj.mx/ojs/index.php/culcyt/article/view/760. Accessed 13 Aug 2020

MacLeod I, Hone G, Farmilo A (2005) The HTA tool [hierarchical task analysis]. IEE and MOD HFI DTC symposium on people and systems—who are we designing for? doi: https://doi.org/10.1049/ic:20050459

Meetings Mexico (2015) Glossary of business tourism [In Spanish]. Mundo Editorial 1–32. In: https://www.meetingsmexico.com/WEB/files/GLOSARIO_DEL_TURISMO_DE_REUNIONES.pdf. Accessed 13 Aug 2020

Moreno García M, Maldonado Macías A, Hernández Arellano J (2018) Hierarchical task analysis, mental load and human error applied to the use of a drug dispenser for older adults. In: López Acosta M, de la Vega Bustillos E, Velarde Cantú J et al. (ed) Research and applications of ergonomics [In Spanish], 1st ed. Pearson, Ciudad de México

Méndez R (2018) Ergonomic postural risk evaluation of workers during the panelized walls installation task in the booths set-up [In Spanish]. Master, Universidad de Guadalajara. https://url2.cl/wyRIN. Accessed 13 Aug 2020

Nussbaum M, Shewchuk J, Kim S et al (2009) Development of a decision support system for residential construction using panellised walls: approach and preliminary results. Ergonomics 52:87–103. https://doi.org/10.1080/00140130802480869

Promann M, Zhang T (2015) Applying hierarchical task analysis method to discovery layer evaluation. Inf Technol Lib. https://doi.org/10.6017/ital.v34i1.5600

RAE (2014) Estand. Diccionario de la lengua española. https://dle.rae.es/?id=Glt0adf. Accessed 10 Aug 2020

Reed Exhibitions (2016) Exhibitor manual [In Spanish]. CWA-Expo Carga E Intralogistics Latin America

Sales M (2002) Pareto chart [In Spanish]. GestioPolis - Conocimiento en Negocios. https://www.gestiopolis.com/diagrama-de-pareto/. Accessed 20 Aug 2020

Sarker S, Chang A, Albrani T, Vincent C (2007) Constructing hierarchical task analysis in surgery. Surg Endosc 22:107–111. https://doi.org/10.1007/s00464-007-9380-z

Stanton N (2006) Hierarchical task analysis: developments, applications, and extensions. Appl Ergon 37:55–79. https://doi.org/10.1016/j.apergo.2005.06.003

Vanexpo (2016) Exhibitor manual [In Spanish]. Expo Eléctrica Internacional 2016

Zegarra R, Andara M (2012) Ergonomic risk analysis through REBA and RULA methods [In Spanish]. UNEXPO 252–259.

Chapter 9
Ergonomic Evaluation of Localized Manufacturing Concerns for Agricultural Appliances in Odisha (India)

Debesh Mishra and Suchismita Satapathy

Abstract In India, the manufacturing of agricultural machineries is undertaken by the village artisans, small-scale as well as large-scale industries. The blacksmiths and carpenters have been fabricating the traditional agricultural appliances. Thus, an ergonomic analysis has been made to review the status and discomfort levels of the village craftsmen manufacturing the agricultural appliances in this paper, so as to plan for any modifications and mechanization of agricultural manufacturing sectors in future.

Keywords Agriculture · Hand-Smith · Ergonomics · Suzanne Rodgers · Strain Index · Discomforts · Odisha

9.1 Introduction

India has witnessed an exceptional growth in agricultural sectors, which has helped to eliminate hunger to self-sufficiency by growing the production of food grains to 208 million tonnes from earlier 51 million tonnes, in addition to surplus for exports. The agricultural machinery adoption growth rate in the country has been achievable because of the local manufactures. The manufacturing of agricultural machineries in India has been done by village-artisans, organized mediums, small-scale industries, as well as large-scale sectors. All the sophisticated machineries like engines, tractors, and dairy equipments are manufactured by the organized sectors, while the blacksmiths and carpenters are the fabricators of traditional agricultural appliances. The earlier mechanization of agriculture in India was significantly influenced by the technical improvements in England. During the later part of the nineteenth century, the horse-drawn and steam-tractors-controlled appliances were imported. When the horse-drawn appliances brought from England were found unsuitable for bullocks and buffaloes, they were properly modified to suit the Indian draught animals by small-scale manufacturers. Thus, an attempt has been made in this study which was

D. Mishra · S. Satapathy (✉)
SME, KIIT Deemed to Be University, Bhubaneswar, Odisha, India
e-mail: ssatapathyfme@kiit.ac.in

A. Realyvásquez Vargas et al. (eds.), *New Perspectives on Applied Industrial Ergonomics*,
https://doi.org/10.1007/978-3-030-73468-8_9

not carried out earlier to make an ergonomic analysis of village craftsmen preparing the traditional hand tools & bullock-drawn implements for agricultural purposes in the province of Odisha in India.

9.2 Literature Review

The committed conventional farmers with a belief about organic production methods to be environmentally harmful or misleading marketing tactics in organic price premiums are less likely to consider organic production as a serious choice (Darnhofer et al. 2005). The power contribution by draught animals in sub-Saharan Africa has been reported to be up to 25% of farm power (FAO 2006). A historical, cultural as well as political connection of Nepal with India have resulted in different patterns investments in farming mechanization (Biggs and Justice 2013). The conventional systems can be managed through less intensive management requirements on larger farms with expected profits higher for the organic systems, with an advantage in profitability which was more for small farms (Delbridge et al. 2013).

Most of the tedious and time-consuming works are performed by farmwomen with traditional methods in varying body postures. An improved sickle was reported of reducing the drudgery level in addition to a higher harvesting efficiency as compared to ordinary sickle (Patel et al. 2015). Singh et al. (2015) have revealed about the domination of traditional tools and implements over the modernized equipment in Wokha district (Nagaland) for all farm activities. It was further suggested for those tools and implements to have further improvement through local artesian along with the farmers to achieve self-reliant in that region. A remarkable diversification has been occurred in the agricultural machinery sub-sector of activities like repairing, manufacturing, and maintenance of farm-implements and machineries (Alam 2005).

Moreover, Alam et al. (2017) have revealed the Bogra region in Bangladesh, as an emerged center for manufacturing of agriculture-related machineries as well as spare parts specially for irrigation pumps, maize-shellers, threshers, and other spare parts for small diesel engines and equipment, etc. and thereby holding a major contribution of the local production. Delbridge and King (2016) have included risk as well as the cost of changeover into a model of the organic implementation decision. Lower changeover and opportunity costs were found to be faced by small farms having more likely to changeover. As energy required for agriculture mechanization comes from farm power sources. The engine power is generally obtained from low-cost tractors and other engine-powered technologies. However, in many developing countries such as China and India, the use of draught animals for farm power is still important (Singh and Zhao 2016). Khapayi and Celliers (2016) have investigated the major limiting factors preventing farmers' progress to commercial agricultural from subsistence farming in the Eastern Cape Province, that included poor physical infrastructures like poor roads, lacking in transportation from the farms to the markets, lacking of marketing skills as well as information, poor infrastructural markets, and higher transactional costs, inadequate land availability for production expansion, lacking in agricultural implements to make production better, poor farm and production management-skills, in addition to lower education levels resulting in a failure to interpret market-related information for production planning as well as marketing.

The "sustainable agricultural mechanization (SAM)" has been recommended for smallholder farmers' producing major food in the world, such that the production will increase by up to 100% by 2050 to nourish the rising inhabitants (Sims and Kienzle 2017). In a study carried out for small farmers in Tanzania, the timely access to adequate quality inputs and machinery supply for farmers was found to be influenced by lack of finance as one of the important barrier (Mdemu et al. 2017). Niewiadomski et al. (2018) have made an attempt to explore different barriers for the effective implementation of the principles of lean management for parts, sub-assemblies in addition to ready-made agriculture-related machinery manufacturers.

Moreover, ergonomics has been regarded as a multidisciplinary approach in science, whose purpose is to modify, design, and improve the conditions of workplaces, products, and the quality of life-promotion. For personnel performance improvement for the workers, the industrial ergonomics gives more emphasis in workplace design (Deborah et al. 1990; Jaffar et al. 2011; Sadeghi Naeini et al. 2014; Karwowski 2006) that may be called as "human-centered" science (Sandom and Harvey 2009). Moreover, there are different aspects of ergonomics in which human as end-clients of the items are examined (Sadeghi Naeini and Heidaripour 2011).

The assessment of the human-beings interaction with the working-systems is done with the help of ergonomics, through introducing some theoretical and practical design for human prosperity advancement as well as overall systems performances (Vieira et al. 2012). Little changes can make critical contrasts and give a more clear comprehension of how the working environment-condition, the board approaches, individual limits, and employment demands affect the security and well-being of people at work (Rodgers 2000). Rucker and Moore (2002) have used the "Strain Index" in a study by considering 28 manufacturing jobs including hose connector as well as chair manufacturing plants. Garg et al. (2007a, b) have revealed that since the development of their method to determine the "Strain Index" in 1995, it has been worldwide used. Further they have suggested the likely future-revisions to the model.

A number of studies have revealed that exposures to high repetitive, higher forces, nonaligned hand or wrist postures as well as hand or arm vibrations by the use of hand tools, are related with a higher-risks of distal upper-extremity symptoms in addition to musculoskeletal disorders (Bernard 1997; Garg and Kapellusch 2011; Harris-Adamson et al. 2011). Batista et al. (2012) have made an attempt to develop an ergonomic-device in order to assist in the maintenance of the units of "Tucuruí-Hydropower" Plant that resulted in reducing the maintenance time, losses in billing and improved the performance by reducing the physical strains for the performances of labors. Meyers et al. (2014) have made an attempt to develop an alternative "Strain Index" risk classification categories. It was suggested to have the "Strain Index" risk category structure to be customized to specific populations. da Silva et al. (2017) have performed an ergonomic analysis in a bicycle industry's workstation of the Industrial Polo in Manaus, by using the "Sue Rodgers" tool for evaluating the degree of risk in the employee's body regions. Garg have proposed the "Revised Strain Index," which is a "distal-upper-extremity (DUE)" physical exposure assessment model on the basis of intensity of exertions, frequency of exertions, duration per-exertion, wrist or hand postures, and the duration of tasks per day.

Nusantara and Suharno (2017) have used three ergonomic tools such as "Rapid Upper Limb Assessment (RULA)," "Manual Task Risk Assessment (ManTRA)," and "Rodgers Muscle Fatigue Analysis (RMFA)" in order to assess the workers' working posture in "CV.Tani Organik Merapi (CV.TOM)" an organic vegetables company. Restuputri and Gangguan (2018) have utilized "Nordic Body Map questionnaires" and analysis of work-risks on hand by "Strain Index" for batik workers in Indonesia. Similarly, the questionnaires for musculoskeletal discomforts have been used and found suitable in different studies (Çakıt 2019; Erdinc et al. 2011; Habibi et al. 2015; Restuputri and Gangguan 2018). Rahma and Faiz (2019) have used three methods such as "Rapid Entire Body Assessment (REBA)," RULA and "Quick Exposure Checklist (QEC)" to analyze the work posture of workers in gamelan craft center, by considering eight jobs divided into two tasks elements: like the dominant task as well as risky task. Ayub and Shah (2018) have used RULA scores sheet, validated version of QEC and Nordic questionnaires to measure the risks for work-related musculoskeletal disorders among the workers of a commercial enterprise. Mishra and Satapathy (2019) have made an attempt to assess the musculoskeletal disorders of Odisha-based farmers in India by using questionnaires as well as RULA method.

9.3 Methodology

In this study, the workers' performing blacksmith jobs were considered followed by the still photography of postures in working and subsequent observation of activities in different operations from Khorda district in Odisha (India). Then, by using the "Ergofellow 3.0" software, the different levels of risks, the classification of those risks, and the discomfort levels involved were identified, so that further changes could be carried out in these task performances. Moreover, a randomly selected worker was considered for the evaluation of different levels of risks and the classification of those risks by the use of "Suzanne Rodgers" tool and "Moore and Garg's Strain Index (SI)" in "ErgoFellow 3.0" software. The "Suzanne Rodgers" tool was used to analyze the muscle fatigue, where three factors were required to be evaluated such as efforts, duration as well as frequencies of any task. In the effort output 1, 2, 3, and 4 were indicated in the first digit for light, moderate, heavy and very hard tasks, respectively. In the duration output 1, 2, 3, and 4 were indicated in the second digit for <6, 6–20, 20–30 & >30 s of tasks performances, respectively. Similarly, in the frequency output 1, 2, 3, and 4 were indicated in the third digit for <1, 1–5, >5–15 & >15 per minutes of tasks performances, respectively. Thus, the final output was obtained in the form of three digits, where the first, second, and third digits indicate efforts, duration as well as frequencies, showing the priorities for any changes, if required in tasks performances (Table 9.1).

Table 9.1 Priorities for changes based on the outputs of "Suzanne Rodgers" tool

Low	Moderate	High	Very high
111	123	223	323
112	132	313	331
113	213	321	332
211	222	322	4xx*
121	231	–	x4x
212	232	–	xx4
311	312	–	–
122	–	–	–
131	–	–	–
221	–	–	–

*x represents any numbers ranging from 1 to 4

Then, Moore and Garg's Strain Index was used, which is the multiplication output of six risk factors associated with task performances. If the value of SI is less than 3, then it is considered to be "Safe." If between 3 and 5, then it is considered to be "Uncertain." If between 5 and 7, then it is considered to be at "Some Risks." And if, the SI value is more than 7, then it is considered to be "Hazardous." Further, the discomfort questionnaire tool available in the "ErgoFellow 3.0" software was utilized to get information from 32 workers performing the blacksmith activities for different agricultural hand tools, at different locations of Khorda district in Odisha (India) about their discomforts during performance of concerned work by considering eighteen different body parts (Fig. 9.10), so that necessary changes or alterations may be carried out in the tasks.

9.4 Results and Discussion

The local blacksmith workers usually found of making the agricultural tools followed by forging metals and using hammering tools to cut or bend the workpieces to the required shapes with sizes as needed for the farmers (Fig. 9.1).

For the evaluation of neck of the worker as shown in Fig. 9.1, the efforts level was found to be moderate, the continuity in effort duration was more than 30 s, and the frequencies in effort were found to be 1–5 per minute showing the three digits output for neck as 242 with purple color. Thus, it indicated a very high priority for change to be occurred in this task performance (Fig. 9.2). In evaluating the back of the worker, the efforts level was found to be moderate, the continuity in effort duration was 20–30 s, and the frequencies in effort were found to be 1–5 per minute showing the three digits output for back as 232 with yellow color.

Thus, it indicated a moderate level of priority for changes (Fig. 9.3). In evaluating the left hand, the efforts level was moderate, the continuity in effort duration was

Fig. 9.1 Blacksmith worker making agricultural tools

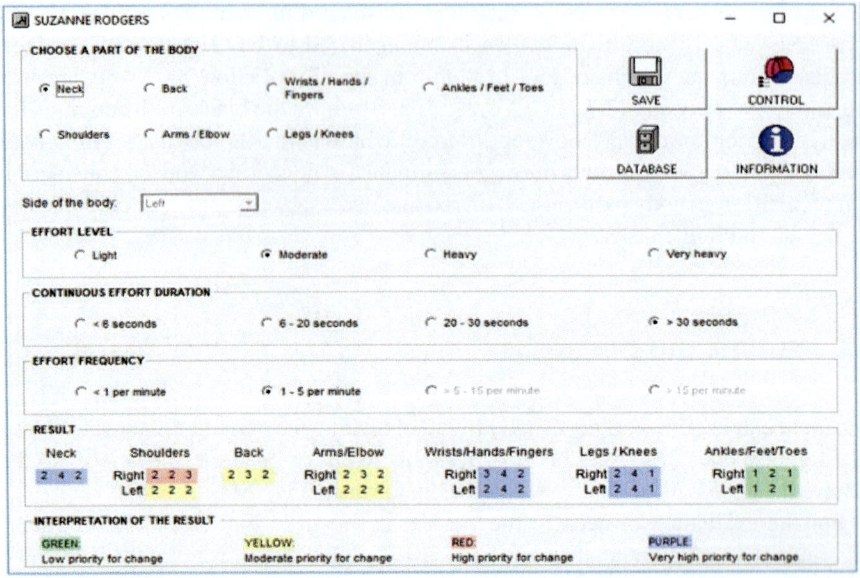

Fig. 9.2 Evaluation of neck of the worker using "Suzanne Rodgers" tool

more than 30 s, and the frequencies in effort were 1–5 per minute showing 242 with purple color as the three digits output for left hand, wrist as well as fingers. Thus, it indicated a very high level of priority for changes (Fig. 9.4a). While, for the right hand, the efforts level was heavy, the continuity in effort duration was more than

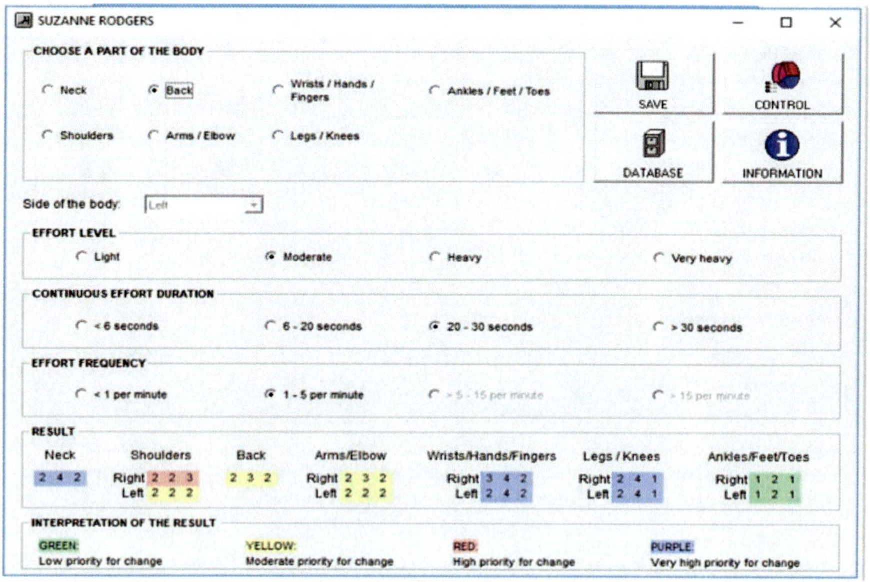

Fig. 9.3 Evaluation of back of the worker using "Suzanne Rodgers" tool

30 s, and the frequencies in effort were 1–5 per minute with 342 as the three digits output having purple color for right hand, wrist as well as fingers.

Hence, it indicated a very high level of priority for changes (Fig. 9.4b). For both the left and right feet, light efforts level, 6–20 s in the continuity of effort duration, and the frequencies in effort of less than one per minute were found showing 121 as the output with purple colors for both left as well as right ankles, feet, and toes. Therefore, this indicated a lower level of priority for changes (Fig. 9.5a, b). Whereas, for the left and right shoulders, moderate efforts level, 6–20 s of the continuity in efforts duration, along with and 1–5 per minute frequencies for left and more than 5–15 per minute frequencies in efforts for right shoulder were observed with 222 as the output having yellow color for left shoulder, as well as with 223 as the output having red color for right shoulder. Hence, for left shoulder it indicated a moderate level of priority for changes (Fig. 9.6a), and for right shoulder a high level of priority for changes was indicated (Fig. 9.6b).

Moreover, for both the left and right arms and elbows, moderate efforts level, the effort frequencies of 1–5 per minute, and 6–20 s in the continuity of efforts duration for the left side and 20–30 s in the continuity of efforts duration for the right side was found showing three digits to be 222 and 232 as the outputs with yellow colors for both. Therefore, this indicated a moderate level of priority for changes for left as well as right arms and elbows (Fig. 9.7a, b). Similarly, a moderate efforts level, more than 30 s in the continuity of efforts duration, and effort frequencies of less than one per minute were found for both the legs and knees showing 241 as the output with

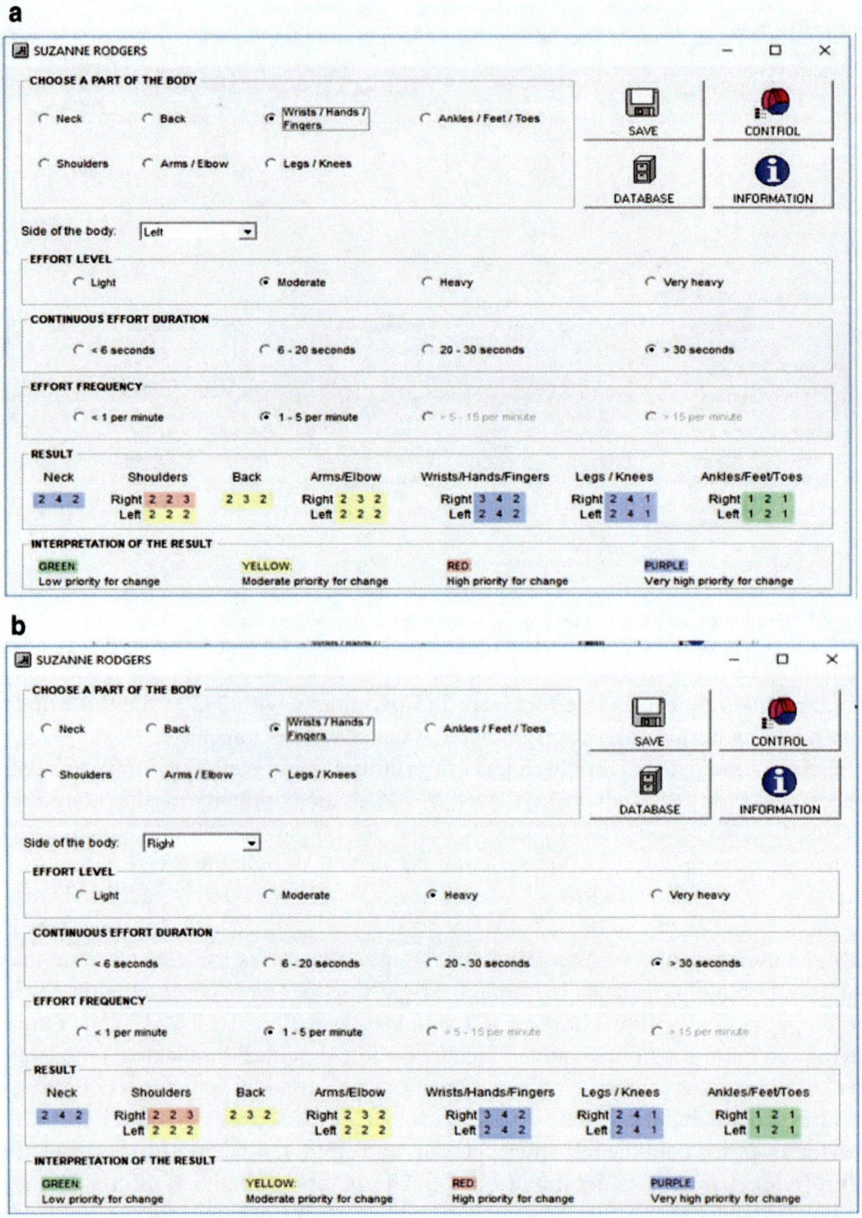

Fig. 9.4 **a** Evaluation of left hand of the worker using "Suzanne Rodgers" tool. **b** Evaluation of right hand of the worker using "Suzanne Rodgers" tool

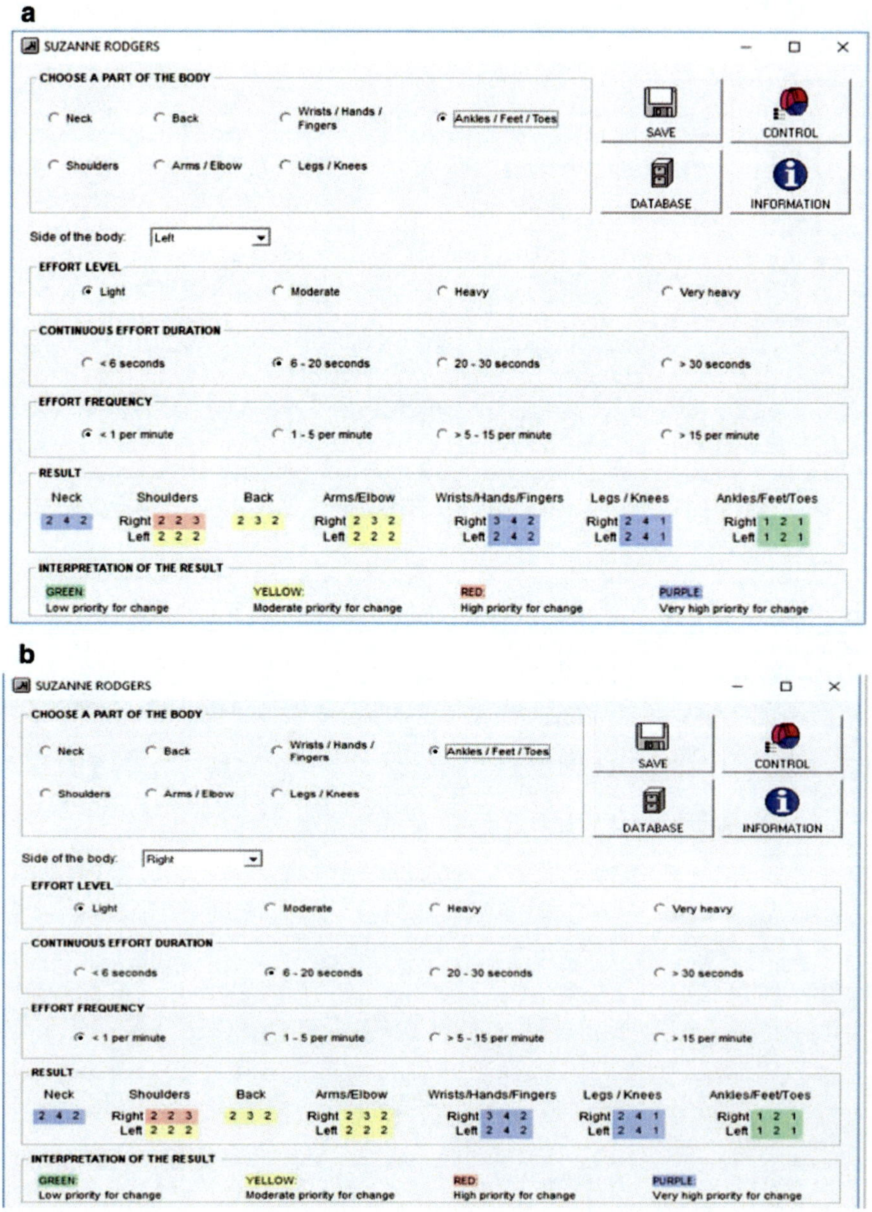

Fig. 9.5 **a** Evaluation of left feet of the worker using "Suzanne Rodgers" tool. **b** Evaluation of right feet of the worker using "Suzanne Rodgers" tool

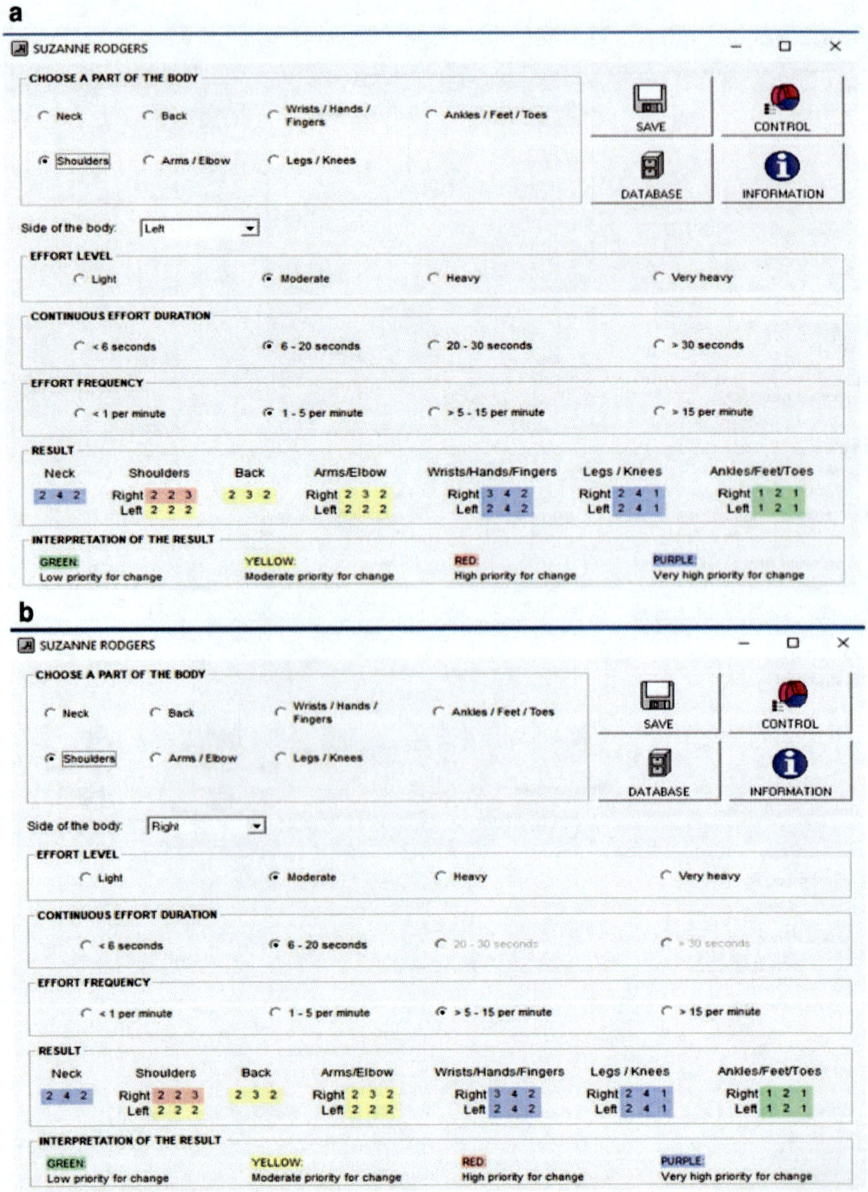

Fig. 9.6 **a** Evaluation of left shoulder of the worker using "Suzanne Rodgers" tool. **b** Evaluation of right shoulder of the worker using "Suzanne Rodgers" tool

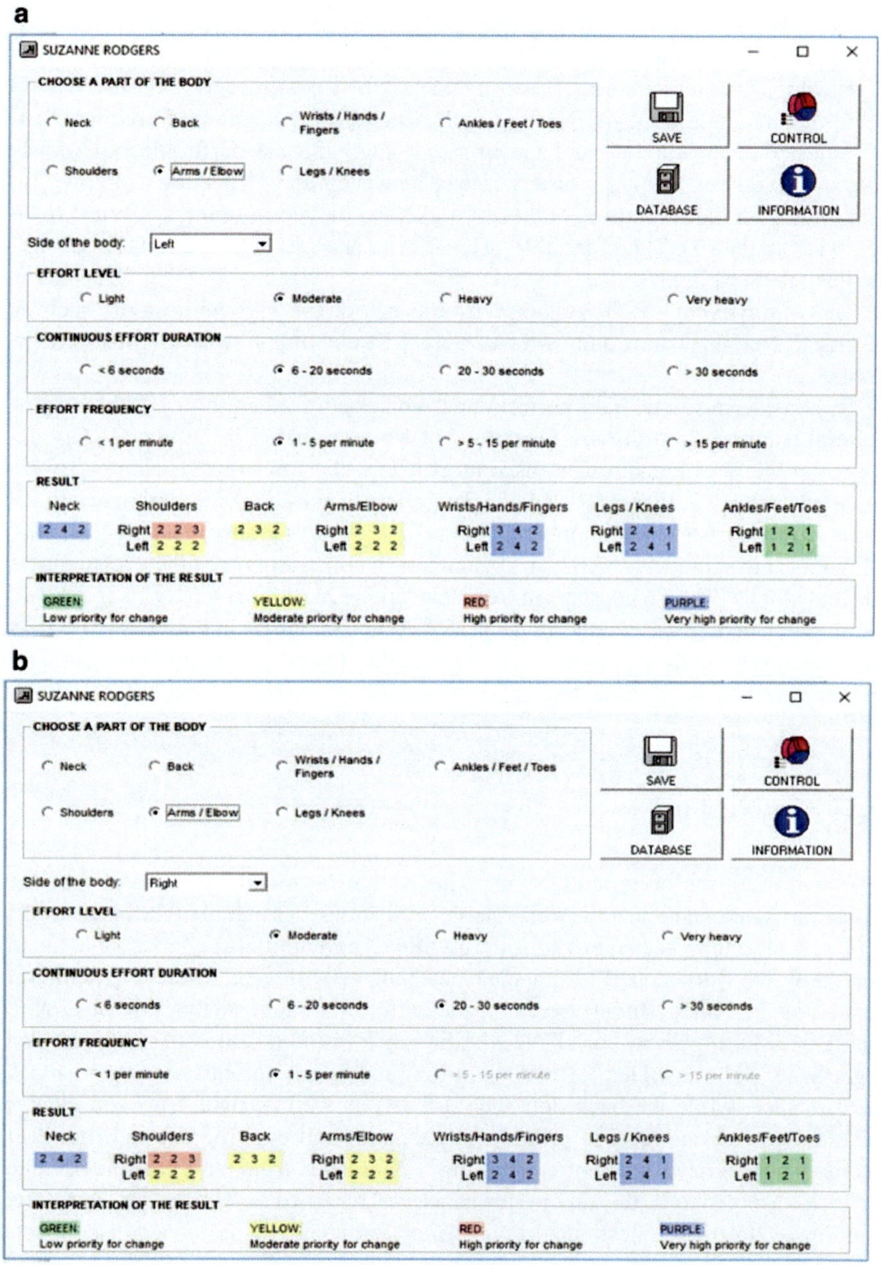

Fig. 9.7 **a** Evaluation of left arm of the worker using "Suzanne Rodgers" tool. **b** Evaluation of right arm of the worker using "Suzanne Rodgers" tool

purple colors for both left and right sides. Hence, a very high level of priority for changes was indicated for left and right legs and knees (Fig. 9.8a, b).

The intensity and duration of exerting the task by the worker were observed to be somewhat hard and of 30–49% of the cycle. The efforts per minute were more than 20, and also the posture of hand and wrist was good, the task performing speed was fast, and the task duration was observed to be between 2 and 4 h per day. Moreover, by multiplying all the six indices of these risk factors, the Strain Index was found to be 15.19 (more than 7). Therefore, this task performance was regarded as "Hazardous" as illustrated in Fig. 9.9.

The summary of results obtained by the use of the ergonomic tools such as "Suzzane Rodgers" tool and "Moore Garg's Strain Index" was as illustrated in Table 9.2.

By considering 32 workers performing the blacksmith activities for different agricultural hand tools, at different locations of Khorda district in Odisha (India) and based on the questionnaires on discomfort levels, the following discomforts were reported (Table 9.3; Figs. 9.10, 9.11): 07 numbers in eyes (21.87%), nil in heads, nil in necks, nil in trapezes, 01 number in thorax (3.12%), 01 number in lumbar (3.12%), 21 numbers in shoulders (65.62%), 20 numbers in upper-arms (62.50%), 23 numbers in elbows (71.87%), 16 numbers in forearms (50%), 21 numbers in wrists (65.62%), 07 numbers in hands as well as fingers (21.87%), 07 numbers in buttocks (21.87%), nil in thighs, 07 numbers in knees (21.87%), nil in lower legs, nil in ankles, and nil in feet as well as toes, respectively.

9.5 Conclusion

The manufacturing of agricultural machineries has been carried out mostly by the village artisans, small-scale as well as large-scale industries in India. The blacksmiths and carpenters have been fabricating the traditional agricultural appliances in India. In this study, the workers performing the blacksmith activities for different agricultural hand tools in Odisha (India) were considered for ergonomic analysis. It was found that a very high priority for changes to be made in the postures of neck, left and right hand, right shoulder, left and right legs and knees, for carrying out this task performance, while the back, left shoulder, left as well as right arms and elbows required a moderate level of priority for changes. And both the left and right feet required a lower level of priority for changes. Moreover, from the Strain Index value of 15.19 (More than 7), this task performance was regarded as "Hazardous" in nature. Further, by the use of questionnaires on discomfort levels, the following discomforts were reported by the workers: 07 numbers in eyes (21.87%), nil in heads, nil in necks, nil in trapezes, 01 number in thorax (3.12%), 01 number in lumbar (3.12%), 21 numbers in shoulders (65.62%), 20 numbers in upper-arms (62.50%), 23 numbers in elbows (71.87%), 16 numbers in forearms (50%), 21 numbers in wrists (65.62%), 07 numbers in hands as well as fingers (21.87%), 07 numbers in buttocks (21.87%),

a

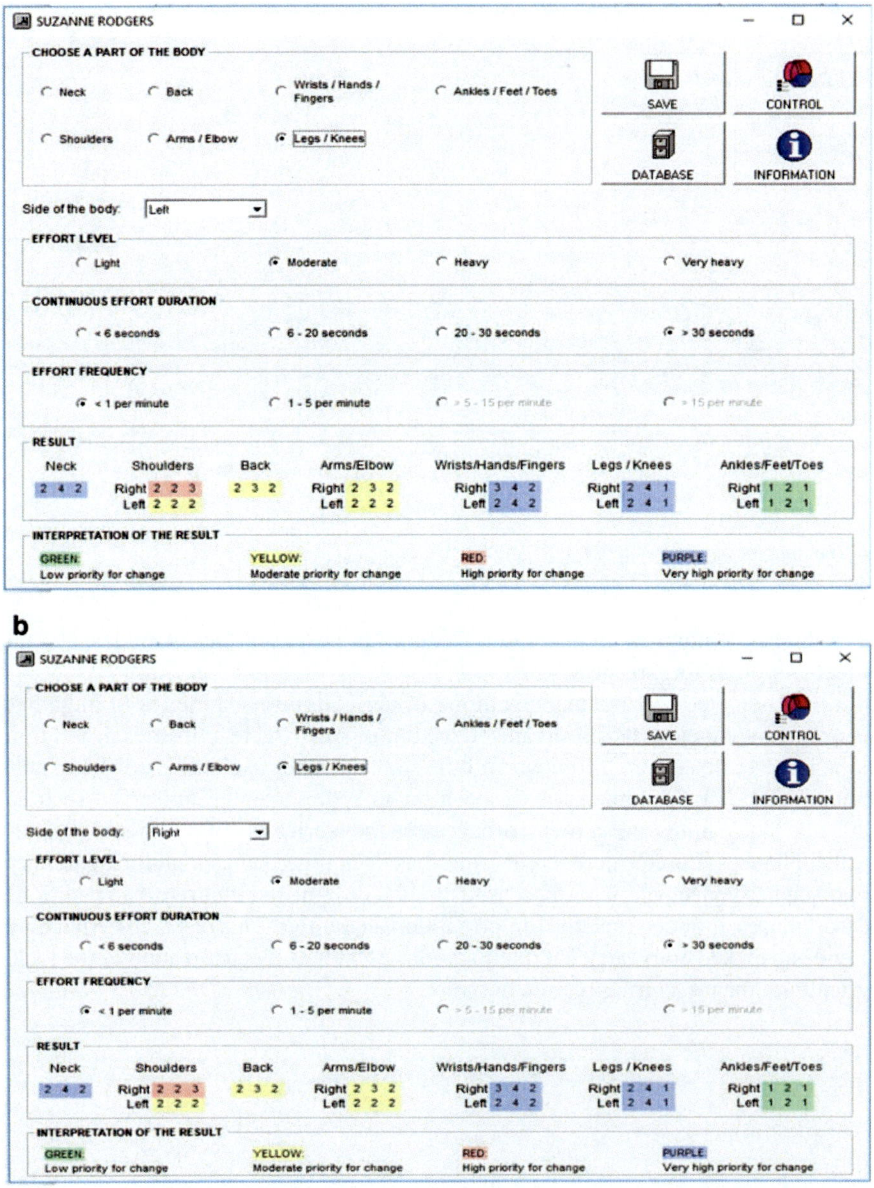

Fig. 9.8 **a** Evaluation of left knee of the worker using "Suzanne Rodgers" tool. **b** Evaluation of right knee of the worker using "Suzanne Rodgers" tool

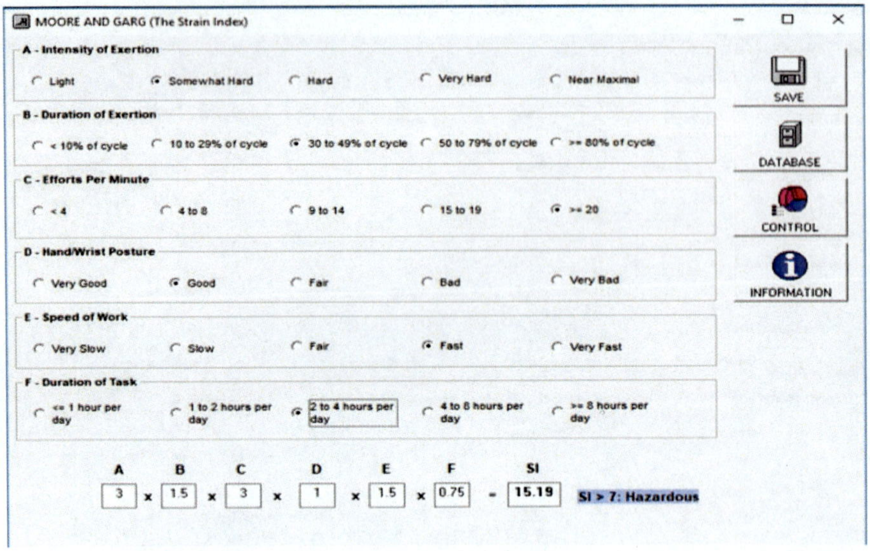

Fig. 9.9 Inputs and output in "Moore and Garg's Strain Index"

nil in thighs, 07 numbers in knees (21.87%), nil in lower legs, nil in ankles, and nil in feet as well as toes, respectively.

It may be noted that the manufacturing of agricultural machineries in India has been done by village-artisans, organized mediums, small-scale industries in addition to the large-scale sectors. Moreover, all the sophisticated machineries like engines, tractors, and dairy equipments have been manufactured by the organized sectors, while the blacksmiths and carpenters have been the fabricators of traditional agricultural appliances. Furthermore, ergonomics has been regarded as a multidisciplinary approach in science, whose purpose is to modify, design, and improve the conditions of workplaces, products, and the life-promotional qualities. This research output may provide an endeavor to carry out modifications as well as mechanization of the local agricultural manufacturing sectors in future.

Table 9.2 Summary of evaluation of worker by ergonomic tools

Evaluation by "Suzanne Rodgers" tool			Evaluation by "Moore and Garg's Strain Index"		
Parameters	Output	Remarks	Parameters	Output	Remarks
Neck	242	Very high priority for change	Intensity and duration of exerting the task	Somewhat hard and of 30–49% of the cycle	Hazardous task
Back	232	Moderate level of priority for changes			
Left hand, wrist as well as fingers	242	Very high level of priority for changes	Efforts per minute	More than 20	
Right hand, wrist as well as fingers	342	Very high level of priority for changes			
Left ankles, feet and toes	121	Lower level of priority for changes			
Left ankles, feet and toes	121	Lower level of priority for changes	Posture of hand and wrist	Good	
Left shoulder	222	Moderate level of priority for changes			
Right shoulder	223	High level of priority for changes	The task performing speed	Faster	
Left arms and elbows	222	Moderate level of priority for changes			
Right arms and elbows	232	Moderate level of priority for changes	Task duration	Between 2 and 4 h per day	
Left legs and knees	241	High level of priority for changes			
Right legs and knees	241	High level of priority for changes	Strain Index	15.19 (More than 7).	

Table 9.3 Workers reported of discomforts in different body parts ($n = 32$)

Body Parts	Number of workers with discomfort levels (%)
Eyes	07 (21.87)
Heads	00 (00)
Necks	00 (00)
Trapezes	00 (00)
Thoraxes	01 (3.12)
Lumbar	01 (3.12)
Shoulders	21 (65.62)
Upper-arms	20 (62.50)
Elbows	23 (71.87)
Forearms	16 (50)
Wrists	21 (65.62)
Hands & fingers	07 (21.87)
Buttocks	07 (21.87)
Thighs	00 (00)
Knees	07 (21.87)
Lower legs	00 (00)
Ankles	00 (00)
Feet & toes	00 (00)

*n = Total number of workers

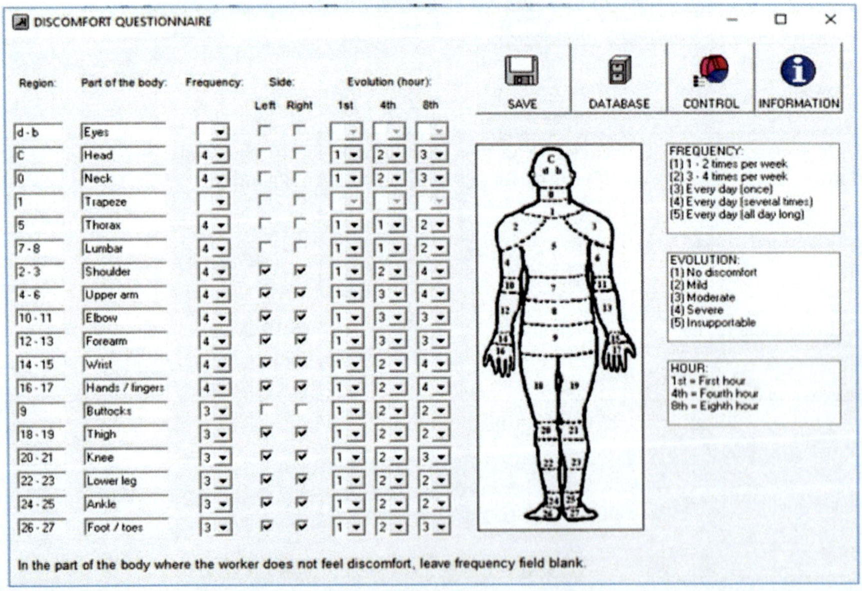

Fig. 9.10 Questionnaires on discomfort levels

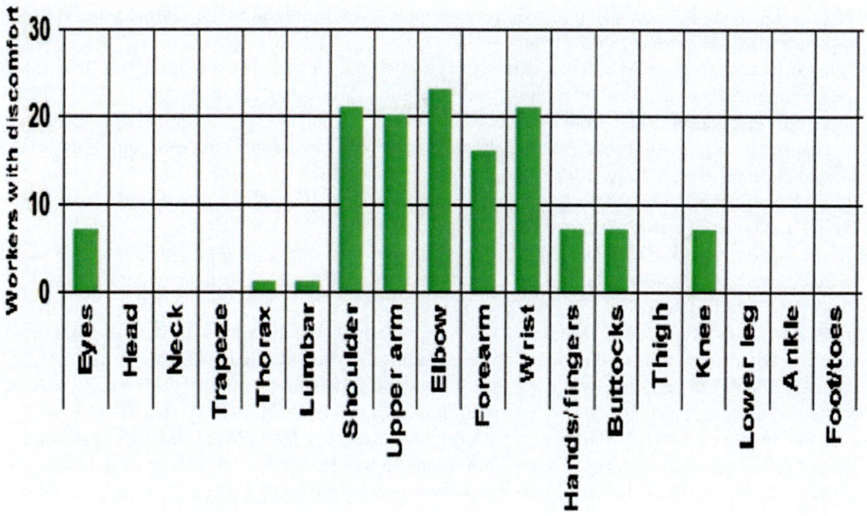

Fig. 9.11 Measured discomfort levels

References

Alam MM (2005) Identification of SMEs and BDS providers and analysis of supply and value chain in agri-machinery sub-sector of Bangladesh. Sponsored by South Asia Enterprise Development Facility (SEDF), United House, 10 Gulshan Avenue, Dhaka 1212, Bangladesh

Alam MM, Khan MIN, Saha CK, Rahman A, Bhuyian MGK (2017) Manufacturing of agricultural machinery in Bangladesh: opportunities and constraints. Agric Eng Int CIGR J 19(1):122–135

Ayub Y, Shah ZA (2018) Assessment of work related musculoskeletal disorders in manufacturing industry. J Ergon 8:233. https://doi.org/10.4172/2165-7556.1000233

Batista IC, Gomes GJC, Teles CS, Oliveira PF, Santos RM, Sassi AC, Sá B, Pardauil AA (2012) Development of an ergonomics device for maintenance of hydraulic generators of Tucuruí hydropower plant. Work 41:5935–5942. https://doi.org/10.3233/WOR-2012-0990-5935

Bernard BP (1997) Musculoskeletal disorders and workplace factors: a critical review of epidemiologic evidence for work related musculoskeletal disorders of the neck, upper extremity, and low back. NIOSH, Cincinnati, OH

Biggs S, Justice S (2013) Rural mechanisation: a history of the spread of smaller scale technology in some Asian countries. Presented in the Asian regional workshop on rural mechanisation: policy and technology lessons from Bangladesh and other Asian countries, planning commission, GoB and BRAC, Dhaka, Bangladesh

Çakıt E (2019) Ergonomic risk assessment using Cornell Musculoskeletal Discomfort Questionnaire in a grocery store. Ergon Int J 3(6):000222

da Silva SM, Worlen Ferreira Gimack WF, Junior JCL (2017) Ergonomic analysis at a workplace of a bicycle company using the Sue Rodgers tool. ITEGAM-JETIA 03(11):13–21

Darnhofer I, Schneeberger W, Freyer B (2005) Converting or not converting to organic farming in Austria: farmer types and their rationale. Agric Hum Values 22:39–52

Deborah ML, Polzella D, Boff K (1990) Human factors, ergonomics, and human factors engineering: an analysis of definitions. CSERIAC Program Office, USA

Delbridge TA, Fernholz C, King RP, Lazarus W (2013) A whole-farm profitability analysis of organic and conventional cropping systems. Agric Syst 122:1–10

Delbridge TA, King RP (2016) Transitioning to organic crop production: a dynamic programming approach. J Agric Resour Econ 413:481–498
Erdinc O, Hot K, Ozkaya M (2011) Turkish version of the Cornell Musculoskeletal Discomfort Questionnaire: cross-cultural adaptation and validation. Work 39(3):251–260
FAO (2006) Agricultural and food engineering services technical report 3. In: Farm power and mechanization for small farms in Sub-Saharan Africa. Food and Agriculture Organization of the United Nations, Rome, Italy, p 65
Garg A, Kapellusch JM (2011) Job analysis techniques for distal upper extremity disorders. Rev Hum Fact Ergon 7(1):149–196. https://doi.org/10.1177/1557234X11410386
Garg A, Moore JS, Kapellusch JM (2007a) The Revised Strain Index: an improved upper extremity exposure assessment model. Ergonomics 60(7):912–922. https://doi.org/10.1080/00140139.2016.1237678
Garg A, Moore JS, Kapellusch JM (2007b) The strain index to analyze jobs for risk of distal upper extremity disorders: model validation. In: Proceedings of the 2007 IEEE IEEM
Habibi E, Taheri MR, Hasanzadeh A (2015) Relationship between mental workload and musculoskeletal disorders among Alzahra Hospital nurses. Iran J Nurs Midwifery Res 20(1):1
Harris-Adamson C, Eisen E, Goldberg R, Krause N, Rempel DM (2011) 1st place, PREMUS best paper competition: workplace and individual factors in wrist tendinosis among blue-collar workers—the San Francisco study. Scand J Work Environ Health 37(2):85–98. https://doi.org/10.5271/sjweh.3147
Jaffar N, Abdul-Tharim A, Mohd-Kamar F, Lop N (2011) A literature review of ergonomics risk factors in construction industry. Proc Eng 20:89–97
Karwowski W (2006) International encyclopedia of ergonomics and human factors, 2nd edn, vol 11
Khapayi M, Celliers PR (2016) Factors limiting and preventing emerging farmers to progress to commercial agricultural farming in the King William's Town area of the Eastern Cape Province, South Africa. S Afr J Agric Ext 44(1):25–41. doi:http://dx.doi.org/10.17159/2413-3221/2016/v44n1a374
Mdemu MV, Mziray N, Bjornlund H, Kashaigili JJ (2017) Barriers to and opportunities for improving productivity and profitability of the Kiwere and Magozi irrigation schemes in Tanzania. Int J Water Resour Dev 33(5):725–739. https://doi.org/10.1080/07900627.2016.1188267
Meyers AR, Gerr F, Fethke NB (2014) Evaluation of alternate category structures for the Strain Index: an empirical analysis. Hum Fact 56(1):131–142
Mishra D, Satapathy S (2019) An assessment and analysis of musculoskeletal disorders (MSDs) of Odisha farmers in India. Int J Syst Assur Eng Manag 10(4):644–660
Niewiadomski P, Pawlak N, Tsimayeu A (2018) Barriers to effective implementation of lean management principles-empirical exemplification in the industry of agricultural machinery. LogForum 14(4):563–576. Accessed on 25 Aug 2019 from http://dx.doi.org/10.17270/J.LOG.2018.308
Nusantara HA, Suharno GTM (2017) Development of working facility to improve work posture at packaging section in organic vegetable industry. In: The 3rd international conference on agro-industry 2016 "competitive & sustainable agro-industry: value creation in agribusiness", KnE Life Sciences: 65–81. https://doi.org/10.18502/kls.v4i2.1658
Patel HS, Kher AO, Bariya MK (2015) Use of improved sickle for drudgery reduction in farmwomen of Gir-Somnath district of Gujarat. J Krishi Vigyan 3(Special Issue):109–112. https://doi.org/10.5958/2349-4433.2015.00047.1
Rahma RAA, Faiz I (2019) Work posture analysis of gamelan craft center workers using quick methods of ergonomic risk assessment. J Phys Conf Ser 1381:012027
Restuputri DP, Gangguan PR (2018) Musculoskeletal Disorder Pekerja Batik Dengan Menggunakan Metode Strain Index. J Teknik Ind 19(1):97–106
Rodgers SH (2000) An ergonomic approach to analyzing workplace accidents. Appl Occup Environ Hyg 15(7):529–534. https://doi.org/10.1080/10473220050028321

Rucker N, Moore JS (2002) Predictive validity of the strain indedx in manufacturing facilities. Appl Occup Environ Hyg J 17(1):63–73

Sadeghi Naeini H, Heidaripour M (2011) Kansei engineering and ergonomic design of products. Int J Occup Hyg 3:81–84

Sadeghi Naeini H, Karuppiah K, Tamrin S, Dalal K (2014) Ergonomics in agriculture: an approach in prevention of work-related musculoskeletal disorders (WMSDs). J Agric Environ Sci 3(2):33–51

Sandom C, Harvey R (2009) Human factors for engineers. The Institution of Engineering and Technology

Sims B, Kienzle J (2017) Sustainable agricultural mechanization for smallholders: what is it and how can we implement it? Agriculture 7:50. https://doi.org/10.3390/agriculture7060050

Singh G, Zhao B (2016) Agricultural mechanization situation in Asia and the Pacific region. Agric Mech Asia Afr Lat Am 47:15–25

Singh LK, Devi SR, Singh MH (2015) Traditional agricultural tools and implements used in Wokha, Nagaland. Indian J Hill Farm 28(1):50–55

Vieira L, Balbinotti G, Varasquin A, Gontijo L (2012) Ergonomics and Kaizen as strategies for competitiveness: a theoretical and practical in an automotive industry. Work 41:1756–1762

Chapter 10
Minimization of Ergonomic Risk in *Autovend*: Case Study—Bread Company

Claudia Yohana Arias-Portela and Ann Godelieve Wellens

Abstract This chapter applies a mapping process and occupational ergonomic analysis to detect the ergonomic interventions required in a distribution process for automatic vending machines. Data collection was achieved by direct measurements such as physical effort, heart and metabolic rate, and observational measurements as OWAS postural risks. These variables were interrelated to observe the correlation between the level of discomfort and cardiovascular effort, and between posture and heart rate. The methodology is developed with data collected from two distribution centers. The results suggest that most work-related accidents occur due to a lack of standardization, process documentation, and good practices. Detected problems regarding occupational profiles, physical and metabolic load above the permissible limits, and postural risks allowed to propose interventions in the distribution system, which directly improve the employee's life quality.

Keywords Metabolic activity · Physical workload · Heart rate · OWAS · Vending

10.1 Introduction

In the system of sales through automatic vending machines operated by various means of payment, also known by the neologisms in English *Autovend* or *vending*, the activities involved in cargo handling and distribution have been the subject of limited research in engineering, ergonomics, and distribution logistics. The vending machine history starts in England around the year 1700, when the first vending machine was created for tobacco distribution inside pubs. In 1902, vending machines started to appear in restaurants in Philadelphia, and by 1905, the US postal service installed stamps vending machines (NAMA 2020).

C. Y. Arias-Portela
Faculty of Engineering, Panamericana University, Augusto Rodín 498, Benito Juárez 03920, CDMX, Mexico

A. G. Wellens (✉)
Departamento de Sistemas, Universidad Nacional Autónoma de México, Av. Universidad 3000, Ciudad Universitaria, Coyoacán 04510, CDMX, Mexico
e-mail: wann@unam.mx

A. Realyvásquez Vargas et al. (eds.), *New Perspectives on Applied Industrial Ergonomics*,
https://doi.org/10.1007/978-3-030-73468-8_10

Between 1999 and 2004, the favorable macroeconomic conditions and the customers' change in eating habits boosted drinks and food sales, which encouraged big companies to establish mechanisms to increase product availability through a variety of channels, among them the vending channel (Moreau 2012). Studies directed by Euromonitor International (2005, 2013a, b) concluded that China and Mexico are the two most dynamic markets in this channel with a compound annual growth rate of 18% and 25%, respectively. The main commercialized products in the vending channel are carbonated beverages (Moreau 2012). The Mexican vending sales sector had a 14% growth in 2008, approximately 3.700 million pesos in comparison with 2007.

Different experiences in the food and product distribution industry, that involved handling operations, have demonstrated that ergonomic interventions have become a primary prevention strategy in the health field. In 1994, the National Institute for Occupational Safety and Health (NIOSH) conducted a study on trucks from the Pepsi-Cola™ Company distribution in Ohio, United States, to evaluate the physical effort as a result of loading and unloading products and to determine if an ergonomic modification could reduce cardiovascular effort (Clark 1996). NIOSH showed a severe impact in the lower back, knees, elbows, and right shoulders produced by unloading products. Furthermore, the study demonstrated that the posture of 33.3% of the employees affected their heart rate and the load and tasks influenced their posture 50% of the time, indicating a correlation between the level of discomfort and the number of loads handled. Other studies, such as the one carried out by Espinosa and Mayor (2010), have shown similar results.

A year later, the Mutua Valenciana Levante institute conducted a study evaluating the occupational risks associated with the physical effort in the food trade sector (García-Molina et al. 2000). The institute determined that 80% of the subjects were at a high risk of suffering neck and shoulder injuries, and almost 50% of them suffering hand-related injuries. Of that 80%, 35% were at high risk of developing a musculoskeletal injury in the dorsolumbar area. Five different risk factors were identified, such as task duration, shelf height, frequency, horizontal distance between the employee and the product, and cargo weight.

The previous studies have shown that the most commonly observed musculoskeletal injuries and levels of discomfort are those of the back (low, medium, and high) and shoulders, because of the lifting load, which makes manual movement subject to high mechanical stresses on the musculoskeletal system (Moore 2009). Studies on the design of the work system in a food distribution center, and on the process of loading and unloading coffee bags, have shown that with interventions in the ergonomic design of the process and work systems, these musculoskeletal injuries and the levels of discomfort can be significantly reduced (García-Acosta and Lange-Morales 2008; Fernández et al. 2001). However, studies on the specific effects of the occupational risks for employees employed in the *Autovend* sector are very scarce.

The present study aimed to reduce occupational risks and improve the life quality of *Autovend* employees, as a vulnerable group within a Mexican bread sales company, based on the results of direct observation, measurement of physiological variables, the

Ovako Working Analysis System (OWAS) method, heart rate measures, metabolic rate, and methods engineering.

The hypothesis of this study is that most accidents in *Autovend* seem to occur due to a lack of standardization in the operation. This can be shown by determining the correlation, on the one hand, between the level of discomfort, productivity, cardiovascular effort, and, on the other hand, the correlation between posture and heart rate. In addition, little documentation was found on best practices in this sector. Based on the diagnosis made, interventions were proposed that seek to commit the company to adapt work to men, which is a key factor for improving occupational indicators, without compromising productivity.

10.2 Literature Review and Theoretical Framework

10.2.1 Ergonomics in Manual Handling

Ergonomics studies the human system and the positive relationship between production and work environment, improving the design of the work system as a result of constructive nonconformity, and meeting the needs of security and hygiene, thus raising qualitative and quantitative productivity levels (Oborne 1988; Solano 1999; Bird et al. 2003; International Ergonomics Association 2010; Solórzano 2012). According to the Occupational Safety and Health Administration (OSHA), ergonomics may help reduce muscle fatigue, increase productivity, and reduce the number and severity of musculoskeletal disorders in an organizational system. Moreover, employers have the responsibility to provide a safe and healthy place for their human capital (Laurig and Vedder 2012).

The manual handling of loads, understanding load as any object susceptible to being moved (HSE 2020), is involved in almost every labor activity. According to NIOSH (Putz-Anderson et al. 1997), 60% of the working population in the world have had a labor-related disease due to the inadequate handling of loads, of which 25% are due to overstraining. Also, it has been reported that 25% of occupational accidents are related to loads handing in the European Union. A study of the Bureau of Labor Statistics (2005) reported that 42% of lost days in an organization may be due to tasks related to cargo handling.

Research on manual handling of loads in Mexico is conducted by the Mexican ergonomic association, in cooperation with the secretary of labor and social welfare; however, most of the literature found in Mexico is based on results obtained from other countries. NOM-006-STPS-2000 and NOM-017-STPS-2008 are some of the norms that rule cargo handling in Mexico (STPS 2008, 2014); they refer to handling and storage of materials as well as security conditions and procedures and safety in the working environment. In 2003, NIOSH published ISO 11228-1:2003(E), the first standardized norm on the subject of cargo handling. This norm states the recommended limits for lifting and manual transport of cargo based on criteria such as intensity, frequency, and task duration (INSHT 2013; Becker 2009).

One of the main occupational risk factors for employees exposed to conditions of inadequate physical loading and overstraining is inadequate postures. These postures may be static or dynamic. According to the NTP 452 norm (INSHT 2009c), one of the main ergonomic correction measures is reducing the static load caused by inadequate postures. This norm takes into account considerations such as the application method, the description of body categories, and registry characteristics. The method resulting from the comparative study applied par excellence for the analysis and evaluation of postural load, apart from offering high analysis reliability, is the OWAS method (Karhu et al. 1977).

10.2.2 Occupational Risk Factors Related to Physical Workload

For the World Health Organization (WHO), the existence of occupational risk factors prevents reaching the highest level of physical, mental, and social well-being of employees in all professions, therefore threatening occupational health (WHO 2007). The physical workload is defined as the set of physical and psychological requirements that employees face throughout the workday (Romero et al. 2011). When addressing the physical workload, the following aspects should be considered:

- Cargo handling often causes back pain.
- Activities that require great physical workload will eventually increase muscle exhaustion.
- Improper work postures contribute to job fatigue and eventually cause serious problems, such as poor blood circulation, spine curvature, or back, neck, and shoulder pain.
- Static workloads increase the heart rate, due to the rushed bombing of oxygen by nutrient requirements on flexed muscles. This could end up being a risk factor for heart conditions and illnesses (INSHT 2015).

The metabolic rate is useful to evaluate the employee's physical workload after finishing a task (Diego-Mas 2015a). Heart rate (HR) measures, in number of beats per minute (bpm), are used to obtain an indirect value for metabolic activity, allowing to determine the increased blood flow demanded (Bernstein 2011). Frimat and Chamoux criteria identify risk factors in terms of work effort. Global heart rate assessment for an 8- or 9-h working day can be carried out with the Chamoux method, whereas evaluation for short phases can be made with the Frimat method (Frimat et al. 1988). These methods, although developed years ago, are still widely used (Manjarres et al. 2020; Mladenovski et al. 2019).

The employed methods use cardiac indicators as heart rates during activity (AHR) and rest (RHR), theoretical maximum heart rate (HRmax.t), which are used to determine the employees' task tolerance and adaptability, expressed as costs:

- Absolut cardiac cost (ACC): ACC = AHR—RHR
- Relative cardiac cost (RCC): RCC = ACC/HRmax.t—RHR
- Heart rate acceleration (HRA): HRA = HRmax.t—AHR

The task's mechanical effort is classified by Frimat between significant and acceptable, with classified heart rates between 20 and 110 beats per minute. Detailed information on the calculations can be found, for example, in INSHT (2009b) and Frimat et al. (1988). Both maximum and allowable HR are considered; the maximum HR can be determined as 200—the age value, and the allowable HR as 75% of the maximum HR value (Billat 2002; Robergs and Landwehr 2002).

Another useful method for the ergonomic analysis of posture load is the OWAS method (Karhu et al. 1977). The OWAS method is based on the task observation and a systematic classification of work postures, consisting of 4 back postures, 3 arm postures, 6 leg postures, and 3 load weights (Diego-Mas 2015b, c). One of four risk categories is assigned (codes 1–4), being 1 the lowest risk and 4 the highest. Each category describes an effect on the musculoskeletal system and corresponding corrective action (Enez and Nalbantoğlu 2019; Riascos et al. 2019).

10.3 Methodology

The global study methodology is presented in Fig. 10.1.

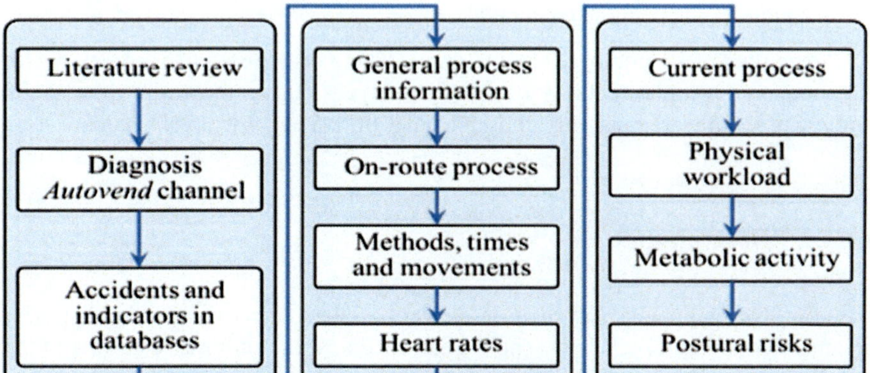

Fig. 10.1 Global study methodology

The main purpose of this study was to develop a diagnosis of the ergonomic, safety, and engineering conditions of the activities that make up the distribution and sale of products from the *Autovend* channel of a bread vending company in Mexico. It evaluates a presumed cause-effect relationship of ergonomic variables and exposure to risk conditions and proposes interventions to decrease risks to the employee. It is based on direct observation, self-reports from employees, interviews, and direct measurement of variables.

The coverage of this *Autovend* channel at the national level is divided into 5 regions: north, northwest, center, metropolitan, and *Bajío*/southwest; each region is divided into sales centers (CeVe). The scope of the study was limited to the metropolitan region, with 97 vendors, because of the high percentage of sales, the number of employees, and, last but not least, the high accident rates compared to other regions. This region has 5 CeVe: *Centro, Ceylan, Iztapalapa, Tultitlán*, and *Naucalpan.* As the number of vendors in each of them is comparable, *Tultitlán* (23 vendors) and *Naucalpan* (18 vendors) were randomly selected for this study. To determine the number of employees to be included in the sample, the following Eq. (10.1) for finite populations is used (Aguilar-Barojas 2005):

$$n = \frac{N z_{\alpha}^{2} pq}{d^{2}(N-1) + z_{\alpha}^{2} pq} = \frac{41 \times 1.645^{2} \times 0.05 \times 0.95}{0.1^{2} \times 40 + 1.645^{2} \times 0.05 \times 0.95} = 10 \qquad (10.1)$$

where:

$N =$ total number of elements in the study population (=23 + 18 = 41).
$z_\alpha =$ 90% confidence value for the normal distribution (=1.645).
$p =$ expected proportion (=5%).
$q =$ $1-p$ (=95%).
$d =$ estimation error of 10%. Being a precision measurement, it depends on the width of the confidence interval.

The number of 10 employees in the sample had to be adjusted to 8, agreeing with the company on the availability of the 8 employees during the entire time of the study; 4 of them were randomly chosen from Tultitlán and 4 from Naucalpan.

10.3.1 Documentation

Based on the company's information on the productive sector, the administrative organization of the canal, work shifts and schedules, the union structure, and the organization and planning of working time, the sampling study could be planned.

In the next step, the company's health and safety department's accident registers, indicators, and disabilities databases were analyzed and documented. Recorded information was complemented with meetings with experts, supervisors, and managers of the *Autovend* channel, in addition to applicable literature reviews. Additionally, the strategic objectives and indicators of the company were documented, in order to establish whether the indicators of the *Autovend* channel are aligned with the corporate strategy.

The health and safety indicators used by the food company are based on the current regulations of the International Labor Organization (ILO) and are recommended by the International Labor Statistics Conference of the ILO (2013). They are as follows:

1. *Frequency index* (*FI*): expresses the number of work-related accidents, reported in a period of 1 (one) year, for every million of employees covered. This is determined by Eq. 10.2:

$$\text{FI} = \frac{\text{Accidents} \times 10^6}{\text{WMH}} \tag{10.2}$$

   where WMH represents the worked man-hours, as calculated by means of Eq. 10.3:

$$\begin{aligned} \text{WMH} = {} & 8 \times \text{Number of employees} \times \text{Total worked days} \\ & - 8 \times \text{Missing days} + \text{Extra time} \end{aligned} \tag{10.3}$$

2. *Severity index* (*SI*): the severity index reflects the number of days not worked due to work incapacity in a year, per million of employees covered. The SI is calculated by applying Eq. (10.4):

$$\text{SI} = \frac{\text{Disability days} \times 10^6}{\text{WMH}} \tag{10.4}$$

3. *Incidence index* (*M*): it expresses the number of injured employees who died due to work activities in a period of one year, for every million of employees covered in the same period. This index is calculated as expressed by Eq. (10.5):

$$\text{M} = \frac{\text{Deceased employees} \times 10^6}{\text{Insured employees}} \tag{10.5}$$

4. *Loss or fatality index* (*FR*): it is similar to the incidence index, but it determines the deceased employees with respect to the total number of cases, or accidents. It is determined by Eq. (10.6):

$$\text{FR} = \frac{\text{Deceased employees} \times 10^6}{\text{Total cases}} \tag{10.6}$$

These measurable values are used by the company's health and safety area to determine compliance with specific business objectives and are included in the annual business report.

10.3.2 Observation and Measurement

A first step of the observation and measurement phase was the general recognition of the process, by means of observation and field visits to the CeVe's chosen in the study. Formats to collect route data were developed; the intention was to recognize the employee's process during a work shift and to describe the corresponding activities. The general steps are summarized below:

1. Collection of CeVe and employee process information.
2. Recording of the employee's personal and demographic information, including name, gender, age, weight, height, and seniority in the workplace.
3. Recording of daily job information, as initial heart rate, the number of vending machines to be supplied during the workday, and use of personal protection elements (PPE).
4. Activity time recording by means of a times-and-movements study. The process was recorded with a camcorder. Corresponding times and heart rate information was registered in predefined formats, as well as information on the loading of the product on the van. Stopwatch times were taken for each operation, monitoring the watch periodically to check for correct operation.
5. Recognition of process operation on route and description of tasks. In this stage, the used method, tools and utensils, times per activity, postures, movement frequencies, number of trays and product quantities handled per day, and repetitive activities were observed, as well as workplace conditions and product handling. Active breaks, rest- and mealtimes, vending machine operation, and employee limitations and capabilities were also observed.
6. The CeVe operation was recorded at the end of the route. A video recording was made, and photographs were taken of the final operations of unloading the van and the administrative operations of delivering documents and sales money.
7. The data from video recording, photographs, and heart rate were downloaded. Every day at the end of the data collection, the data taken in the operation was stored, thus avoiding loss of information and creating a backup for each employee.

10.3.3 Process and Tasks Analysis

Process analysis was implemented to detect the critical activities, susceptible to intervention. These activities are the ones that generate high mechanical demands and long exposure times. The statistical analysis was carried out with the SPSS® software. The analysis of the process allows uncovering the adopted postures, the frequency of movements and the repetitiveness of the task, as well as to establish the physical load and the metabolic consumption, based on the heart rate and the workload.

The analysis of the sales process was done through flow diagrams, time, methods and movements studies, direct observation, and data measurement. Each employee was accompanied on their route for a full working day. The field monitoring areas were the CeVe and the programmed route. The flowchart allowed to express the sequence of steps carried out to provide the vending service, as well as to identify problems and causes of possible inefficiencies. From the duration of the tasks, the most frequent and long-lasting activities were determined.

In a second step, the physical workload was measured. The employee's HR was measured with a personal watch, to determine the activities with the greatest physical workload and the metabolic consumption of each activity. Using an infrared device, the collected information was transferred to the polar.com software and was analyzed with the *polarpersonaltrainer* application (app transitioned to the Polar Flow platform on December 31, 2019; https://www.polar.com/en/pptclosed). For each employee, data on the duration of recorded activity, average heart rate, maximum heart rate, calories, distance traveled, average speed, and training load was analyzed. The software presents the heart rate zones where the employee stayed the longest, as shown for an example in Fig. 10.2.

In Fig. 10.2, the Y-axis represents the HR in units of beats per minute, and in the X-axis the elapsed time in h/min/s. The graph shows the behavior of the employee's HR during the sales activity to be observed on the red line, denoting maximum and minimum HR for the operation. In the lower part, the HR zones in which the employee accumulated more time in operation are presented. In this example, the employee was observed to be during 51% of the time in zone 2 (average HR between 110 and 127 heartbeats per minute). The *polarpersonaltrainer* data and the *Ergonautas* software (Diego-Mas 2015b, c) were used to obtain the Chamoux and Frimat criteria and evaluate task difficulty.

Finally, posture load analysis was done. The sales process was divided into key activities that were repeated in each work cycle by the employee, and each of these

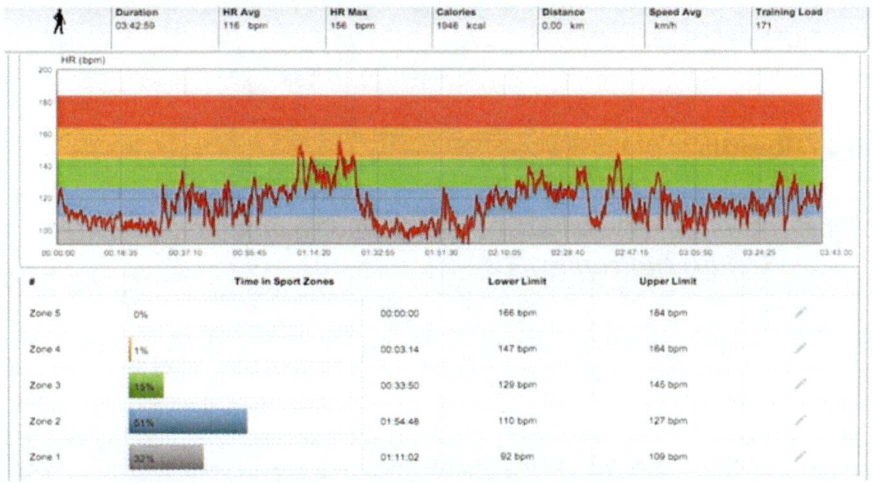

Fig. 10.2 Example of an individualized training and heart rate report

activities was recorded for each of the 8 employees in the sample. Each task was observed for 20 min and was divided into 30-s intervals to capture the positions taken by the employee. In parallel, the activity carried out by the employee was recorded in formats, to later cross-analyze between the performed activity and the adopted posture. The four tasks that by direct observation were determined as critical and repetitive in each cycle were:

- Product loading in van,
- Displacement with load in trolley,
- Order picking in van,
- Vending machine order assortment.

The present study focuses on these four tasks because they represent 44% of the employee's daily activities and 55% of the entire working day; consequently, interventions that reduce occupational risks in these tasks will be analyzed and selected.

10.3.4 Interventions and Expected Benefits

Based on the summary of the risk and engineering problems found in the observation and measurement stage, intervention proposals were generated at both the process and the human level. In conjunction with the CEO of the health and safety area, the observed problems and risks were prioritized to select those that required immediate intervention, always considering the company's corporate strategy. At the process level, the participation of the employees was highly valued, sharing with them ideas about the possible optimal solution to be developed. The generated ideas were validated in the workplace and shared with some other employees to see if the proposed solution had functional feasibility.

10.4 Results

10.4.1 Vending Channel Description and Occupational Health Indicator

The food company in study's *Autovend* channel has been dedicated to providing distribution service of snacks, food, snacks, juices, soft drinks, and coffee through vending machines for more than 15 years. Currently, it has more than 30,000 vending teams operating in the market and it performs more than 150 million transactions annually. In 1999, understanding that customers require a complete and easy-to-manage vending service and a comprehensive product offering, the *Autovend* channel of the company expanded its offer and incorporated not only snacks but also cold beverages (soft drinks and juices) and in 2001 it integrated the coffee product line, through a joint venture with a leading company in this category. In 2005, *Autovend* vending machines for this company were located in more than 64 cities.

The vision, mission, and values of the *Autovend* channel, aligned with the food company's business strategy, are represented in "developing the value of the food company's brand, providing a service that provides the greatest satisfaction to customers and consumers, seeking the highest efficiency per machine with an appropriate distribution cost and being the leading operator in Latin America of vending machines in reliability, recognition, and preference of customers and consumers."

According to the data reported by the safety and health area in the 2012 year-end report, the company's sales channel showed a decrease in the occupational accident indicator compared to 2011. This indicates positive interventions and the existence of control systems in terms of safety and health. However, when the restructuring of the sales area was carried out within the company, and *Autovend* became an independent area, the accident indicator for this individual area turned out to be high (Corporate Company 2012).

10.4.2 Sales Process Analysis

In a first stage, the analysis of the current process was carried out. Each CeVe has a weekly routing plan that assigns each employee to an assortment area with a specific number of vending machines and a sales goal. The employee is in charge of the assortment area and one or more routes, which means he must cover all the vending machines in the area, with different visiting frequencies and sales targets.

After a training period, the employee starts the regular operation in his route, if available. If no routes are available, he needs to provide assistance to other route operators. A daily route comprises on average 12 vending machines, and visits are programmed to minimize travel times. The orders placed for each vending machine are known before the planning and are summarized in a sales ticket, so the employee knows at the beginning of each week the amount and type of products requested for the route, as well as the number of sales points or vending machines that must be visited and supplied. The product delivered by the dispatch area is left in steel cages with 34 horizontal spaces, to place up to two trays in each space. Each tray is delivered according to the seller's order, each with a different type of product. In the loading area, in the back of the van, the employee verifies the delivered product against the provided sales ticket.

If the quantity and type of products delivered by the dispatch area match the sales ticket and the products are not damaged or opened, the employee can start the loading of products in the van; if the quantities do not check, the employee makes a verbal request to the supervisor to add or remove products from the order. Also, in the case of products that are damaged or have open packaging, a request for a product change is made, and the supervisor is informed of the problem. This can take between 15 and 25 min, depending on the quantity and type of reported product.

The most frequent failures are due to poorly sealed packaging or leaking beverages. After loading the van, the employee starts supplying the vending machines on his route with the required product, as well as collecting the sales revenue, returning to base, counting the stock and product for return, to end his workday.

Figure 10.3 graphically represents the activities of a complete work shift. The four most frequent tasks, as mentioned above, are shown in blue.

In Fig. 10.3, HH = Hand Held, SP = Sales Point, CP = Computer, and VM = Vending Machine.

Although annual courses are taught on manual handling of loads, employees lack a general manual on occupational health, ergonomics, and conditions of cargo handling, resulting in a variety of work strategies. To move the products, the employee

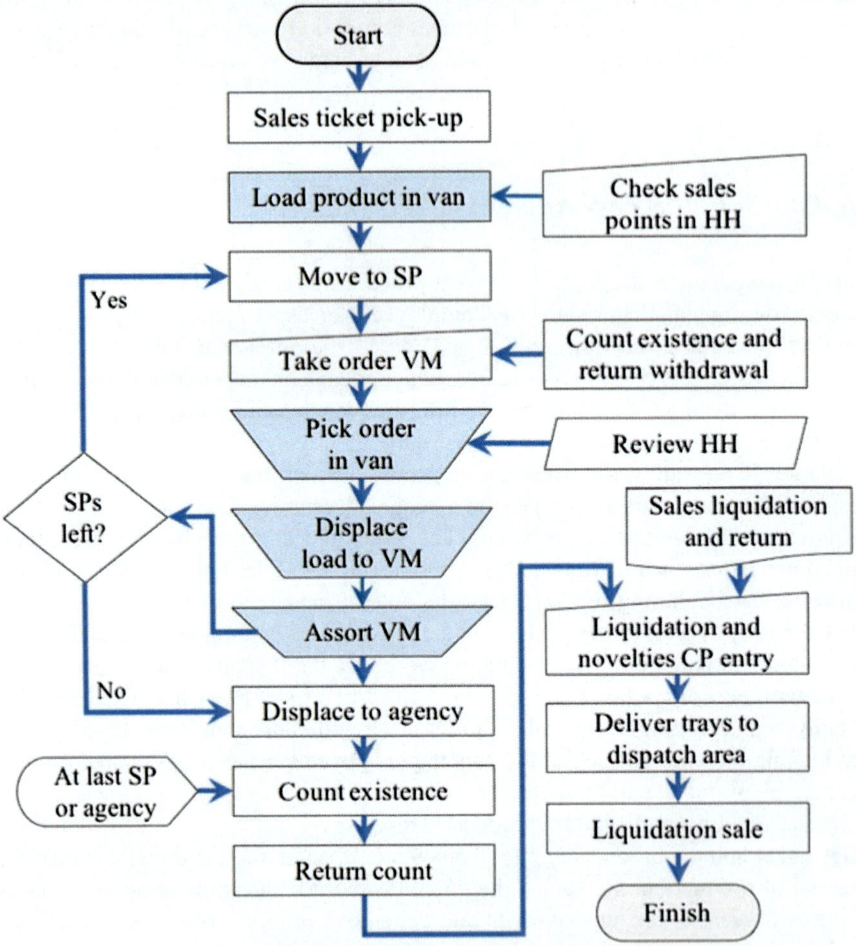

Fig. 10.3 Flow chart for vendor activities in the *Autovend* channel

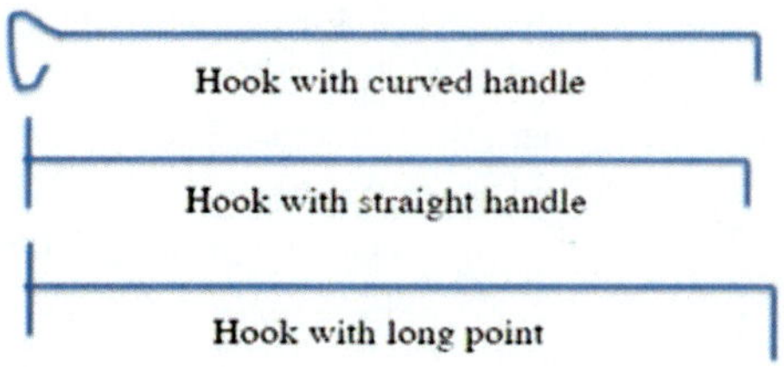

Fig. 10.4 Types of hooks for pulling and pushing trays

mainly uses two tools. The first one is a steel hook used to pull the required tray through the rails to the employee (Fig. 10.4). The hooks are made by the employees themselves and allow to reduce the picking time by up to 2 min per product, according to the data taken by the times-and-movements study.

The other tool is a hand trolley, also known in Latin America as *diablito*, with 2 wheels at its base, and handles on the top for steering. It has a shelf-type base at the bottom, to place objects in a certain order and facilitate the movement of products from the van to the vending machine.

In a second stage, the analysis of physical workload was carried out to determine the difficulty of the sales task of the *Autovend* channel. The results of the evaluation are based on the determination of the values of maximum HR and allowable HR per employee, in order to have a reference framework for the evaluation and control limits of the variable. Table 10.1 presents the maximum and allowable HR values for the selected employee sample.

The results show that the mean maximum HR is between 176 and 189 bpm, while the mean allowable HR is between 132 and 142 bpm. Based on these values, the global hardship of the sales task can be classified as *difficult*, according to the NTP 323 standards (INSHT 2009a). Likewise, since it is greater than 110 bpm, cardiac demand is classified as *important* according to the Frimat criteria.

The analysis suggests interesting differences between men and women. Although not many women work in this profession and there was only one in the selected sample, this woman presented mean HR values lower than average. Her HR during activity was below 91 bpm, and the HR at rest did not exceed 60 bpm, compared

Table 10.1 Maximum and allowable HR values

Employee	Age	Initial HR (bpm)	Maximum HR (bpm)	Allowed HR (bpm)
1	31	90	189	142
2	31	94	189	142
3	35	58	185	139
4	44	84	176	132
5	38	68	182	137
6	37	118	183	137
7	37	62	183	137
8	38	85	182	137

to men who reach HR during activity of up to 156 bpm and HR in rest up to 118 bpm. The observed results also indicate that overweight or obese employees tend to have a higher average active HR than normal-weight employees. This correlation is presented in Fig. 10.5.

Individual scores and categories of the physical workload in the Frimat and Chamoux analysis were compiled in Fig. 10.6 and Table 10.2.

The workload categories correspond to the ones presented in INSHT (2009b). The mean score was 20.8 for the Frimat analysis, while for the Chimoux analysis it was respectively 16.8 for ACC or the individual tolerance of the employee in a certain task, and 16 for RCC or the adaptation of the employee to the job.

When comparing the assessment data from Table 10.2, it can be observed that for 63% of the employees the effort at work is more grievous when evaluating short phases of the working day (Frimat), than the evaluation of the entire working day

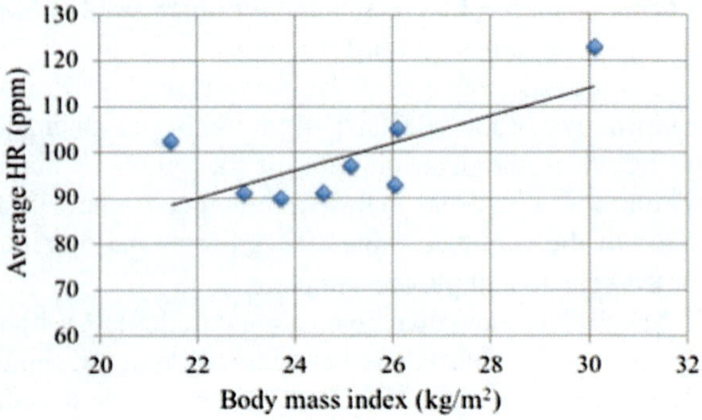

Fig. 10.5 Average HR versus body mass index correlation

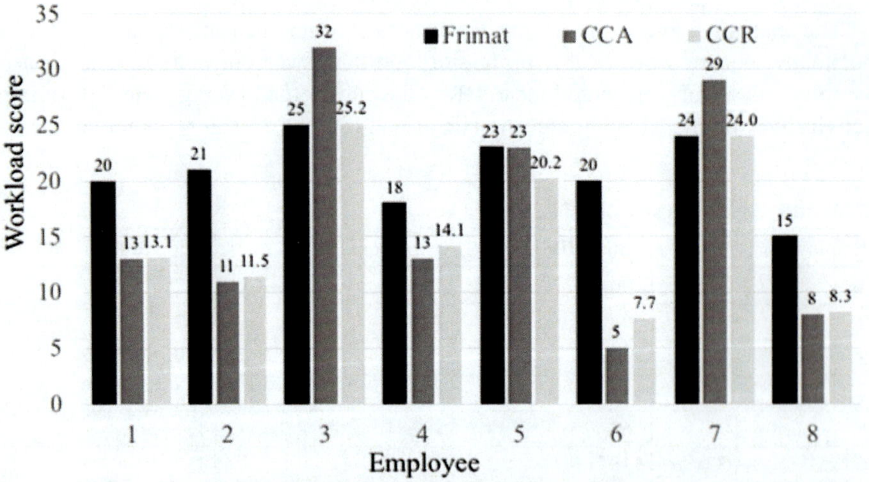

Fig. 10.6 Frimat and Chamoux methods valuation

Table 10.2 Frimat y Chamoux evaluation of the individual physical workload

Employee	Activity	Short intervals		8-h workday			
				CCA		CCR	
		Score	Category	Score	Category	Score	Category
1	Assorting	20	Grievous	13	Light	13.13	Light
2	Order picking in van	21	Hard	11	Light	11.46	Light
3	Order picking in van	25	Extremely hard	32	Heavy	25.2	Moderate
4	Order picking in van	18	Bearable	13	Light	14.13	Light
5	Order picking in van	23	Very hard	23	Moderate	20.18	Moderate
6	Order picking in van	20	Grievous	5	Very light	7.69	Very light
7	Moving with load	24	Very hard	29	Moderate	23.97	Moderate
8	Order picking in van	15	Bearable	8	Very light	8.25	Very light

(Chamoux), and that for the remaining 37% the effort in the entire working day is more painful than evaluating short phases of the day. It is noticeable that employee 3, corresponding to the women in the sample, had the highest scores in both short interval and 8-h workday evaluations, indicating that the job might not be designed for female co-workers. However, her scores are very similar to the scores for employee 7, an overweight person. In general, overweight persons seem to show more difficulties during specific short intervals but can cope appropriately with the load for the complete workday.

As shown by the results, the heart rate of the employees was observed to be excessive in some operations. This can be attributed to the lack of standardization in operations, not adapting the work environment to the employee to minimize risks and apparent ignorance of minimum guidelines for manual handling of loads.

Finally, for the postural workload analysis with *ergotools*-OWAS, the same four process phases were considered: product loading, product transfer, order picking in the van, and order assortment in the vending machine, as shown in Fig. 10.7.

Each posture observed in the abovementioned tasks is assigned a posture code consisting of four digits. The first digit corresponds to the position of the back, the

(a) Product loading (b) Product transfer

(c) Order picking inside van (d) Order assortment in VM

Fig. 10.7 Process phases considered in the OWAS analysis

second to the posture of the arms, the third to the posture of the legs, and the fourth to the handled load, as shown in Fig. 10.8 for a specific example.

A total of 56 postures for different back, arms, and legs positions, as well several load conditions, were observed during the 4 tasks. Summarizing the individual posture codes, a global OWAS risk category is determined for each observed posture, taking into account the relative frequency of each individual posture adopted per evaluated body part (Diego-Mas 2015c); this information is presented in Table 10.3. As shown in this table, 53.6% of the postures are classified as having a risk category of 3 and 4. This reveals that the employee's posture during his shift can have extremely damaging effects, requiring immediate corrective action.

Of the 56 observed postures in the four most frequent tasks, the most critical posture is a bent back with a twist, both arms down, legs maintained on flexed knees, and carrying a load between 10 and 20 kg. It is evaluated with risk 4, as shown in Table 10.4. This critical posture is performed 35.7% of the time during the workday.

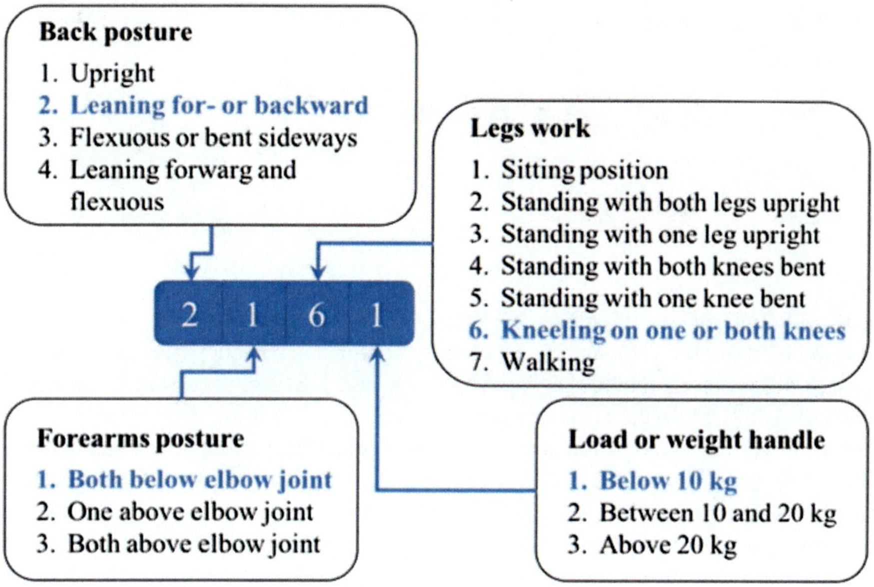

Fig. 10.8 OWAS work-posture coding. Adapted from Özkaya et al. (2018)

Table 10.3 Observed global postures by risk category

Risk category	Effects on the musculoskeletal system	Posture incidence (%)
1	Normal posture without harmful effects	26.8
2	Posture with possibility of causing damage	19.6
3	Posture with harmful effects	28.6
4	Posture with extremely damaging effects	25.0

Table 10.4 Most critical posture observed

	Back	Arms	Legs	Loads
Code	4	1	4	2
Posture	Twisted bent back	Both arms down	Over bent knees	Between 10 and 20 kg
Risk	4			
Frequency		35.7%		

Due to the postural load during the handling of trays, products, drinks, and loading tools, the back is the part of the body that suffers the most during the operation, both for the magnitude (weight of the load) and for the repeatability or frequency of the

Table 10.5 Body part risks

Body segment	Risk 4	Risk 3	Risk 2	Risk 1
Back	11.1%	29.9%	34.7%	24.3%
Arms	0%	20.0%	0%	80%
Legs	9.0%	0%	21.4%	69.6%

task. Although most of the observed postures during the four analyzed tasks are of low and medium risks, the incidence of risky postures is considerable; 29.9% of the back postures correspond to risk 3 and 11.1% to risk 4. The legs present 9% postural loads of risk 4, while the arms present a risk of 3 in 20% of the times (Table 10.5). The arms were found to be the part of the body subject the least serious risks. During 80% of the time, arm postures are in risk category 1, expecting thus fewer long-term health effects than for the other studied body parts.

Figure 10.9 shows the back, arms, legs, loads, and forces assessment made with OWAS; the colors used in the pie graphs correspond to the severity of the risk associated with each posture.

The employee keeps his back bent during 36% of the time, and during another 34% of the time, the back is bent with a turn, presenting risks 2 and 3, which require immediate corrective action. The tasks where these postures occur are the loading of the van and the picking of products in the van. The greatest risk in the short term occurs for postures with a twisted back (5% incidence).

The employee's arms are kept down 57% of the time. During 27% of the time, one arm is down and the other is raised, presenting risk 2, and the need for corrective actions in the near future. The tasks that accompany these positions are the assortment of the vending machine and the product picking in the van. The greatest risk occurs when both arms are raised (16% incidence), and it occurs in medium-height employees who, to reach, must raise their arms or rise to reach trays and products. This posture is associated with a probability of increased risk in the arms in the short term.

The employee's legs are kept on one bent knee 21% of the time of his workday and, during 20% of the time, the employee stretches one of his legs while the other is flexed, thus unbalancing his weight. The highest risk occurs when maintaining a standing position for long periods of time (incidence of 25%), presenting risk 2 of corrective care in the near future. The most critical postures occur during the assortment of the vending machine, the loading of the product in the van, and the product-picking inside the van when the employee climbs up to reach the upper trays.

The load handled by the employee is between 10 and 20 kg, sometimes exceeding the 20 kg limit, imposed by the Mexican National College of Ergonomics (CNEM 2001). This presents risk 2 of corrective care in the near future. Critical activities in this regard are the displacement with the load, the loading of products in the van, and the picking of products in the van. The greatest risk occurs when the employee must

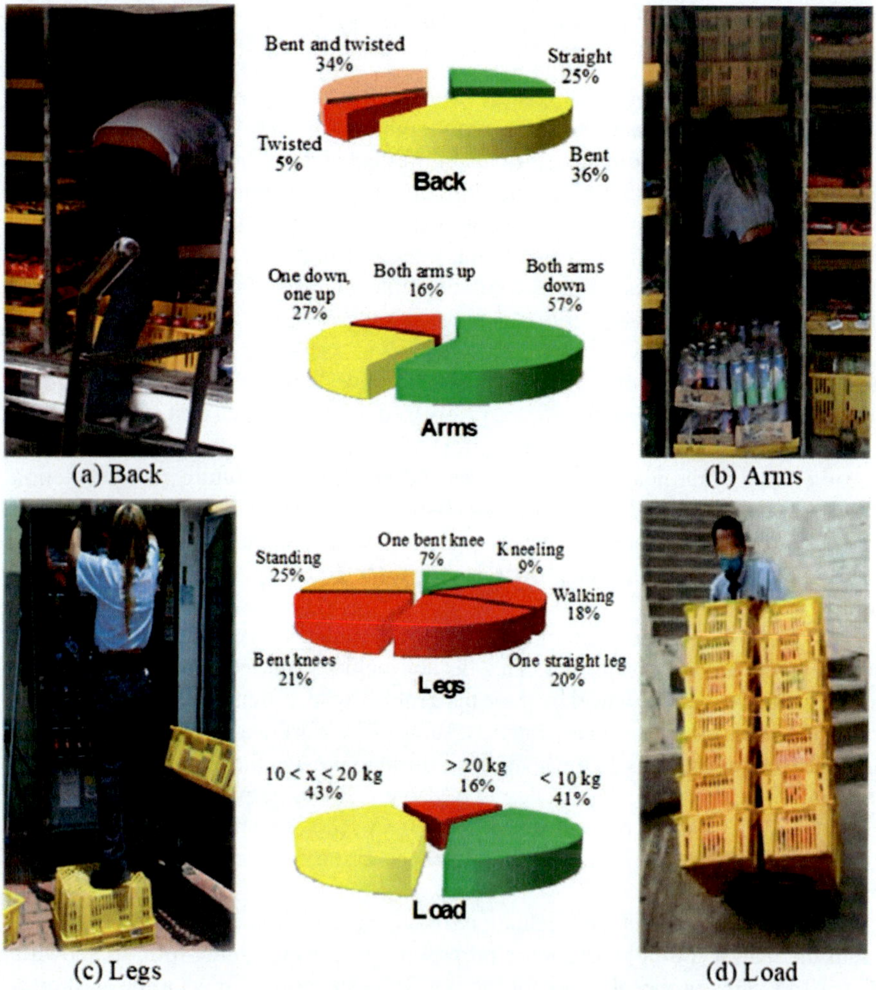

Fig. 10.9 Back, arms, legs, and load posture assessment

move the load (43% of the time); it can increase considerably because employees must enter with their loads to companies without access to freight elevators or ramps, or even where the access of the hand trolley is not allowed.

Although roughly the same back, arms, legs, and loads postures are observed for both men and women, the woman in the sample presented a higher incidence of postural risk due to relatively heavy loads and the height of the vans, both designed for male co-workers.

10.4.3 Interventions and Expected Benefits

The interventions proposed to reduce the risk to the process and to the human being have been designed based on requirements for continual improvement, the changing needs of the company, the capabilities, and limitations of the employees of the *Autovend* channel, as well as their life quality. In addition, in all recommended interventions, synchrony between ergonomics and engineering has been sought.

Experiences in the field of ergonomics suggest that interventions must first be developed at the process level, to continue at the environmental level and finally at the human level (Quintana 1999). This order ensures a less intrusive effect on the employee because the work environment adapts to the person and not the other way around. In joint work with the company, the proposals for interventions mentioned in Table 10.6 have been analyzed and prioritized, proposing in a first stage measures 1A, 1B, and 2. The remaining interventions (4, 5 and 6) were evaluated and put into consideration by the health and safety area of the company for future implementation.

Proposals 1A and 1B consist of a standard hand trolley with ergonomic support, and easy to use in the loading operation, as well as the location of the hand trolley in a designated fixed place; both proposals are presented in Fig. 10.10. and Tables 10.7 and 10.8.

The proposal for the loading tool (1A) contemplates a modified standard hand trolley to go up and downstairs. This design includes the replacement of, on both sides, the current single wheel by three hard rubber wheels that allow fluid movement both on flat surfaces and when there are steps (The Workplace Depot 2021). It also includes retaining hooks on the loading platform to prevent the products from sliding when moving the trays. The trolley is foldable and can be placed in the suggested bracket on the door of the van (proposal 1B). A recommendation is also provided for a polyethylene bench ramp, which is edged at the ends to prevent the cart from tipping over.

Proposal 1B includes a reorientation of the van door. Instead of opening the van from the rear, a sliding side door is proposed that goes up to the roof, as shown in Fig. 10.11. This proposal seeks for the travel distance of the trays to be as short as possible. In this way, the employee extends his arms as little as possible to remove the trays, making the loading and unloading task comfortable and with low postural risk for the back, arms, and legs. The tray pulling time will also be shorter. Currently, the drag distance of the tray is between 0.5 and 5 m when using the hook, but with the lateral discharge of the product, the maximum drag distance decreases to 1.70 m. The savings per distance traveled is over 66%, which is directly reflected in operating times and postural comfort.

Considering the costs expressed in Table 10.8, the yearly benefit ascends to $26,665 MXN − $8,676 MXN = $17,989 MXN per employee, corresponding to a 67.5% saving due to risk reduction.

In terms of productivity, understood as the production capacity per unit of time, in the current situation an employee with an observed average working day of 529 min (8.8 h) ends up investing 20.4% of his time in loading and unloading the hand

Table 10.6 Proposed interventions to decrease ergonomic risks

Nr	Intervention level	Recommendation	Advantages and disadvantages
1A	Process	Standardized hand cargo trolley for general operation	• Increases the product volume handled by the employee • Reduces the potential factors for cumulative trauma • The cost might exceed the budget recommendation
1B	Process	Secured hand cargo trolley support on the outside of the van door	• Reduces back and shoulder biomechanical efforts • Reduces heart rate, fatigue, and uncomfortable postures • Reduces cycle times • Reduces the risk of theft of the hand cargo trolley • Might be inappropriate for non-folding trolleys
2	Process	Design of a tray with wheels and a brake on a sliding train inside the van, for handling beverage cargo	• Increases the product volume handled by the employee • Reduces cycle times • Reduces the potential factors for cumulative trauma • Risk of product damage if more product is handled than allowed by specification
3	Process	Standardized hook to pull and push the trays, with a protective handle	• Tool standardization • Less risk of scratching and friction due to use of the hook without palmar protection • Reduces cycle times
4	Human	Document the standardized task procedure	• Allows a better work planning • Disclosure of appropriate methods to carry out activities and tasks
5	Human	Use of recorded videos on good labor practices	• Task standardization for new employees • Improves procedures for handling loads, postures, frequencies, and weights

trolley, and the remaining time allows him to supply about 12 vending machines. After implementing the intervention proposal, the employee is expected to spend only 7% of the time loading and unloading the hand trolley (a decrease of 66%). With this gain in time, the employee could supply around 2 more vending machines per day.

As a second proposal, a tray with wheels and a brake was designed on a sliding train inside the van, for handling the load of beverages. It allows the employee a

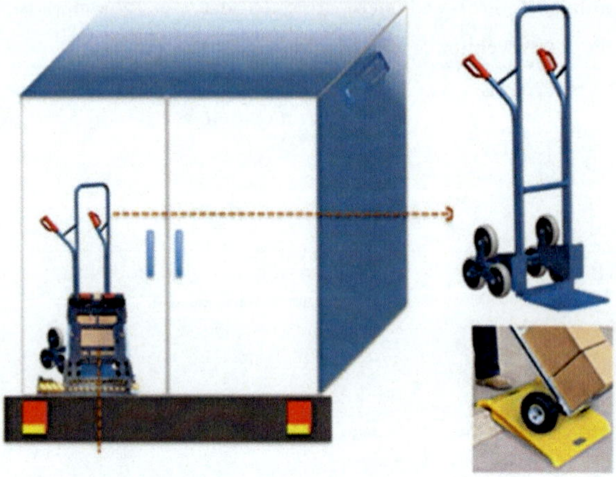

Fig. 10.10 Design of proposals 1A and 1B Trolley adapted from The Workplace Depot 2021)

safer and more effective work in the handling of beverages inside the van, both in the loading of the van and in the assortment of the order for each vending machine.

It also eliminates the forced postures of bent backs with a twist, arms up, and hyperextension of the legs in the operation. Additionally, it helps to reduce the force of the displacement of beverages with the arms and the hook. Tables 10.9 and 10.10, as well as Fig. 10.12, present the description and benefits of the proposal.

A pilot test was carried out in the *Autovend* channel to simulate intervention 2. The test consisted of asking the employee to simulate the operation of loading and unloading the order of soft drinks or juices for a customer with a pilot version of the sliding train. *Autovend* maintenance personnel installed 4 wheels on a closed beverage tray, as well as 2 rails on the bottom of the van. After checking its correct functioning, an operator used about 15 min to get trained in how to use the hook together with the rolling tray. After that, he simulated the loading and unloading operation for about one hour, obtaining a time gain of 44% over the system in use. The aim of this intervention proposal was to increase the comfort of the employee at work, shortening the travel distance of the tray as much as possible, so that the employee extends his arms the shortest possible distance to remove the trays. The savings per distance traveled is over 66%, directly reflected in the operating times.

The average and maximum number of beverage trays used during this study were respectively 3 and 6. Only on one occasion 6 trays were used, this being the maximum number expected in regular operation. Considering three trays on each installed lane, the intervention proposal contemplates equipping 2 sliding trains and 6 wheeled trays in each van. With the costs expressed in Table 10.10, the system would have a total cost of \$6152 MXN per van. Considering a 3-year lifetime for the rail system, and an 80% decrease in the number of events, the annual benefit amounts to \$26,665 MXN − \$7,384 MXN = \$19,281 MXN per employee, corresponding to a 72.3% saving due to risk reduction. If both proposals are implemented, both obtaining a risk reduction of 80%, the expected annual benefit per employee will increase to a 75.8% cost saving.

Table 10.7 Description of the proposed standardized hand trolley and door support

Situation	Description		
Problem	Apparently, most of the accidents in the *Autovend* channel happen due to a lack of standardization in the operation		
Involved activities	Van loading, vending machine assortment, and product picking in the van		
Outcomes	Process characterization	Physical load	Postural load
	• Trolley is stored inside the van in free space • Difficulties with loading and unloading the trolley • The trolley (±15 kg) can be manipulated on average 35 times/day	• Between bearable and extremely hard for short intervals (Frimat) • Between light and moderate for 8-h workdays (Chamoux)	• Postural risks: 2 and 3 • Corrective care needed in the near and immediate future
Graphical representation	Current situation: Hand trolley in free space	Proposed situation: Anchored hand trolley	

Table 10.8 Expected benefits of intervention proposals 1A and 1B

Factor	Current situation	Expected situation
Loading/unloading time of hand trolley	± 9 min/event	± 3 min/event
Nr of hand trolley loads per day	35	20
Postural risk	• Postural risk 3. Harmful postures and effects on the musculoskeletal system	• Postural risk 1. Without harmful effects on the musculoskeletal system
	• Postural risk due to repetitiveness in the extension and flexion of the trunk when raising/lowering the trolley	• Right back and semi-bent legs, both arms down. Maximum loading distance from door to floor 60 cm
	• Mechanical risk (low back pain) due to excessive force when loading/unloading the hand trolley	• Hand trolley released with only one button
Physical load risk (task's difficulty)	Frimat: Very hard Chamoux: Moderate	Expected load: Light
Possible risks	• High operation times	• 66% time-saving in trolley manipulation
	• Occupational injuries	• Decreases mechanical risk
	• Uses valuable loading space	• Does not use cargo space
	• Possible cargo damage	• Decreased cargo damage with the standardized hand trolley
Associated costs	• Annual work disability due to low back pain in Mexico: 10 days per employee, on average (all sectors)[a] • Average cost of disability due to low back pain: $25,433 MXN* per acute event[b] • Cost for days lost at work: on average $1,232 MXN per event[a]	• HTS Ultrarack support: $18,233 MXN,[c] 10-y estimated lifetime • 2 × 3-wheel hand trolley: $6,752 MXN,[d] 8-y estimated lifetime • Annual trolley maintenance: 10% of the purchase cost • Annual disability cost: 5,335 MNX, considering an 80% decrease in events

[a]From IMSS (2018), [b]From Ponce (2013) and Covarrubias-Gómez (2010), [c]From HTS Systems (2020), [d]From the Workplace Depot (2021)
*MXN = Mexican pesos

Currently, on a common route of 12 vending machines (a 529-min or 8.8-h workday, on average), an employee uses 20.4% of his working time as beverage picking time. In terms of productivity, after implementing this proposal, the employee is expected to save 33% of this picking time, allowing him to supply around one more vending machine per day.

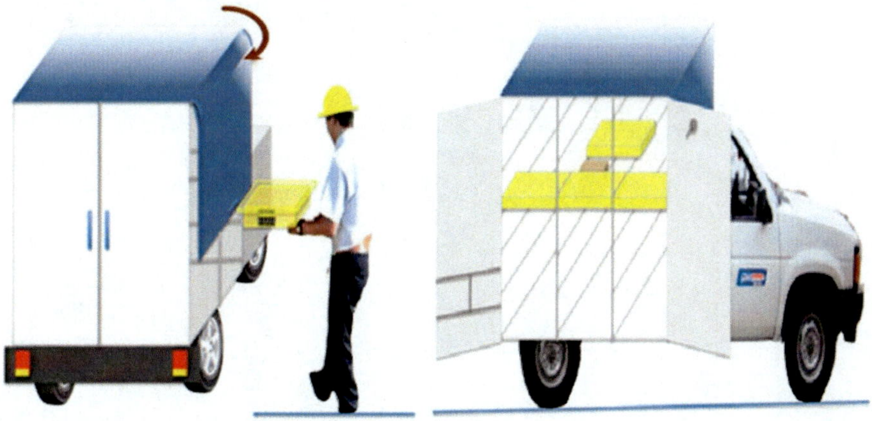

Fig. 10.11 Change of orientation in the van area for product unloading

It should be noted that not only process modifications will improve job safety; the latter also depends on the commitment of the company who must provide the employee with the minimum elements of personal protection for the operation.

10.5 Conclusions and Recommendations

The work of selling through vending machines constitutes an activity of high occupational demand and ergonomic risk. The interventions that can be proposed at the process level and at the employee level to reduce risk must satisfy the elimination of the problem at its roots, investing capital and specialized labor in the search for effective high-impact interventions within the channel. This research develops a structural methodology of diagnosis, measuring, control, and implementation of strategies in a sales process through vending machines, in order to achieve positive interventions that allow increased productivity at the level of control and personnel, as well as monitoring variables that influence the comfort of the workstations.

When characterizing the process and working methods, the main problems were found to be probably the lack of standardization in the process, the lack of a procedures' manual, and the documentation of good practices. This could be evidenced in the mechanical aids that the employees have developed for the operation, such as hooks and attachments for the trolley. These have been developed without the technical supervision that allows standardization. Sales and distribution activities en-route were measured, this being the most complex stage of the study due to the long hours that employees handle, in strenuous working conditions.

Regarding demographic variables, the population appears to be a young workforce. The active occupational life within the *Autovend* operation appears to be short-lived. Due to forced labor during the active occupational life, most employees

Table 10.9 Design of a sliding train in the van, for handling beverage cargo

Situation	Description		
Problem	Apparently, most of the accidents in the *Autovend* channel happen due to a lack of standardization in the operation		
Involved activities	Van loading, product picking in the van, and vending machine assortment		
Outcomes	Process characterization	Physical load	Postural load
	• Maintained postures in extremities • Risk of low back pain due to flexion with twisting of the back • Hyperextension of arms • Risk of accidents due to slips, trips, and falls	• Between bearable and extremely hard for short intervals (Frimat) • Between light and moderate for 8-h workdays (Chamoux)	• Most critical postural risks: 4 • Adopted 50% of the time • Corrective care needed in the immediate future
Graphical representation	Current situation: Product loading and unloading	Proposed situation: Using a sliding train	

Table 10.10 Expected benefits of intervention proposal 2

Factor	Current situation	Expected situation
Beverage picking time	± 9 min/event	± 6 min/event
Postural risk	• Postural risk in loading and picking operations: 3 • Mechanical risk in lower and middle back • Risk of low back pain due to flexion with a twist • Maintained postures in extremities and hyperextension of arms to reach/move products in van • Overexertion while lifting and displacing loads > 20 kg, or 10–20 kg with a bent back • Risk of accidents due to slips, trips, and falls	• Loading/picking operations risk: 1 • Mechanical risk minimization: right back, arms down, on semi-bent legs, load between 10 and 20 kg • 36% less time of exposure to occupational risk • Risk minimization in upper and lower extremities: they are no longer exposed to forced postures • Minimization of load handling overstrain, due to only pulling the loaded tray onto the undercarriage • Risk minimization of incident and accident, as the employee does not lean on the ceiling of the truck
Physical load risk	• Frimat score: between grievous and extremely hard • Chamoux score: between moderate and heavy • Average heart rate above allowable values	• Average HR during loading and order picking activities in the van is expected to decrease as overexertion while loading (cause of HR peaks) is avoided • Expected savings in metabolic consumption over 30%
Process risk	• No standard location of the beverages in the van	• Standardized location of beverages in the van
	• Increased probability of battered drinks or poorly sealed packaging	• Minimization of product damage and corresponding rework
	• Longer operation times due to the manual handling of heavy trays with beverages	• Shorter operating times. Savings of approximately 44% in van loading and picking times
Associated costs	(Same associated costs as stated in Table 10.8) • Average cost of disability due to low back pain: $25,433 MXN per acute event[b] • Cost for lost workdays: on average $1,232 MXN per event[a]	• Sliding wheels: twin black plastic wheels with bolt hole: $18 MXN each. 4 wheels per tray = $72[c] • Galvanized sliding rail with brake. $286 MXN/m * 5 m* 2 rails (length of the van) = $2,860 MXN. It can be produced in the existing maintenance workshop of the *Autovend* channel[d]

[a]From IMSS (2018), [b]From Ponce (2013) and Covarrubias-Gómez (2010), [c]From ULINE (2020), [d]From the company's maintenance department

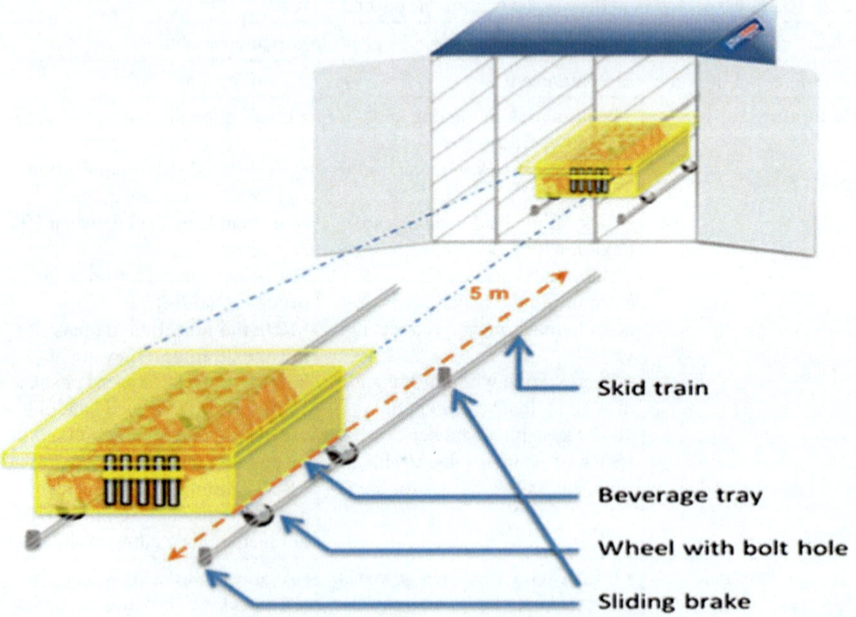

Fig. 10.12 Design of a tray with wheels and brake on a sliding train inside the van for the handling of beverage cargo

leave these types of jobs before the age of 45. Another demographic conclusion is that there are more men in this type of job, which is perhaps designed for the male gender. However, this does not prevent it from being carried out by a woman. The estimated caloric expenditure of the working day is up to 3,500 kcal/day. Nevertheless, 50% of the analyzed employees are overweight, probably related to poor eating habits during the operation, few food-breaks during the day, as well as the lack of physical activity and water consumption.

The physical workload values showed that the labor of employees in the *Autovend* channel is classified as critical since it presents a cardiac demand over 110 bpm. Some employees even exceed the maximum allowed heart rate. The average heart rate can also increase due to exhaustion in the last hours of the workday. In addition, a positive correlation was observed between body mass index and heart rate.

Although for complete workdays the Chamoux criteria indicate a bearable workload, the Frimat criteria show heavy to extremely heavy physical workloads for short periods. The greatest physical workload for short intervals is the order picking in the van, which accounts for 75% of the activities with the greatest impact on physical workload.

In terms of postural loading, the most critical activity is the loading and unloading of products from the van, especially canned beverages, where 37.51% of employees have a high risk of musculoskeletal injuries. Likewise, it was evidenced that the parts of the body with the greatest ergonomic risk after a full work shift are the middle back and arms, due to the high repetitiveness, the weight of the load, and the postures maintained when loading and unloading beverages. The most critical posture of the sales activity is the back bent with a twist, arms down, on bent legs and load between 10 and 20 kg. The relationship between postural load and physical workload was evidenced because posture directly affects the heart rate and therefore metabolic consumption, and the weight of the load influences the posture adopted and maintained by the employee.

Based on the most frequently observed postures and risks, several interventions could be proposed to the company for immediate implementation, among which a standardized hand trolley to facilitate unloading and displacement of the products, secured trolley support on the outside of the van, a reorientation of the van doors and a sliding train to diminish the pulling distance when unloading beverages. Proposed medium-term interventions include the design of a standardized hook to pull the products and trays for unloading, the documentation of standardized work procedures, and the recording of videos to transmit good work practices to the employees in the *Autovend* channel.

References

Aguilar-Barojas S (2005) Formulas for the calculation of the sample in health research [In Spanish]. Salud en Tabasco 11(1–2):333–338. ISSN: 1405-2091. Retrieved from https://www.redalyc.org/pdf/487/48711206.pdf. Sept 29, 2020

Becker J (2009) ISO 11228 standards in manual handling of loads [In Spanish]. XV Congreso Internacional de Ergonomía SEMAC, México

Bernstein D (2011) Evaluation of the cardiovascular system: history and physical evaluation [In Spanish]. In: Kliegman RM, Stanton BF, St. Geme JW III, et al (eds) Nelson, Textbook of Pediatrics, 19th edn, Chap 416. Elsevier Saunders, Philadelphia, PA

Billat V (2002) Physiology and training methodology [In Spanish]. Editorial Paidotribo. A & M graphic, España, Barcelona

Bird FE, Germain GL, Clark MD (2003) Practical loss control leadership. Det Norske Veritas; Inc., Duluth, GA

Bureau of Labor Statistics (2005) Lost worktime injuries and illnesses. Retrieved from https://www.bls.gov/news.release/archives/osh2_03302005.pdf. Sept 8, 2020

Clark B (1996) Ergonomic intervention for the soft drink beverage delivery industry. CDC, National Institute for Occupational Safety and Health, Atlanta, USA

CNEM (2001) Technical standard NT-CNEM-001. Maximum allowable limit for weight loads per person (Mexico) [In Spanish]. Colegio Nacional de Ergonomía en México A.C. Retrieved from the CNEM database, Sept 8, 2020

Corporate Company (2012) 2011 corporate health and safety report [In Spanish]. Author, México, Ciudad de México

Covarrubias-Gómez A (2010) Low back pain: a public health problem [In Spanish]. Revista Mexicana De Anestesiología. 33(1):106–109

Diego-Mas J (2015) Estimation of metabolic rate [In Spanish]. Ergonautas, Universidad Politécnica de Valencia. Retrieved from https://www.ergonautas.upv.es/herramientas/tasamet/tasamet.php. Sept 8, 2020

Diego-Mas J (2015b) Assessment of physical workload using heart rate [In Spanish]. Ergonautas, Universidad Politécnica de Valencia. Retrieved from https://www.ergonautas.upv.es/herramientas/frimat/frimat.php. Sept 8, 2020

Diego-Mas J (2015c) Postural assessment using the OWAS method [In Spanish]. Ergonautas, Universidad Politécnica de Valencia. Retrieved from https://www.ergonautas.upv.es/metodos/owas/owas-ayuda.php. Sept 8, 2020

Enez K, Nalbantoğlu SS (2019) Comparison of ergonomic risk assessment outputs from OWAS and REBA in forestry timber harvesting. Int J Ind Ergon 70:51–57

Espinosa J, Mayor A (2010) Diagnosis and proposal for improvement in the ergonomic conditions of order enlistment and validation activities, in the distribution of products of a mass consumption company in the city of Bogotá [In Spanish]. Bachelor thesis. Pontificia Universidad Javeriana Colombia

Euromonitor International (2005) Vending in emerging countries. Reino Unido, Londres. Retrieved from https://blog.euromonitor.com/2005/07/vending-in-emerging-countries.html. Sept 8, 2020

Euromonitor International (2013a) Market Research for México. Reino Unido, Londres. Retrieved from https://www.euromonitor.com/mexico?id=2&sortby=5&pagesizes=10. Sept 8, 2020

Euromonitor International (2013b) Vending in México. Reino Unido, Londres. Retrieved from https://www.euromonitor.com/vending-in-mexico/report. Sept 8, 2020

Fernández A, Hernández R, Otero S (2001) Diagnosis and pilot design of ergonomic controls for the process of storage and dispatch of coffee bags in the Almacafé warehouses in the city of Bogotá [In Spanish]. Bachelor thesis, Pontificia Universidad Javeriana, Colombia

Frimat P, Amphoux M, Chamoux A (1988) Interpretation and measurement of heart rate [In French]. Revue De Medicine Du Travail XV(4):165

García-Acosta G, Lange-Morales K (2008) Macroergonomic study of food sector company distribution centres. Appl Ergon 39(4):439–449

García-Molina C, Page del pozo A, Tortosa-Latonda L, Moraga-Maestre R, Ferraras-Remesal A (2000) Assessment of occupational risks associated with physical loads in the food and beverage trade sector [In Spanish]. Instituto de Biomecánica de Valencia, Mutua Valenciana Levante. ISBN 84-95448-00-9

HSE (2020) Manual handling at work. HSE books, Health and Safety Executive, United Kingdom

HTS Systems (2020) Order form. Pennsylvania, EU. Retrieved from https://www.handtrucksystems.com/order.htm. Sept 8, 2020

IMSS (2018) More than 300 thousand consultations for low back pain in 2017 [In Spanish]. Press release. Archivo No. 246/2018 [Online]. Retrieved from https://www.imss.gob.mx/prensa/archivo/201810/246#:~:text=En%202017%2C%20el%20Instituto%20Mexicano,especialidad%20de%20Traumatolog%C3%ADa%20y%20Ortopedia. Sept 8, 2020

INSHT (2009a) NTP 323. Determination of energy metabolism [In Spanish]. Instituto Nacional de Seguridad e Higiene en el trabajo. Retrieved from https://www.insht.es/InshtWeb/Contenidos/Documentacion/FichasTecnicas/NTP/Ficheros/301a400/ntp_323.pdf. Sept 8, 2020

INSHT (2009b) NTP 295. Assessment of physical load by monitoring heart rate [In Spanish]. Instituto Nacional de Seguridad e Higiene en el trabajo. Retrieved from https://www.insht.es/InshtWeb/Contenidos/Documentacion/FichasTecnicas/NTP/Ficheros/201a300/ntp_295.pdf. Sept 8, 2020

INSHT (2009c) NTP 452. Evaluation of working conditions: postural load [In Spanish]. Instituto Nacional de Seguridad e Higiene en el trabajo. Retrieved from https://www.insht.es/InshtWeb/Contenidos/Documentacion/FichasTecnicas/NTP/Ficheros/401a500/ntp_452.pdf. Sept 8, 2020

INSHT (2013) Manual handling of loads. Snook and Ciriello tables. ISO 11228 standard [In Spanish]. Retrieved from https://www.insst.es/documents/94886/509319/SyCISO+11228.pdf/a1838f7f-6592-4d68-b91f-fd9495895ea2. Sept 8, 2020

INSHT (2015) Work postures—risk assessment [In Spanish]. Instituto Nacional de Seguridad e Higiene en el trabajo. Retrieved from https://www.insst.es/documents/94886/96076/Posturas+de+trabajo.pdf/3ff0eb49-d59e-4210-92f8-31ef1b017e66. Sept 8, 2020

International Ergonomics Association (2010) Definition and domains of ergonomics. Retrieved from https://iea.cc/what-is-ergonomics/#:~:text=Ergonomics%20(or%20human%20factors)%20is,system%20performance%20(ratified%20by%20the. Sept 8, 2020

ILO (2013) Manual for the use and interpretation of labor statistics [In Spanish]. International Labor Organization. Retrieved from https://www.ilo.org/wcmsp5/groups/public/---americas/---ro-lima/documents/publication/wcms_216075.pdf. Sept 8, 2020

Karhu O, Kansi P, Kuorinka I (1977) Correcting working postures in industry: a practical method for analysis. Applied Ergonomics 8(4):199–201

Laurig W, Vedder J (2012) Ergonomics—tools and approaches [In Spanish]. INSHT, Spain. Enciclopedia de salud y seguridad en el trabajo OIT, Chp. 29:2–40

Manjarres J, Narvaez P, Gasser K, Percybrooks W, Pardo M (2020) Physical workload tracking using human activity recognition with wearable devices. Sensors 20:39

Mladenovski O, Achkoski J, Goleva R (2019) System development for monitoring physiological parameters in living environment. In: Ganchev I, Garcia N, Dobre C, Mavromoustakis C, Goleva R (eds Enhanced living environments. Lecture notes in computer science, vol 11369. Springer, Cham

Moore A (2009) Assessment of risk factors for development of work-related musculoskeletal disorders (RSI). Appl Ergon 25(3):157–164

Moreau R (2012) Vending in emerging countries. Market Research Findings, United States. Retrieved from https://www.marketresearchworld.net/content/view/157/77/. Sept 8, 2020

NAMA (2020) National s. History of Convenience Services. Retrieved from https://www.namanow.org/convenience-services/history-of-convenience-services/. Sept 8, 2020.

Oborne D (1988) Ergonomics in action. The adaptation of the work environment to man [In Spanish]. Trillas, México, México D.F.

Özkaya K, Polat O, Kalinkara V (2018) Physical workload assessment of furniture industry workers by using OWAS method. Ergon Open J 11(1):11–19

Ponce M (2013) Institutional cost of the patient with temporary incapacity for work due to mechanical low back pain [In Spanish]. Bachelor thesis (Medicine), Universidad autónoma de Queretaro, Querétaro, México

Putz-Anderson V, Bernard BP, Burt S, Fairfield-Estill C, Fine LJ, Grant KA, Gjessing C, Jenkins L, Hurrel J, Nelson N, Pfirman D, Roberts R, Stetson D, Haring-Sweeney MH, Tanaka S (1997) Musculoskeletal disorders and workplace factors: a critical review of epidemiologic evidence for work-related musculoskeletal disorders of the neck, upper extremity, and low back. DHHS-NIOSH, Document 97B141.

Quintana L (1999) Determination of the maximum acceptable weight of lifting, carrying, pushing and pulling loads for male employees in Colombia. PhD Project, University of Houston. Houston

Riascos C, Lopes L, Amaral L, Merino G (2019) Human factors at work: OWAS application for identification of musculoskeletal disorders in a maintenance assistant. Hum Factors Des 8(16):82–104. https://doi.org/10.5965/2316796308162019082

Robergs R, Landwehr R (2002) The surprising history of the "HRmax=220-age" equation. J Exerc Physiol 5:2. Retrieved from https://www.researchgate.net/publication/237258265_The_surprising_history_of_the_HRmax220-age_equation. Sept 8, 2020

Romero M, Álvarez C, Prieto A (2011) Evaluation of the physical workload, by monitoring the heart rate, in nursing assistants of a municipal geriatric residence [In Spanish]. Revista de Enfermería del Trabajo 1(4):193–202. Retrieved from https://dialnet.unirioja.es/servlet/articulo?codigo=3868157. Jan 12, 2021

Solano J (1999) Ergonomics and productivity [In Spanish]. Revista Industrial Data 2:48–50

Solórzano O (2012) Ergonomic risk assessment in manual handling of loads in laundry plant operators [In Spanish]. Master thesis, Instituto Politécnico Nacional, México

STPS (2008) Official Mexican Standard NOM-017-STPS-2008. Personal protective equipment—selection, use and management in the workplace [In Spanish]. Secretaría de Trabajo y Previsión Social, México

STPS (2014) Official Mexican Standard NOM-006-STPS-2014. Materials handling and storage—occupational health and safety conditions [In Spanish]. Secretaría de Trabajo y Previsión Social, México

The Workplace Depot (2021) Webpage. Retrieved from https://www.theworkplacedepot.co.uk/

ULINE (2020) Webpage Distribuidora Uline. Retrieved from https://es.uline.mx/

WHO (2007) Health issues—occupational risk factors [In Spanish]. World Health Organization. Retrieved from https://www.who.int/topics/risk_factors/es/. Sept 8, 2020

Chapter 11
Use of QUITE Method to Improve the Productivity and Quality of Manufacture Process in a Textile Industry

Carlos Raúl Navarro González, Yanet Villareal González, Pedro Alberto Escárcega Zepeda, Ana Laura Sánchez Corona, Juan Gabriel López Hernández, Verónica Arredondo Robledo, Elizabeth Romero Samaniego, and Ismael Mendoza Muñoz

Abstract This investigation evaluates the use of ergonomic methods in a textile industry installed in Mexicali city, with the objective of optimizing the manufacturing processes. This city is in the northwest of Mexico, and it is considered an industrial zone, being very important in the economic development of Baja California State, together with Tijuana city. The analysis described cases of 100 people evaluated on a production line, where ergonomic design factors of industrial equipment and machinery, work method and aspects that generated a change in attitude of workers in the evaluated textile industry were analyzed. The study was highly relevant, based on the correlation analysis of the three factors mentioned, and was made to improve the working conditions of the evaluated personnel and reducing the number of accidents and health symptoms. The majority of evaluated people presented musculoskeletal disorder (MSD), that suffered of discomfort and pain, essentially in head, neck, shoulders and spine. The textile industry where was the investigation, fabricates work uniforms to other industrial plants of the Mexicali city. This generated the need to apply the Quality Improvement Through Ergonomics (QUITE) method to improve the quality of the manufactured products and avoid the presence of workers with discomfort or pain in the mentioned body parts. With this method, a reduction in

C. R. Navarro González (✉) · A. L. Sánchez Corona · J. G. López Hernández · V. Arredondo Robledo · I. Mendoza Muñoz
Departamento de Ingeniería Industrial, Universidad Autónoma de Baja California, Mexicali, BS, Mexico
e-mail: cnavarro51@uabc.edu.mx

Y. Villareal González · P. A. Escárcega Zepeda
Departamento de Ingeniería Industrial, Tecnológico Nacional de México, Instituto Tecnológico de Mexicali, Mexicali, BS, Mexico

E. Romero Samaniego
Departamento de Ingeniería Industrial, Tecnológico Nacional de México, Instituto Tecnológico de Ensenada, Ensenada, BS, Mexico

A. Realyvásquez Vargas et al. (eds.), *New Perspectives on Applied Industrial Ergonomics*,
https://doi.org/10.1007/978-3-030-73468-8_11

times and movements was observed to increase the number of manufactured products and immediate delivery to customers. The investigation was made from 2017 to 2019.

Keywords Ergonomic methods · QUITE · Textile industry · Productivity · Quality control

11.1 Introduction

Currently, all industries work very hard on the topic of occupational health, so all companies must have a department dedicated to this activity to reduce the presence of accidents or health symptoms that generates a severe damage in workers of industrial plants. On some occasions, a situation of serious discomfort or pain can occur, and cause a temporary or permanent disability of a worker, causing a person not to return to work throughout their lives. This topic became very relevant from the beginning of the twenty-first century in the industry in any industrial activity, because the number of injured workers increased considerably in the last 10 years (Carrasco Martinez 2010). In the last 20 years in Mexicali, the occurrence of accidents and diseases for industrial activities increased in the majority of industrial plants located in this city with at least a dozen cases per month.

This happened in the development of activities of the manufacturing areas essentially with inadequate operations and without adequate ergonomic methods and tools, with major occurrence in manual operations and with minor presence in automatized functions. From the knowledge of this problematic situation, the representatives of the owners of the companies, managerial personnel, mainly from the area of management, accounting and human resources, and the supervisory personnel took care and took actions to avoid the high turnover of personnel that was presented by this type of incident. This was made, because some specialized workers suffered accidents or diseases and was necessary inability them, and also was required collocate new and inexpert people in these operations. In addition to pay the health recuperation of injured workers, it was necessary to pay great money to Mexican health authorities for labor fine risks.

One of the relevant factors considered in the industrial activities is that Mexicali city is an arid zone with extreme temperatures (in summer until 45 °C and in winter near to 0 °C), being an important factor in the comfort of workers, to elaborate industrial operations. This was considered from the 2000 to this date, as an aspect of interest in the health occupational analysis by the World Health Organization (WHO), which is a part of the ergonomic topics. This aspect is very relevant in the manufacturing areas of industrial plants, to conserve the good health of workers (López Torres et al. 2012). One of the aspects of importance for 20 years, in the occupational health care, has been the analysis of ergonomic methods, evaluating their application in industrial processes to reduce occupational risks. This has supported in the decrease in workers with discomfort and pain in some parts of the human body, such as the head, back, shoulders, hands, arms, legs, feet, and spine, and possibly some severe complication in some internal organ.

That is why analyses have been developed in manufacturing processes in manual and automated operations, evaluating the times and movements of workers in each industrial activity, to determine if there is any type of risk that causes a negative effect on the health of the personnel who work in industrial plants. That is why the evaluations in the industrial processes include activities that analyze factors of the elaboration of operations to determine if any of these may cause a risk in the health of the workers in the manufacturing area. In addition, the productivity and quality indices of the manufactured products are evaluated, balancing these two factors with the manufacturing processes of manufacturing personnel activities.

A method that considers these three factors (occupational risk, productivity, and quality) is the so-called Quality Improvement Through Ergonomics (QUITE), which has had great relevance in the industry in any zone of the world. The QUITE method evaluates various ways of developing an activity to select the most appropriate one, safeguarding the integrity of the workers who work in the manufacturing areas, with the objective of reaching the goals of productivity and quality in the manufacture of products.

This is achieved based on the use of the continuous improvement with the adequate tool as the Ishikawa diagram, being the most optimal to reduce time and movements, labor risks and the generation of errors by the human factor (Ozalp and Oguzhan 2008; García et al. 2011). In the ergonomics analyses, six factors are mainly evaluated with the Ishikawa diagram, being (1) the attitude of the workers, (2) operation of industrial machinery, (3) work environment, (4) material used in industrial processes, (5) working method, and (6) measurement process. Once these six aspects have been analyzed and improvements have been made with ergonomic methods for the reduction of occupational risks, three factors are most frequently presented in ergonomics evaluations, which are the measurement process, work method, and attitude of the workers.

For this reason, in this investigation, all aspects were evaluated, especially those three that presented the most incidents in ergonomic studies, and are illustrated in Table 11.1. In this investigation, were evaluated the six ergonomics aspects with a brainstorm with its advantages and disadvantages showed in Table 11.1, and was made a pilot test in a manufacturing line of the textile industry evaluated.

Table 11.1 shows the three most important ergonomics factors that were applied immediately in the evaluated manufacturing line, to reduce the rates of workers with discomfort and pain in the body parts mentioned above. The Ishikawa diagram was used for its ease and clarity in the evaluated information, with which the main advantages and disadvantages of the three aspects evaluated in Table 11.1 were determined, being an important statistical tool of continuous improvement (Watson 2004; Drury 2010).

Table 11.1 Analysis of ergonomic factors

Characteristics	Ergonomic factors		
	Measurement process	Work method	Attitude of workers
Advantage	Need specialized industrial equipment, machinery and systems, specialized workers with good attitude to generate an optimal yielding	With the adequate method, people worked with good attitude and generates an optimal yielding	People worked in normal periods and overtime with a good attitude with high production yielding
Disadvantage	Specialized training to workers to handle specialized industrial equipment, machinery, and system to avoid errors	Long training time is required to avoid errors	People worked in normal periods with low productive yielding with bad attitude

11.1.1 Problem Statement

In the last ten years in the city of Mexicali, the trend of the increase in the personnel turnover in the industry has been for the lack of application of ergonomic methods. And also, it has been observed a decrease of the workforce in industrial plants with the objective to save costs, forcing a single person of industrial operations to work with more than two functions and forcing their parts of the human body such as joints, upper and lower limbs, in addition to the spine, and as a first consequence the fatigue mental and physical and the second suffering constantly discomfort and pain in some parts of the body. In some industries of Mexicali, these events are not analyzed, until a worker specialized in some type of industrial operation suffers continually from some discomfort or pain. This causes that worker does not report to work and it is necessary to require untrained personnel in that area, where that position was vacant and the errors in the manufacturing processes are made, generating economic losses.

This was the reason to do this project in a textile industry of Mexicali city, where the manager and supervision personnel not considered this important aspect from 2015 to 2017 until the investigation was began, and around 30% of workers visited to medical services in inside of this industry, and some workers visited the Mexican health institutions located outside of the company evaluated, suffering of discomfort and pain the upper and lower extremities for a long time, until they stopped working. Before knowing the problematic situation, the directive, managerial and supervisory personnel of the evaluated company considered this type of industry as an industrial process without major difficulties for the development of manufacturing operations, so they constantly modified some activities at various manufacturing stages to increase the productivity, only in the last five years, this industrial plant began to present an increase in accidents and health symptoms of its workers, particularly personnel specialized in certain types of industrial operations, and the directive, managers and supervision personnel was concerned for this.

11.1.2 Ergonomics in the Industry

Ergonomics is considered as a discipline that has the objective of evaluating the way of elaborating operations of any type of work, be it commercial, industrial, or even in activities as school and domestic type. The analysis of this topic is to analyze mainly industrial-type operations and to improve them to avoid or reduce occupational risks, which both affect the economy of workers, as they are sometimes in conditions of partial or total disqualification in some type of limbs, joints, or organs of the human body (Chandra et al. 2009). This discipline associated with the area of health can evaluate physiological, psychological, anatomical aspects, including analyses of the capacities of workers in an industry, as well as to analyze work areas and methods applied to manufacturing processes so they are the adequate with a good work environment, and enough for optimal processing of industrial operations. By having the right working conditions, the activities with a high productive performance are developed, with which, the industrial plants of any type of activity present profits that are later remunerated in our country with generation of bonus as extra money to workers. In change, if industries have inadequate facilities in manufacturing processes, either manual or automated, the methods are very rigid, can cause discomfort or some situation of pain in the limbs, joints, or organs of the human body of the workers, being a concern on the part to directive, manager, and supervision personnel for the high costs generated by the medical attention to the workers of each company, whether foreign or national, and mainly the fines from the health sector of the Mexican government. For this reason, for the development of ergonomics analysis in the industry, various types of ergonomic interventions that improve working conditions are considered, which are explained below (Ahram and Jang 2014):

- Physical ergonomic interventions. They are evaluated to develop structures, systems, or devices, partially (making some adaptation to an existing process) or total (fully carrying out the necessary adaptation in an industrial activity), with the aim of improving working conditions. This type of intervention is very useful because industrial design specialists of structures as arm, hand, and body rechargers. Also, these specialized people have abilities to develop support assemblies to facilitate the operations of the workers and installation of equipment and systems within reach used for the industrial processes, supports to improve the ergonomic conditions in the manufacturing processes of industrial plants. These specialized adaptations of industrial structures, systems, equipment, and machinery, adapted to the hands, wrists, arms, shoulders, back, legs and feet, are very important to avoid discomfort and pain in the head, neck, upper and lower extremities, and spine. The objective of this function is to achieve the positioning of the joints and extremities of the workers in a correct way so that they can carry out their activities in the most comfortable way possible. In this study, this ergonomic intervention was applied in a worktable in a manufacturing line of a textile company, a process of placing a thread in a garment, without having a structure to recharge the arm, and was designed by industrial experts an adaption to put adequately the arm with a simple and a low-cost structure, and to be used by various people and for a long time. With this process, any kind of discomfort and pain of workers was avoided.

- Institutional ergonomic interventions. They are developed as part of an activity called labor gymnastics, which supports to avoid repetitive functions for long periods of time that cause discomfort to the personnel who work in industrial plants or even the risk of a serious accident or diseases. In most of the industrial plants of the Mexican Republic, being foreign and national, the directive, managerial, and supervisory personnel are concentrated in achieving the objectives of productivity and quality, without considering that the workers tend to generate a mental and physical fatigue. This has led to saturating the effort of the personnel that works in the manufacturing areas with rude and repetitive activities, without thinking that it is necessary at least on certain occasions of each working day, to apply some activity that allows the mind and both lower extremities and upper hands and neck, may have a moment to rest and after a period. Once the mind and body parts have rested, industrial operations can be regenerated with greater energy and a positive attitude on the part of the workers. This is known as occupational gymnastics, which is performed in some companies, mainly in the states of Sonora and Chihuahua but not in Mexicali, achieving that the personnel who work in manufacturing areas, in three daily sessions of 10 minutes, separate from their work area. With music emitted by horns in various areas of the industrial processes, a stage of stretching and limb movement begins and whoever wants to speak quietly, sing or shout does so only at that moment, to obtain a moment of relaxation that helps to motivate workers to return to their jobs with good spirits and a positive attitude. This has generated good operating performance and improved productivity and quality in manufacturing processes, and occupational accidents and illnesses were avoided due to negative attitudes. In this investigation was proposed an evaluation to the industrial company, to prepare special methods to have the good work conditions, to administrative, managers, supervisors and workers in manufacturing areas.
- Cognitive ergonomic interventions. It is analyzed to assess the capabilities of industry workers and thereby design structures, equipment, and systems appropriate to their knowledge and skills with the aim of obtaining the best industrial operating performance. This type of evaluation is carried out so that the industrial design specialists elaborate the structures or suggest the systems, equipment, or industrial machinery, for the work personnel, depending on the weight, height, length of limbs and mental abilities to develop the industrial operations. This type of ergonomic intervention is of great importance to obtain maximum operational performance from workers in manufacturing areas.
- Training interventions. They are evaluated to train new personnel entering an industry, which generates an initial expense for the company on the part of the worker, and which is very necessary to optimize activities that generate rapid operational performance and avoid accidents and some health symptoms. This type of analysis corresponds to human resources personnel specialized in industrial operations, where it is necessary to evaluate the experience, skills, and attitude of the workers to be hired or the personnel that may be relocated to other areas for

personal improvement or needs due to disabled workers. These types of functions must be performed in all industrial plants, to ensure optimal working conditions for workers, and have the least number of accidents and illnesses from manufacturing activities. Only, in the last 10 years, most of the times, industries have selected experienced personnel so as not to generate training expenses but the cost of specialized and experienced people is major than workers without experience. This increases the costs of industries, and contract only workers without experience in a period of emergency, when specialized people of the manufacturing areas are injured, and for this reason, this type of ergonomic intervention is not applied properly.

- Multifaceted ergonomic interventions. They are developed to avoid fear and mistrust in the use of industrial equipment, systems, and machinery, which generates a rapid development of the capabilities of workers in an industry. One of the relevant aspects in the elaboration of industry activities is having the confidence to carry out the operations without any fear. Sometimes for the lack of training, some people of a large part of the personnel that work in the manufacturing areas develop a fear and distrust to operate some type of instrument, structure, system, equipment, or industrial machinery. These workers think that they can damage it and generate considerable expense to the industry where they work. It is for this reason that this type of intervention is of great interest, to develop confidence in workers, to eliminate those fears, which do not allow them to develop their skills and abilities to optimal operational performance and thereby generate a negative attitude to time to work, sometimes even missing their place of work or resign. These ergonomic interventions are very important in industrial processes and are described in Table 11.2 with principal advantages and disadvantages.

Table 11.2 shows the five ergonomic interventions, illustrating the principal advantages and disadvantages in each intervention and observing the principal causes of the presence of workers injured and angry with his functions in the manufacturing areas of industries.

11.1.3 Ergonomics in Manufacturing Processes

It is considered that the ergonomics applied to the industry dates to the end of the nineteenth century, where the risks suffered by workers of industrial operations were increased due to the accelerated number and speed of movements, being a serious problem since that time by the increase of industrialization (Harshal 2017). The accelerated rate that began in industrial plants of that time led to the presence of a greater number of risks in work activities and, therefore, people who worked in industries with faster and more harsh operations began to suffer from accidents. As time went by, the industries required more advanced technology, so that automated systems began to be used, with faster manufacturing periods, which caused the implementation of more sophisticated methods and the need for specialized personnel to avoid

Table 11.2 Ergonomic intervention in industrial plants

Factors	Ergonomic interventions				
	Physical	Institutional	Cognitive	Training	Multifaceted
Advantage	Support to improve the labor conditions with new structures and adaptations	Generates a moment of relaxation after moments of mental and physical fatigue in workers who work in industries	Originates reliability on personnel to evaluate the capacities and abilities of the workers are properly evaluated to design the structures and industrial machinery properly and obtain optimum operating yielding	Generates reliability in personnel of human resources, supervision, and managerial personnel to train new people and workers of the companies, to have an optimal operating yielding	Support to avoid the fear to operate industrial systems
Disadvantage	It necessarily has experts in industrial design to improve the industrial processes	It necessarily controls the periods of time and the times in which the workers rests and avoid be deconcentrated and occurs errors	It necessarily has experts in industrial design to improve the industrial processes	Is necessary expert personnel of human resources, supervision, and management areas	Is necessary a psychologist or personnel expert in attitudes

that occupational hazards would occur more frequently. Currently, manufacturing processes in industrial plants of any type of industrial process are divided into operations with automated control systems and manual activities in production lines or manufacturing cells. Regarding ergonomics, in 1890, occupational risk analysis was started at a first congress held in the capital of France (Paris), where from that date, operations with a high risk of suffering an accident or some disease (Imada 2007). Subsequently, events continued to be developed to continue analyzing manufacturing methods, with the aim of reducing occupational risks and thereby lowering the costs of medical care that concerned senior managers in various types of industries.

11.1.4 *The QUITE Method in the Industry*

In the field of industry, various factors are contemplated, such as the hiring of trained personnel and the purchase of specialized equipment and machinery, as well as the use of optimal industrial engineering strategies to increase the productivity and quality indices of the manufactured products (Ozalp and Oguzhan 2004). But the human factor which must be viewed with great value in sometimes is forgotten by the management and supervision personnel, to the extent that they are forced to carry out activities in industries, with a high level of risk of suffering any health condition. This is a principal reason that was considered a specific area to support to this problematic situation and was applied to improve the labor conditions, being the ergonomics. This area has been contemplated from a lot of years with specific analyses and it is considered in the development of specialized procedures or standards that greatly support workers not to suffer from any discomfort or pain that generated a severe situation in their health (Axelsson 2000).

That is why ergonomics methods have been developed that have largely supported the prevention of occupational hazards and based on that the QUITE method has been applied for 20 years, which has been of great relevance in the areas of manufacturing of industrial plants (Drury 2010). The QUITE method, by its initials, means an analysis of quality and continuous improvement based on the use of ergonomic methods, which are very useful in evaluating manufacturing processes. The development of research for 20 years of the QUITE method has provided the necessary instruments and strategies for the reduction of accidents or the generation of diseases in work activities. Every day in all companies worldwide, industrial operations are evaluated, especially the way of preparing activities. The evaluations consist of determining if the industrial operations, whether manual or automated, are the most adequate to avoid the creation of a situation that causes certain types of discomfort or pain in the upper or lower extremities, which have health complications at future, until they become disabled either temporarily or permanently (González et al. 2003).

11.1.5 *History of the Textile Industry*

Since primitive times, man has sought methods to transform materials into clothing, sewing the skins of animals he hunted, to later develop fibers with the skins and parts of animals, in addition to vegetable products to create flax, wool, and silk (McIntyre 2004). Subsequently, animal skins and plant products that were uncomfortable and heavy were replaced with materials with which strong and intertwined threads were developed, generating the fabrics and with it the beginning of the textile industry, which is of great importance today. Because the fabrics made in the early days disintegrated, it is very difficult to know exactly when this process started and ended, being the origin of the spinning and texture.

According to experts on the subject, linen was created in the stone age, wool in the bronze age, and silk in the time of ancient China, around 5000 years ago (Udale 2014). The sewing machine was a great team that supported increasing productivity levels, by having faster manufacturing processes than making them by hand. It was not until the year of 1830 that Walter Hunt developed a double pint mechanism to develop sewing activities, to interweave threads. For the year 1846, Elias Howe invented the first sewing machine with a needle near where the stitch was made on the fabric to be manufactured, to have a double stitch and make fabrics with great resistance (Baugh 2011). In 1851, Isaac Singer built an improved sewing machine from Howe's to make clothing faster and more efficient.

These types of sewing machines are currently being used with great popularity and whose surname has the name of a company recognized worldwide (Hencken 2010). The textile industry was one of the first activities to be industrialized with mechanized systems, being an important source of job creation, where manual activities began to be changed to processes with machines that replaced operations carried out by workers. Currently, there are different types of sewing machines, according to each garment to be manufactured, the productivity and quality indices and the ergonomic methods used, to improve the working conditions of the personnel who work in the areas of textile industry manufacturing.

A priest named Edmund Cartwright invented a sewing machine in 1785 and perfected it, and with some businessmen who were his partners in the early nineteenth century built the first factory in Great Britain, having about 400 steam operated mechanical looms. These machines no longer relied on water mills to operate and could be installed anywhere the businessmen wanted, this being a great advantage (McIntyre 2004). In 1801, José Jacquard invented in France a sewing machine to make printed fabrics, with perforated patterns to form drawings, giving a great turn to the textile industry, with the decoration factor. This machine was to develop silk fabrics, supporting the mechanization process, reducing the cost of this material, and increasing the great use of wool and linen. In 1830, there were 100,000 mechanical looms, being large machines that operated in industries. In the early 1800s, they were working in spinning operations and some 250,000 in fabric processes.

Once the railway began operations in 1830, the textile industry expanded greatly (Udale 2014). The textile industry is of great importance worldwide, for developing processes with fibers (natural or synthetic), to manufacture threads and fabrics, with which thousands of products necessary in everyday life are manufactured, especially the clothes we use daily. This type of activity generates many jobs directly and indirectly and requires processes with industrial equipment and machinery and manual operations (Karthik and Gopalakrishnan 2014). The textile industry mainly depends on four sectors, being livestock, agriculture, industrial chemistry, and petrochemicals, where most of the raw material comes from, which has the objective of generating threads and fabrics with which the confession of clothing of any type of material and for an infinity of types of clothing. Due to the enormous technological advance of this industry, more functions began to be developed for other types of industries that required less manufacturing time due to the need to demand products for the world population (López Torres et al. 2011).

11.1.6 Occupational Health in the Textile Industry

As the textile activity is one of the industries of great relevance in the world and there are around 20 companies in the city of Mexicali (INDEX 2019), and being very important in the global market, a research located in Mexicali, where steps of manufacturing processes of a company that manufactures vests for personnel in the training process of any commercial and industrial activity were evaluated. It was found that the managerial and supervisory personnel thought that it was not necessary to take care of the personnel who work in this company where the investigation was carried out, considering that anyone could carry out the activities of the injured people.

For this reason appeared extra costs occurred due to the increase of accidents and health symptoms, which were generated by the medical care in social security institutions in this city which began to concern the managers of this evaluated industry. It was observed that since 2015, the number of workers with MSD (Muscle Skeleton Disorders) conditions increased, in activities carried out in the manufacturing processes and sometimes requiring urgent medical attention and for long periods of time, with excessive expenses on medicines supported financially by the company analyzed (Ahram and Jang 2014).

The Mexican health institutions detected that the cases of hospitalization of workers of this company increased significantly. In other foreign and national companies installed in this city, the rate of workers injured was increased, being a great concern of the industries. What was most worrying was that these situations occurred in foreign companies, and some foreign companies did not want to take responsibility for this serious situation, and when evaluating expenses for disabled people, it could generate a tendency to withdraw from the country, unexpectedly. This would leave personnel who worked for several years in a disabled manner in damaged limbs, spines, or organs of the human body, and without work. Also, foreign industries would leave these persons with some type of temporary or permanent disability, and sometimes until the occurrence of deaths due to the lack of application of adequate methods to avoid accidents or illnesses due to occupational risks.

The lack of application of adequate ergonomic methods in the evaluated industry generated that expert workers suffered from some health symptom, such as headache, neck, shoulder, and hip pain; and with this the company would have to do without inexperienced personnel from other areas or hire inexperienced people, causing human errors due to not having the capacity and ability to operate industrial equipment and machinery, and thereby decrease the quality of manufactured products. In addition, these persons can suffer more quickly from the aforementioned health conditions. All personnel who work in an industry must quickly understand the changes required in the manufacturing processes, which is why this causes, in most cases, resistance to these changes, primarily due to the psychological factor (Oron and Hancock 2017). The industry evaluated generated more expenses for not properly analyzing occupational risks and preventing negative effects on the health of workers.

11.1.7 Risk Factors in the Industry

Proposals for improvement in industrial plants by managerial and managerial personnel are highly relevant in manufacturing processes, where in recent years the need to increase productivity and quality indices has been generated, which is why on occasions rough and repetitive activities are defined for workers in manufacturing areas, without adequate working conditions and the necessary tools and safety equipment (Aven 2015). In the research carried out in the textile industry of this arid city, this situation was observed and the recommendations for the development of activities with the ergonomic methods necessary to protect the safety of the worker were elaborated, showing an improvement in the processes in stages industrialists of the evaluated manufacturing area.

11.1.8 Diagnosis of Work Risks with Ishikawa Diagram

In the industry, different types of evaluations are prepared, with which relevant information is generated that helps determine the possible causes of problematic situations and continuous improvement processes (Aven and Zio 2013). One of the analysis tools that is widely used in manufacturing areas is the Ishikawa diagram, with which the events that are occurring are observed in detail through an evaluation of six main factors that have great effects in obtaining the levels of optimal productivity and quality and the reduction or elimination of accidents and health symptoms generated by work activities (Khan et al. 2015). The six factors that were evaluated for this study are explained below:

1. Attitude of the worker (AW). It is an important aspect, because based on this, when it develops manufacturing activities in any type of industry, and with good attitude of workers and with a positive way, the daily objectives are met and are achieved the manufacture of products efficiently, and in the adequate time of delivery to the client, on the days proposed during the sales contracts. This supports the conduct of industrial operations, with enthusiasm reflected in efficient productivity and quality indices and a zero-accident trend.
2. Operation of industrial machines (OIM). It encloses the systems, equipment and industrial machinery, being relevant in the aspect of having the confidence to operate properly and considering the necessary care of accessories, parts and operation in general, with the application of corrective maintenance activities, preventive and predictive. Operating the industrial machines correctly, with the confidence of using the functions properly, has the supervisory personnel preparing continuous improvement operations, without the need to keep an eye on the operating personnel and generating stress on them, by constantly supervising, and trying to avoid accidents or health symptoms due to occupational risks.

3. Work environment evaluation (WEE). It is an important factor in the development of any activity, where both the noise levels of industrial machines are evaluated, as well as having music in manufacturing processes, which not all industries in the city of Mexicali consider to be beneficial, according to experts from industrial operations and human resources personnel, and managers. In addition, the communication environment that exists between operational and supervisory personnel is analyzed, which sometimes becomes tense, due to the logistics and achievement of objectives for productivity and quality in manufacturing areas.
4. Material used in industrial processes (MUIP). It is an aspect that must be evaluated, even though many times, it is not considered by the managerial and supervisory personnel, who consider that it is necessary to work with the materials that are received from the parent companies, whether foreign and nationals. Only it has happened on occasions that certain types of materials can affect the operational performance of workers in industrial processes, because they are difficult to manage in activities with fast operating rates and mistakes can be made or even suffer from some work accident.
5. Working method (WM). It is considered together with the attitude of the worker as a very relevant factor, because, based on the way of developing manufacturing operations, an attitude or negative is generated in the operating personnel. This is because, if there are repetitive, rough, or complex functions, it can cause mental and physical fatigue quickly, causing the personnel who work in industrial operations to perform these functions in a bad mood, causing so much damage to their personal integrity such as parts, by-products or finished products, reducing productivity and quality levels, and increasing accident rates and health symptoms. In this part was considered the labor relation between workers with the supervision personnel and managers, to obtain the goals proposed.
6. Measurement process (MP). It is considered in the analysis of work risks, to design and use new structures and industrial systems to avoid discomfort and pain in workers; also generate a bad attitude in workers and originate errors in the measurements of dimensions or functionality of by-products and products outside the limits or ranges allowed and cause a failure in the final production lines, with the client who makes the sales contracts to the industries or with the user who is very excited to purchase some type of product. The analysis of ergonomics factors mentioned was evaluated with the Ishikawa diagram (Watson 2004). This evaluation process was carried out to determine the possible causes that originate the development of ergonomics evaluations (García et al. 2011). Next, the six factors known as causes are represented in an illustration, explained with one of the many effects that are caused by the mentioned causes. At each stage of an industrial process, an Ishikawa analysis diagram can be drawn to determine possible factors influencing in the productivity and quality levels and in the presence of accidents of diseases by labor activities, representing in the Ishikawa diagram a problematic situation (PS). The information is described in Fig. 11.1.

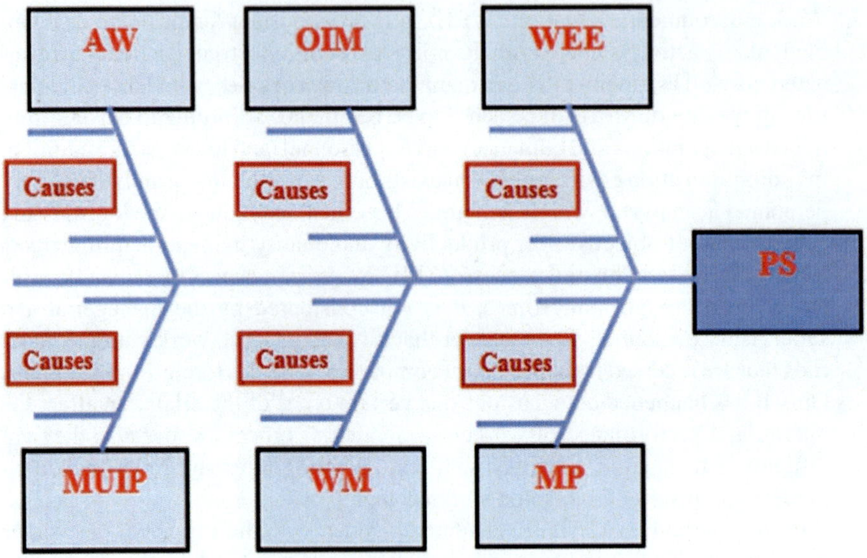

Fig. 11.1 Risk analysis evaluated by Ishikawa diagram

11.1.9 Muscle Skeletal Disorders (MSD)

Risk factors can be evaluated in any type of activity, essentially in industries, where they occur more frequently in manufacturing areas by elaborating repetitive, rough, and complex operations that are part of the industrial processes and with operations carried out manually or with systems, devices, equipment, instruments, and industrial machinery. This has generated great concern in the supervisory, managerial, and executive personnel of companies located in the city of Mexicali, where not all industrial plants carry out the analyses with ergonomic methods and this generates large expenses that must be carried out before the health institution of Mexico, either from foreign or national companies. This type of actions generates discomfort and pain for workers of any type of industry located in this desert city and anywhere in the world, causing musculoskeletal disorders (Bevan 2015).

The workers most likely to suffer from an accident or health symptoms are those who have less experience for making decisions sometimes in a hasty way, without measuring the consequences. Musculoskeletal disorders are mainly generated in work activities, essentially in manufacturing processes, where operations, in addition to being complex and repetitive, most of the time, do not train work personnel and improperly develop their functions with a high risk of an accident or health symptom (Da Costa and Vieira 2010). These disorders have a negative effect on the shoulders, neck, back and upper and lower extremities, generating from slight damage to the joints and tissues of the human body, to severe discomfort, which may sometimes require surgical interventions. The main causes of the generation of musculoskeletal disorders are mentioned below (Punnett and Wegman 2004; Monroe 2000):

a. Handling of high weight of materials, subproducts, and products. This type of activity generates discomfort or damage due to inappropriate positions when lifting and/or moving an object, especially when they are heavy, even when it is at short distances.
b. Repetitive or unnecessary movements. This aspect is evaluated because there are recurring operations or unnecessary movements, which in sometimes are used to increase productivity, but are not necessary to use this type of movements, without considering the consequences that they may cause in the workers.
c. Inappropriate postures. They mainly cause neck and back discomfort due to inadequate structures (systems or places to sit) or work methods that are not suitable for workers without evaluating their height, weight and abilities and causing slight or serious discomfort.
d. Vibrations of industrial systems. They are analyzed because sometimes vibrations are generated from instruments, equipment, or industrial machinery that cause discomfort due to repetitive and rapid movements, causing the chairs where the workers are sitting or the areas where they must carry out their duties to move, being a bit annoying at the beginning and with a greater negative effect on the health of the personnel who work in industries after long periods of exposure to vibrations.
e. Poor lighting. It must be evaluated to avoid damage to the sight of workers or health symptoms in the neck that sometimes must generate unnecessary movements such as bending down continuously.
f. Uncomfortable climate. It is an important factor in the elaboration of activities, because based on this aspect the good or bad attitude of the workers when performing their functions depends, and with it the optimal operational performance or failures and error generation.
g. Work at high speed. A rigorous analysis is carried out in specialized operations that require a high level of speed, where sometimes, due to inadequate decision-making or not knowing when to perform its function, accidents or minor or serious health symptoms originate.
h. Being seated, standing or in the same position for long periods. This aspect in few companies carries out a detailed analysis, so having a single position for long periods of time essentially causes damage to the health of workers in the upper and lower extremities, neck and back.

The causes described briefly are those that occurred in the textile industry evaluated most frequently and occur in the manufacturing processes of the majority of industries located in the Mexicali city, where the directive, managerial, and supervisory personnel sometimes do little; they are not aware of the costs generated by fines before the health sector of institutions of the Mexican government. In this investigation, in each evaluated cause that caused discomfort and health symptoms in the workers, continuous improvements were analyzed, shown in Table 11.3, with the use of the appropriate tools and work methods.

Table 11.3 Causes and improvements of MSD

Causes	Factors			
	Improvements	Devices	Methods	Tools
a	Could propose an automatized system to support in the heavy operations to avoid pain or discomfort in workers	Electronic system with automatized functions	Elaborate a capacitation to use the automatized system	Could use Ishikawa diagram to evaluate the six factors of manufacturing processes
b	Could propose an automatized system to support in the heavy operations to avoid pain or discomfort in workers	Electronic system with automatized functions	Elaborate a capacitation to use the automatized system and in the manual operations was used the labor gymnasts	Could use Ishikawa diagram to evaluate the six factors of manufacturing processes and brainstorm
c	Could propose adequate structures to adapt to workers and realize its functions correctly	Adequate structures and systems in manual operations	Elaborate a capacitation of workers in the new industrial structures and systems	Could use Pareto diagram to evaluate the principal improvements to the bad positions
d	Could propose damping sponges to dampen vibrations	Automatized system with an indicator of light detect vibrations	Rigid inspection of the supervision personnel of vibration areas	Could control graphics to measure vibrations with control graph and Ishikawa diagram
e	Could propose an automatized system to detect the adequate light intensity in the manufacturing processes	Electronic system to detect low, adequate and high intensity of light	Rigid inspection of the supervision personnel of light zones	Could use control graphics to measure vibrations with control graph and Ishikawa diagram
f	Could propose an automatized system to detect the adequate climatic conditions in the manufacturing processes	Electronic system to detect low, adequate and high levels of heat and cold	Rigid inspection of the supervision personnel of manufacturing processes	Could use control graphics to detect variations in the climatic conditions with control graph and Ishikawa diagram
g	Could propose an automatized system to detect velocity of the manufacturing processes	Electronic system to detect low, adequate and high levels of velocity of the manufacturing processes	Rigid inspection of the supervision personnel of manufacturing processes	Could use control graphics to detect variations in the climatic conditions with control graph and Ishikawa diagram

(continued)

Table 11.3 (continued)

Causes	Factors			
	Improvements	Devices	Methods	Tools
h	Could propose adequate structures and systems in the manufacturing processes	Electronic system to detect level of position or body of workers in the manufacturing processes	Rigid inspection of the supervision personnel of manufacturing processes	Could use notch sensors in the body of workers as analysis to detect the adequate positions

Table 11.3 shows the causes and improvements of MSD generated in this company analyzed, obtained the information with the visits to medical services in inside and outside and in the manufacturing processes, using the Ishikawa diagram as the principal tool of continuous improvement utilized in this investigation. With the application of ergonomic methods, the ergonomic conditions were improved and generated an increase in productivity and quality, in addition to the reduction of accidents and health symptoms in workers in the manufacturing areas analyzed.

11.1.10 Occupational Health Regulation Standards

There is an occupational health law that is regulated by the Secretaría del Trabajo y Prevision Social (STPS) and Secretaría de Salud (SSA) in the Mexican Republic, and must be applied in all industries located in our country. These regulations have the function of protecting the workers of foreign and national industrial plants, and its norms are in the Diario Oficial de la Federacion de Mexico, decreed and published from some years ago (DOF-Mexico 2019). At the global level, occupational health assessments have been prepared, generating the necessary norms and laws that support having the optimal conditions for industrial workers. In Mexico, the STPS and SSA elaborate and regulate the industrial operations, as well as was observed in the textile industry evaluated, being very important in the occupational health and showed in Table 11.4.

All Mexican regulations described in Table 11.4 were used in this investigation, to maintain the good labor conditions in the textile industry evaluated.

11.1.11 Manufacture Process

They are mainly of two types, one of which is manufacturing in a specific way with activities carried out manually and the other is when they are carried out with systems, equipment, and industrial machinery with automated operations. Based on the mentioned types, they are carried out in various stages, evaluating each of

Table 11.4 Mexican regulations by the STPS in textile industries

Regulations	Concepts	
	Title	Objectives
NOM-001-STPS-2008	Constructions to industries	Application in areas of labor centers about work conditions and labor safety
NOM-004-STPS-1999	Machinery safety devices and systems	Application in protection systems and safety devices in machinery and equipment used in the workplace
NOM-006-STPS-201	Material handling and storage	Application for safety and health conditions at work activities
NOM-034-STPS-2016	Access and development of activities of workers with disabilities	Application in labor centers with security conditions for access and development of activities for people with some type of disability
NOM-010-STPS-1999	Chemical contaminants	Application in the labor centers of the safety and hygiene conditions where use of chemical contaminants is generated
NOM-011-STPS-2001	Noise	Application in the workplace of the safety and hygiene conditions where noise is generated
NOM-015-STPS-2001	High or low thermal conditions	Application in work centers with safety and hygiene conditions with high or low thermal conditions
NOM-024-STPS-2001	Vibrations	Application in work centers with the conditions of security and hygiene evaluating the levels of vibrations
NOM-025-STPS-2008	Illumination	Application in work centers with safety and hygiene conditions analyzing lighting levels
NOM-017-STPS-2008	Personal protection equipment	Application in the workplace of the selection, use, and management of personal protection
NOM-019-STPS-2011	Safety and hygiene commissions	Application in work centers of the constitution, integration, organization, and operation of safety and hygiene commissions
NOM-030-STPS-2009	Preventive health and safety services	Application in work centers of health and safety services and commissions
NOM-035-STPS-2018	Psychosocial risk factors	Application in the workplace of psychosocial risk factors at work, as well as risk identification, analysis and prevention

(continued)

Table 11.4 (continued)

Regulations	Concepts	
	Title	Objectives
NOM-036-SSA2-2002	Prevention and control of diseases	Application in work centers analysis of vaccines, toxoids, sera, antitoxins, and immunoglobulins in humans

the activities carried out in each phase, with visual observations and specialized measurement instruments (Alvarez et al. 2015). The objective of the manufacturing processes is to have materials, whose characteristics are modified, mentioned below are some examples such as the size, representation or shape of the material, weight, height, and thickness essentially.

The transformation of materials (also called raw material) is according to the type of characteristic to be modified, the systems that support material changes, the quantity of products to be manufactured and specialized personnel to carry out industrial operations. Depending on the type of material to be transformed, there are various manufacturing processes, from casting, cutting, folding, deformation, gluing, heat treatments, chemical processes, and molding, mainly (Karthik and Gopalakrishnan 2014). This occurs in all industrial plants, and in companies dedicated to textile activities, which use industrial equipment and machinery and trained personnel to essentially operate sewing machines, specialists in textile materials and the manufacture of garments to be manufactured, to raise productivity and quality levels and problem solving.

Table 11.5 shows the main manufacturing processes and their essential characteristics, where they are observed and obtained information that represents the conditions of each industrial operation and being related to each process. It is shown in all processes with manual operations, low speed process activity, the requirement for specialized personnel, the need for a complex process, and long manufacturing time. The other representations are from the various types of industries where manual operations are applied. For automated operations, high-speed activities, having specialized personnel, a simple process and short manufacturing times are illustrated in all processes. The other representations show the types of industries where automated operations are applied.

11.1.12 Quality Control

It is a very important stage in any industrial manufacturing process, where the functional characteristics of the manufactured products are evaluated, which analyzes the levels of productivity and quality. The quality control process has become a fundamental step in industrial plants, where all functions are evaluated, which are part of the competitiveness regulations of the world market (Alvarez et al. 2015). Each product contains the specifications according to its functionality and adaptability

Table 11.5 Main industrial manufacturing processes

Process	Features	
	Manual Operations	Automated Operations
Chemical process	LSP, SP, CP, LMT, CHI, MDI, MTI, OI	HSP, SP, SP, SMT, CHI, MDI, MTI, OI, PI
Cut	LSP, SP, CP, LMT, AI, EI, MTI, TI	HSP, SP, SP, SMT, AI, COI, EI, MTI, OI
Deformation	LSP, SP, CP, LMT, AI, MTI, TI	HSP, SP, SP, SMT, COI, EI, MTI, OI
Folding	LSP, SP, CP, LMT, AI, MDI, PI, TI	HSP, SP, SP, SMT, AI, MDI, PI, TI
Gluing	LSP, SP, CP, LMT, COI, MDI, PI, TI	HSP, SP, SP, SMT, COI, FI, TI
Melt	LSP, SP, CP, LMT, AI, EI, MTI, OI	HSP, SP, SP, SMT, AI, EI, MTI, OI
Molding	LSP, SP, CP, LMT, AI, EI, MDI, MTI	HSP, SP, SP, SMT, AI, EI, MDI, MTI, PI
Thermal treatments	LSP, SP, CP, LMT, CHI, MDI, MTI, OI, PI	HSP, SP, SMT, CHI, MDI, MTI, OI, PI

LSP = Low speed processes, HSP = High speed processes, SP = Specialized personnel, SP = Simple processes, CP = Complex processes, SMT = Short manufacturing time, LMT = Longer manufacturing time, AI = Aeronautical industry, CHI = Chemical industry, COI = Communications industry, EI = Electronic industry, FI = Food industry, MDI = Medical industry, MTI = Metallic industry, OI = Oil industry, PI = Pharmaceutical industry, TI = Textile industry

to customers and users, who are primarily responsible for evaluating this factor of importance in manufacturing areas (Besseris 2014). In industries with high technology, from simple manual operations to various automated systems are used to control product specifications.

Quality control not only focuses on the manufacture of products but on the scope of companies dedicated to providing services of any kind. Sometimes this aspect is evaluated with market analysis with surveys or functionality tests. With this activity, which consists of evaluation with manual and automated instruments, equipment and industrial measuring machines, continuous improvement tools are also used, with the Ishikawa diagram, brainstorming and control charts being the most used in this study. With this type of quality tools and automated systems in industrial operations, you can ensure that manufacturing processes are carried out correctly, avoiding the generation of errors. Manual and automated measurement systems can show the

Table 11.6 Automatized systems and improvement tools used in the investigation

Quality Tools	Systems	
	Manual	Automatized
Ishikawa diagram	Evaluate information of measurements of manual operations using manual instruments to describe causes of defective products with a low velocity	Evaluate information of measurements of automatic operations using automatized instruments to describe causes of defective products with a high velocity
Control graph	Evaluate information of measurements of high and low data of manual operations using manual instruments to describe causes of defective products with a low velocity	Evaluate information of measurements of high and low data of automatized operations using automatized instruments to describe causes of defective products with a high velocity
Brainstorming	Evaluate brainstorming to improve the manufacturing processes of manual operations in industries	Evaluate brainstorming to improve the manufacturing processes of automatized operations in industries

numerical data of the manufacturing processes of industries, which can support to elaborate the required adjustments and quality tools for process control. An analysis of literature (Alvarez et al. 2015) of the mentioned is showed in Table 11.6.

Table 11.6 represents the principal continuous improvement tools used in previous studies in industrial operations, indicating the importance of these tools in the evaluation in the manufacturing processes.

11.2 Methodology

11.2.1 Materials and Instruments

Once the problematic situation that arose in the textile industry where the research was carried out was detected, the appropriate activities for the analysis were designed with various materials and instruments that were used in each phase of the study. The material resources of the company evaluated that were considered from the beginning to the end of the study, and those provided by the three participating educational institutions, supported to realize the activities. The company provided specialized measurement equipment to obtain numerical data of the specifications of the manufactured products and mathematical values of the observation of times and movements of the industrial operations realized by the workers of the analyzed industrial plant, to evaluate if they were made correctly, and with this avoid occupational risks. In addition, the company supported with two desktop computers for the capture of numerical data of the evaluations made in each step of the manufacturing process.

The investigation personnel provided two laptop computers to have the supporting information, in addition to carrying out between four researchers the mathematical analysis with different sections of the information, and thereby obtaining the results analysis of what happened in the process as quickly as possible of the manufacturing processes. The computer programs required for the research were MATLAB (Magrab et al. 2011) and Excel (Alexander et al. 2016), which were provided and used by the personnel of the participating educational institutions. With these programming systems, the capture of the numerical data and its evaluation was prepared to determine what happened at each stage of the industrial processes analyzed. The instruments used to obtain numerical data that were organized and represented in tables and graphs were used based on the six work risks evaluated with the Ishikawa diagram mentioned above. Each risk is described below with the instrument used:

1. **Attitude of the workers**. During the development of the study, a survey was applied to the manager of the evaluated manufacturing area and another to the supervision personnel, in addition to those applied to the 100 workers individually, who participated in the research to obtain information on the activities realized. The survey of the manager and supervision personnel was to know their sensitivity and awareness in the development of industrial operations and the worker surveys were to evaluate the main ergonomic factors such as the activity of measuring variables in the manufacturing process, as well as evaluate the working methods at each stage and determine the main causes that generate different attitudes of the workers, when elaborating their functions in the evaluated industry. These three ergonomic factors were the most relevant of the six analyzed.
2. **Operation of industrial machines**. Workers were working with industrial equipment, machinery, and systems, and some of these were uncomfortable to work, as the Garnett machine that is used in the textile industry. This is a big machine that handles large materials, in addition which it contains various electromechanical mechanisms that few specialized people know how to repair this machine and elaborate its corrective, preventive, and predictive maintenance. Workers of this textile industry evaluated were complained about the materials used in the manufacturing, because the materials got stuck in some mechanisms of the machine, due to their great size and the machine stopped constantly, and of the manufacturing speed of this machine. In this phase was used a high-quality stopwatch H5671, with a countdown stopwatch, step marker, and programmable memory, to measure the velocity of this big machine in each step.
3. **Work environment**. In this phase, five variables that are part of the work environment were evaluated, being the levels of noise, vibration, lighting, climate (temperature and humidity), and air pollution (chemicals). The instrument used to analyze noise levels, according to the NOM-011-STPS-2001 standard (NOM011 2001), was the personal dBadge2 noise dosimeter with a range of 55 dB to 140 db. For the evaluation of vibration according to NOM-024-STPS-2001 (NOM024 2001), in equipment, machinery and industrial systems, the LCD vibration meter basic model with speed ranges from 0.1 m/s to 199.9 m/s

was used and a displacement from 0.001 mm to 1999 mm. The lighting analysis according to NOM-025-STPS-2008 (NOM025 2008) was carried out by the backlit luxmeter model MXFLU-001-001. The humidity and temperature measurement were prepared according to the NOM-015-STPS-2001 (NOM015 2001) standard with a CW-T20WIFITH model datalogger wifi thermohygrometer with temperature ranges from −40 °C to + 80 °C (± 0.3 °C) and humidity from 0 to 100% RH (± 2RH%) and with wireless Internet connection to send data quickly and easily to a computer and be analyzed. The contamination levels according to NOM-172-SEMARNAT-2019 (NOM172 2019) were evaluated with a multiple gas monitor that detects oxygen, carbon monoxide, hydrogen sulfide, sulfides and nitrogen oxides. Grainger Model 337WY5.

4. **Material used in industrial processes**. An analysis of the materials used in the manufacturing process was carried out, as well as their texture and methods of handling and manipulation to prevent workers from suffering from any health symptoms. In this phase, no instrument was used, only observations from the company manager and supervisory personnel.
5. **Working method**. An analysis of the materials used in the manufacturing process was carried out, as well as their texture and methods of transport and manipulation to prevent workers from suffering from any health symptoms. In this phase, no instrument was used, only observations from the company manager and supervisory personnel. In this phase, notch sensors were used to evaluate different positions of the workers and were analyzed according to the activities made to determine if it was necessary to modify any operation with a work method that supports avoiding the generation of accidents and diseases in the industry.
6. **Measurement process**. This activity was elaborated with specific instruments to determine the amount of fabric to be used for the manufactured products, using the Ohaus Scout Model H-7294 electronic scale with a platform size of 6.7 × 5.5 and a precision capacity of 2.2 grams, and utilized to measure the grammage of textile materials and to know the quantity required for each product manufactured. Another measuring instrument used was the Mitutoyo H2781 digital caliper with a resolution of 0.0005¨ and a range from 0¨ to 0.47¨, to measure the adequate thickness of the fabrics to be used and of the manufactured products. The third instrument used was a high-quality stopwatch H5671, with a countdown stopwatch, step marker, and programmable memory, to measure time and movements, and velocity of industrial machines to fabricate in each step.

11.2.2 Method

The investigation consisted of five fundamental phases to obtain the most relevant information that would support the rapid detection of the causes of the increase in accidents and diseases in workers of this textile industry. The steps are explained next:

a. **Evaluation of activities that generate occupational risks**. In this phase, an analysis of six ergonomic factors to determinate the correlation of the principal causes evaluated and generate labor risks with the highest incidence in the evaluated textile industry was made. This was made with the standards of ISO 14000 (ISO 14000 2015), of ergonomics (ISO/TC159/ SC1, ISO/TC159/SC3, ISO/TC15/SC4 and ISO/TC159/SC5) and the OHSAS (Occupational Health and Safety Assessment Series) as the OHSAS 18:000 (OHSAS 18:000 2007).
b. **Analysis of occupational health**. An assessment of visits to the doctor of workers inside the evaluated company and to public health institutions in Mexico was made. Accident and disease rate were evaluated from the start of the investigation (2017) to the improvement stage (2019), while the study analysis was continued in observation.
c. **Assessment of MSD cases**. The health symptoms in various parts of the human body were determined, being the back, head, neck, hands, shoulders, legs, feet, and spinal column; where this information was presented, which was obtained from visits by workers to doctors inside the company analyzed and to public health institutions in Mexico, evaluating the degree of discomfort and pain of the MSD.
d. **Evaluation of ergonomic factors**. The ergonomic factors mentioned in Table 11.1 were evaluated in two ways (measurement processes and work method) with observations at detail in industrial equipment, machinery, structures, and systems. This supported largely with new design, fabrication and made a test of the new measurement functions, elaborated by the supervision and manager personnel. The attitude of all people of the industry evaluated was analyzed with observations of the work methods of manufacturing and with a questionnaire of 20 questions, which were associated with the presence of accidents and diseases of workers and related to the presence of errors and with the productivity and quality indices of the manufactured products were evaluated.
e. **Analysis of productivity and quality indices**. The productivity and quality indices of the products manufactured from 2017 to 2019 were analyzed, correlating them with ergonomic interventions. This phase was developed according to the ISO (International Standardization Organization) standards as the ISO 90001 (ISO 9001 2015).

In this investigation, some improvements were made in the industrial processes of the evaluated manufacturing area and with it and be attended the claim of certain personnel involved in these optimization operations, which were mainly by the increase in the productivity and quality indices of the manufactured products. The modified activities were to elaborate the operations in an easier and more comfortable way, that even knowing the workers that it benefited them, they continued elaborating the activities with the previous methods due to the habit to make. At first, resistance was generated by workers in the manufacturing area, and as they understood the new ergonomic methods, they were using them.

In this study, the inductive method was used, according to the requirement of the development of the activities in the manufacturing processes, with seven essential actions to follow mentioned below:

1. The evaluations of the industrial processes of each stage based on observations.
2. The registration of numerical data.
3. The analysis of values obtained from the observations and with the mentioned instruments.
4. The classification of the numerical information obtained.
5. The elaboration of hypotheses.
6. Validation of the information obtained.
7. The development of continuous improvement to avoid recurrence of errors or failures.

These seven steps were applied to this research in order to achieve the reduction of workers affected in health by industrial processes (Álvarez et al. 2015) with activities without the required tools and equipment and very rigid (such as the handling of large rolls of cloth without considering their weight and method of transport within the evaluated textile industry). In addition, a valuation of the lack of application of ergonomics evaluation techniques (in the sense of lack of posture analysis for the development of industrial operations, which can generate discomfort or pain in any part of the human body) was prepared (Harshal 2017). The two essential continuous improvement tools used were the Ishikawa diagram and the control chart for the analysis of numerical and textual information, with which the industrial engineering and ergonomics methods used were determined. The reference values used were validated with the safety and environmental analysis standards, as well as the ISO and OSHAS standards, as well as the manufacturing specifications based on the needs of the clients. The research was developed in a manufacturing line that consists of five stages (Udale 2014), to serve the experimental process, explaining immediately:

1. Reception of raw material (linen, silk, cotton, silk, nylon, polyester, cellulose, and rayon, as the most important). The material is received and handled with the required security procedures.
2. Spinning process. The threads and fibers are made according to the manner and products to be manufactured.
3. Product design. A pilot test of the product to be developed is designed and elaborated.
4. Finish. Certain characteristics of the threads or fabrics of manufacture are modified, with the purpose of improving their aesthetics or functionality.
5. Preparation. Garments are developed with sewing machines.

The textile industry evaluated contains five production lines (line A, line B, line C, line D, and line E) and this experimentation process was prepared in a preliminary way in line A, to be applied to other manufacturing lines, following the same methods and criteria evaluated in this investigation for line A. The main objective was the reduction of occupational risks and with it visits to the doctor of the workers of this company and unnecessary expenses of disability, as well as the hiring and training of new personnel or workers who work in this industry. Once the analyses were prepared, we proceeded to evaluate solution proposals that would support the main objective mentioned in this paragraph. This pilot test was replicated in other manufacturing lines of the company where the investigation was made. The characteristics of communication, adaptability, climate variations and risk situations must always were considered in an essential way, to have workers who had the confidence to make the activities in a comfortable way. Industrial designers of structures of recharges, backrests, support to place hands, made these structures for the comfort of workers, as developed in this research, regarding the postures and breaks of operating personnel. The structures were made according to the evaluations of time and movements of the workers in each step.

11.3 Results

The development of the study included analyses and evaluations that showed relevant information of what was happening in the textile industry evaluated, which are shown in the following sections. The manufacturing processes evaluated with ergonomic methods are of great relevance, because with the analyses carried out, information was obtained that supported the improvement of the attitude of operating personnel and of all areas of the industry evaluated and with this an increase in the level of quality and productivity in industrial operations. The workers showed greater commitment in their functions by having suitable working conditions for the development of manufacturing processes. An adequate adaptation was achieved among the operational and supervisory, and managerial personnel, and with it a good harmony and adaptability of the personnel in their work areas. This analysis was of great importance for the evaluated industry, which supported generating savings in costs of accident prevention and increased productivity and quality. In the process of the study, the industrial activities carried out by the workers were evaluated and when observing that in certain stages, some type of support was required to have the correct posture and carry out the activities with greater activity, they were designed, manufactured, and tested automated structures and systems. One of these was an electronic device that automatically stopped a machine to prevent the material from falling to the ground when the operator did not have the time to load the material at another stage of the process, or when the worker felt tired and only saw drop the material on the floor of the machine. This electronic device did not have the machine.

11.3.1 Evaluation of Labor Risks

In the evaluations was observed that when the working conditions are adequate, high-level operating returns are obtained and the productivity and quality indices of the manufactured products are high, and with this there is a positive effect on the finances of the companies in this arid city with the aspect of profit sharing. This information is expressed in Table 11.7, showing a correlation analysis of the causes that can generate labor risk with levels from 0 to 100%, being represented at the beginning of the study (in February of 2017), indicating the possibility that an evaluated industry, workers have suffered from an accident or suffered from a disease depending on the industrial operations. In Fig. 11.2, it is observed by color level, the evaluation of the six ergonomic factors analyzed.

Table 11.7 Analysis of ergonomic factors

Month	Year		
	2017	2018	2019
	Factors		
January	Au, Bh, Cu, Dh, Eh, Fa	Ac, Be, Cc, De, Ee, Fh	Ac, Be, Cc, De, Ee, Fh
February	Au, Bh, Cu, Dh, Eh, Fa	Ac, Be, Cc, De, Ee, Fh	Ac, Be, Cc, De, Ee, Fh
March	Au, Bh, Cu, Dh, Eh, Fa	Ac, Be, Cc, De, Ee, Fh	Ac, Be, Cc, De, Ee, Fh
April	Au, Bh, Cu, Dh, Eh, Fa	Ac, Be, Cc, De, Ee, Fh	Ac, Be, Cc, De, Ee, Fh
May	Au, Bh, Cu, Dh, Eh, Fa	Ac, Be, Cc, De, Ee, Fh	Ac, Be, Cc, De, Ee, Fh
June	Au, Bh, Cu, Dh, Eh, Fa	Ac, Be, Cc, De, Ee, Fh	Ac, Be, Cc, De, Ee, Fh
July	Ac, Be, Cc, Dh, Ee, Fh	Ac, Be, Cc, De, Ee, Fh	Ac, Be, Cc, De, Ee, Fh
August	Ac, Be, Cc, Dh, Ee, Fh	Ac, Be, Cc, De, Ee, Fh	Ac, Be, Cc, De, Ee, Fh
September	Ac, Be, Cc, Dh, Ee, Fh	Ac, Be, Cc, De, Ee, Fh	Ac, Be, Cc, De, Ee, Fh
October	Ac, Be, Cc, Dh, Ee, Fh	Ac, Be, Cc, De, Ee, Fh	Ac, Be, Cc, De, Ee, Fh
November	Ac, Be, Cc, Dh, Ee, Fh	Ac, Be, Cc, De, Ee, Fh	Ac, Be, Cc, De, Ee, Fh
December	Ac, Be, Cc, Dh, Ee, Fh	Ac, Be, Cc, De, Ee, Fh	Ac, Be, Cc, De, Ee, Fh

AW = Attitude workers; B = Operations of industrial machines; C = Work environment; D = Materials used in industrial processes; E = Working method; F = Measurement process; Ac = Workers comfortable; Au = Workers uncomfortable: Be. Easy to make operations; Bh = Hard to make operations; Cc = Workers comfortable; Cu = Workers uncomfortable; De = Easy to handle material; Dh = Hard to handle material; Ee = Easy to make operations; Eh = Hard to make operations; Fh = Happiness for values in ranges; Fa = Annoyance for values in ranges

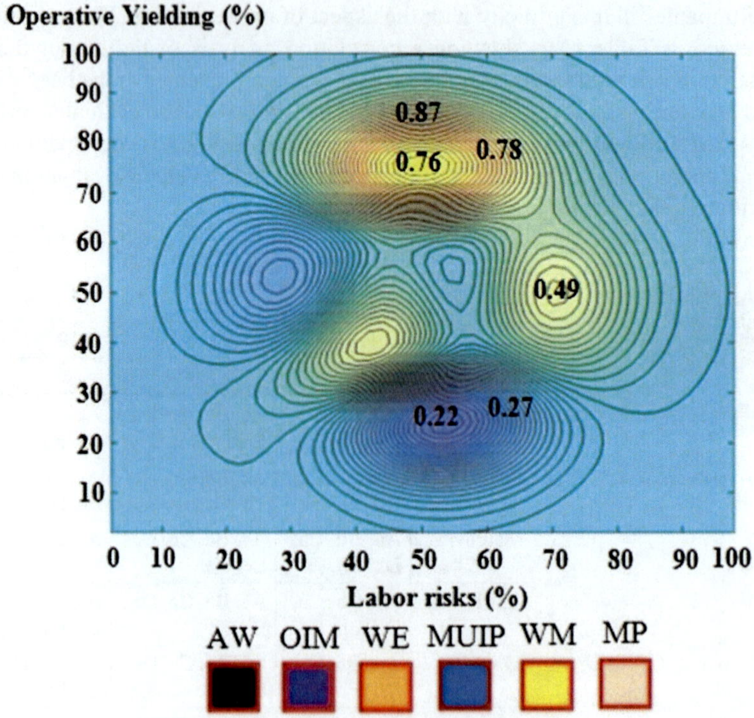

Fig. 11.2 Grade of occurrence of ergonomic factors in 2017

Table 11.7 shows an analysis as textual information of the evaluations made of the six ergonomics factors, where it is observed that in the first six months of 2017, dissatisfaction levels are illustrated in all the evaluated factors. This analysis considered both manager, supervisory personnel and especially of the workers who elaborated the industrial operations in the evaluated company. From the seventh month of 2017 to the last month of 2019 in which the investigation ended, indices of satisfaction were presented that led to obtaining results that decreased the levels of occupational risks and increase of productivity and quality. Figure 11.2 illustrates an incidence analysis of the six occupational risks evaluated, indicating the indices in which these six analyzed are presented, with their percentages of occurrence.

In Fig. 11.2, AW = Attitude workers; OIM = Operations of industrial machines; WE = Work environment; MUIP = Materials used in industrial processes; WM = Working method; and MP = Measurement process.

The percentage values were obtained by evaluating the presence of each occupational risk on a weekly basis, obtaining percentages as reference values, whose data were obtained from constantly carried out surveys. Workers graciously agreed to provide the information for the analysis of survey information. As can be seen in the graph, the attitude of the workers was a very important aspect when preparing the activities, showing a 0.87 level of incidence. Values close to 1 mean that they have a

Table 11.8 Analysis of visits to medical attention in indoor of textile industry (2017–2020*)

Characteristics	Year							
	2017		2018		2019		2020*	
	I	O	I	O	I	O	I	O
Workers with experience	67	34	49	23	18	14	13	10
Workers without experience	103	53	69	36	29	24	10	6

I = Indoor; O = Outdoor; *It is an analysis projection

high intensity of occurrence. This aspect was taken as occupational risk when certain workers of the evaluated company thought negatively and elaborated the activities without following the safety procedures. The next work risk by occurrence indices was the work method with a percentage value of 0.78, followed by the work environment with 0.76. Subsequently, the measurement process was presented with a 0.49. The last two aspects evaluated represented the materials used in industrial processes and the operation with industrial machines or equipment with levels of 0.27 and 0.22, respectively. Based on this, solution proposals were made.

11.3.2 Occupational Health

With the development of the investigation, relevant information was obtained, which supported the supervisory, and managerial personnel to reduce costs, due to the prevention of accidents and diseases of workers in the areas of manufacture.

The quantity of visits to the doctor inside the company and in the public health was an important factor in this study, being a relevant element of the occupational health of this investigation developing an analysis of this aspect from 2018 to 2019 and with a projection of analysis of 2020. This represents the Table 11.8.

Table 11.8 illustrates the rates of visits to the doctor both inside the company and outside in public health institutions in Mexico, observing that visits inside were greater than those outside. In addition, Table 11.8 represents the rates of inexperienced workers with higher levels of visits to the doctor, than the personnel with experience of manufacturing activities, because the lack of experience generated nervousness in the workers and developed the activities without following the procedures that avoided occupational risks.

11.3.3 MSD Analysis

In this investigation an evaluation of information obtained from the medical personnel inside the company, as well as from outside the Mexican public health institutions were made. In this section was analyzed, cases of discomfort and pain in the upper and lower extremities, and to determine in which parts of the human body the highest degree of health symptoms was generated. The main areas of the body that suffer from discomfort or pain are the upper and lower extremities, as well as the neck, head, and back. The presence of MSD generated personnel change in essential areas at the time that workers required medical attention or temporary disability. This rotation of internal personnel generated the need for training the personnel of the evaluated company or new hires that required rapid training, so that production stopped at times or in desperation of the manager and supervisory personnel, they put to work manufacturing personnel without having the experience. This was what generated the errors and caused failures in the machines of the analyzed industry. The percentage of these changes was 44% in 2017, 31% in 2018, 22% in 2019.

The maximum errors that caused defective products and failures in the industrial equipment and machinery were 66 and 33 in 2017, 48 and 26 in 2018, 28 and 12, by each worker without training and experience in each step. The two principal diseases occurred by the presence of MSD were cervicalgia and epicondylitis with a frequency of 14 every month of cervicalgia and 12 of epicondylitis in 2017, 9 every month of cervicalgia and 7 of epicondylitis in 2018 and 6 every month of cervicalgia and 4 of epicondylitis in 2019. The workers without experience visited to doctor with discomfort and pain in indoor of the industry evaluated were 170 and in the Mexican health institutions considered as the outdoor visits were 87 in 2017; 118 in indoor and 59 in outdoor in 2018 and 63 in indoor and 38 in outdoor in 2019, observing a reduction of cases in last two years for the applications of ergonomic methods. The principal parts of the human body that suffer were discomfort and pain in back, hands, head, neck, and shoulders, representing by degree of occurrence in Fig. 11.3.

Figure 11.3 shows the analysis of 2017 of the visits to medical attention, which indicated that the major values with an incidence of 54-27 (being the first value the indoor visits to doctor (in the industry evaluated) and the second value the outdoor visits to Mexican health public institutions); followed back by 46-23, head with 33-17, shoulders with 26-12, and finally hands with 13-8. To represent these numerical data from 2017 to 2019 the analysis was made, which is represented in Table 11.9.

11.3.4 Evaluation of Ergonomic Factors

When a disability occurred, the management and executive personnel of the industries were concerned when considering the high costs that are generated before Mexican health authorities, whether they are foreign or national companies, for the fines before the Mexican health authorities and for the rehabilitation of the workers. This is why workers had to be trained in the correct way to avoid the presence of accidents or diseases. The evaluation of the ergonomic interventions is showed in Table 11.10, and its analysis indicates that at high levels of the attitude and comfort was varied. The evaluation indicated that with good relation between manager, supervision personnel and workers of industrial operations, the attitude and comfort of them was good. That is why the constant supervision of these ergonomic factors evaluated in this textile

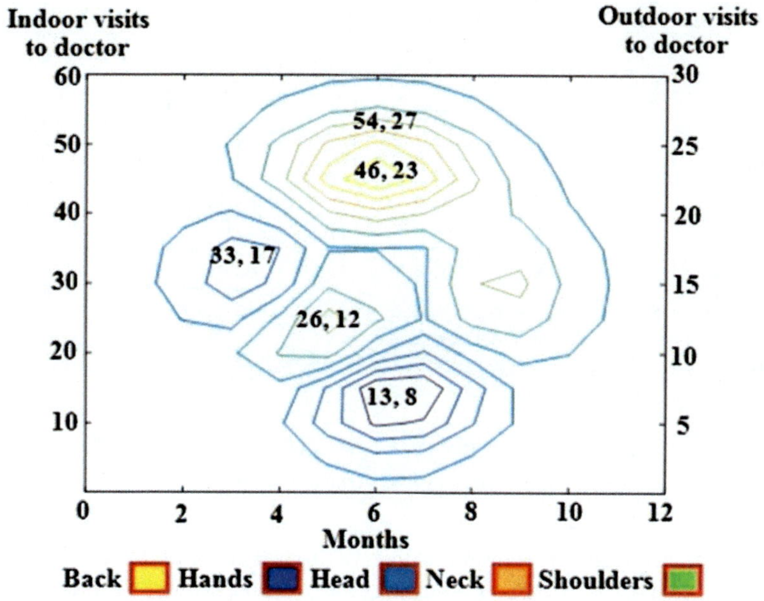

Fig. 11.3 Grade of occurrence of discomfort and pain in upper parts of the human body (2017)

Table 11.9 Evaluation of MSD symptoms (2017–2019)

Year	2017		2018		2019	
MSD Symptoms	Indoor	Outdoor	Indoor	Outdoor	Indoor	Outdoor
Neck	54	27	37	18	16	13
Back	46	23	29	14	13	10
Head	33	17	24	11	9	7
Shoulders	24	12	18	9	6	5
Hand	13	8	10	7	3	3
Total	170	87	118	59	47	38

I = Indoor; O = Outdoor

industry, and the development of continuous improvement processes, was important, generating good work environment, being a relevant factor in the productivity and quality indices of the manufactured products.

Table 11.10 shows an analysis of ergonomic interventions, where attitude and comfort levels were evaluated in workers with and without experience. It is observed in said table that at the beginning of the investigation there was a hostile attitude of the personnel who worked in the manufacturing areas and did not have the comfort to carry out their activities. As the study progressed, the attitude and comfort indices went from the bad level to the regular level and to good level.

Table 11.10 Evaluation of MSD symptoms (2017–2019)

Year	2017		2018		2019	
Ergonomic interventions	WE	WWE	WR	WWE	WE	WWE
Physical	B, B	B, R	R, R	R, B	G, G	E, G
Institutional	G, R	G, B	R, G	B, R	G, R	R, G
Cognitive	B, G	R, B	B, R	R, G	R, G	G, G
Training	G, R	G, R	G, R	G, R	G, R	G, G
Multifaceted	R, B	B, B	G, G	G, G	G, R	G, R

WE = Workers with experience; WWE = Workers without experience
Attitude level: G = Good, R = Regular B = Bad
Comfort level: G = Good, R = Regular B = Bad

11.3.5 Correlation of Levels of Productivity and Quality with Ergonomic Interventions

A correlation analysis of ergonomic interventions and productivity and quality indices in the period of the investigation was made, which is represented in Table 11.11. The ergonomic factors using the Ishikawa diagram were analyzed, observing an influence in the manufacturing variables, due to levels of productivity.

As can be seen in Table 11.11, the six ergonomic factors indicated a relation with these aspects with the industrial factors mentioned, which presented the highest incidence, and which were complaints from workers in the manufacturing area.

11.4 Conclusions

The human factor is of great importance in the elaboration of industrial operations in any type of manufacturing area, both in manual and automated processes. In this investigation, it was observed that this factor influenced to a great extent in the productivity and quality indices due to the attitude of the workers as an ergonomic aspect evaluated. The QUITE method supported the generation of solutions to avoid occupational risks and increase productivity and quality that were challenges for managerial and supervisory personnel. With this method was reduced the levels of labor risks and the workers injured, decreasing the costs in the industry evaluated. With the applied methodology, very relevant information was obtained, and based on this, the solution proposals were evaluated and developed in the stages and areas where they were required.

The methodological process was carried out with the greatest number of variables involved, with specialized measurement instruments and expert company personnel. Limitations of lack of financial support for the preparation of the research were presented, so some solution proposals, such as developing innovative automated

Table 11.11 Analysis of ergonomic aspects with the Ishikawa diagram in textile industry (2017–2019)

Aspects	Analysis					
	1		2		3	
	CA	CO	CA	CO	CA	CO
AW	Bad structures in industrial operations	Bad operation development	Strong supervisor calls	Bad communication with the supervisor and other workers	Low salary without bonuses of productivity, quality and assistance	Miss work or retire from work
OIM	Bad adjust in mechanisms of industrial systems	Presence of defective products	Lack of personnel training for the operation	Presence of defective products	Personnel with bad attitude	Presence of defective products
WEE	Bad relation between supervisor and workers	Bad operation development	Bad climatic conditions	Bad attitude and presence of defective products	Industrial operations made in a forced way	Miss work or retire from work
MUIP	Defective material	Bad operation development	Use of materials with labor risk	Bad operation development	Use of materials with labor risk without protection	Miss work or retire from work
WM	Incorrect method	Bad operation development	Complex method difficult to elaborate	Bad operation development	Lack of support with the work method	Bad operation development
MP	Measurement instrument without misadjusted	Presence of defective products	Incorrect instrument to measure	Presence of defective products	Lack of support with the instrument to measure	Bad operation development

CA = Cause, CO = Consequence

systems to improve the comfort of the works to make his functions and the operation of machines and industrial processes, will make in a next stage. In addition, the information of the company was limited as of 2015, because it was not captured and organized.

A future stage is to obtain financial financing to develop systems that detect problematic situations of occupational risk, since sometimes the personnel of the industries focus on the indices of productivity and quality, but not on the personnel who make the industrial operations. With the analysis of the main ergonomic components, aspects of interest were generated that the evaluated textile company did not consider, mainly with the treatment of workers in the manufacturing area.

The development of the research helped to have a vision with a greater focus on the human factor, which should be supported for the development of industrial operations, with the necessary elements that originated a better attitude of the operating personnel and with the tools for continuous improvement, to increase the quality and productivity indices, which was a concern of the supervisory, and managerial personnel of the evaluated textile industry. In all the analyses carried out, the participation of the workers in the manufacturing area was considered, achieving progress with the application of ergonomic methods, considering their skills and qualities, to develop activities in a more comfortable, agile and excellent way. Attitude of workers in manufacturing areas.

The evaluations show how important the human factor is in the productivity and quality indices, which generated gains in the evaluation period and the necessary improvements were developed in detail to continue increasing the percentage of profits and reduce the levels of occurrence of accidents and diseases in workers of this company. In addition, shared analyses of industrial processes are being prepared on a regular basis, by evaluated industry personnel and scientists from the areas of ergonomics and industrial engineering, with the aim of following up on any type of failure and nonconformity of operating personnel.

The objective is to have a plant of operational and specialized personnel in various industrial operations of the analyzed textile company, to replace the necessary workers without generating defective products or causing accidents. or illnesses in the personnel of the manufacturing area. This was carried out with the aim of safeguarding the health in the psychological field of the workers and having great harmony when developing industrial operations, which lead to increasing productivity and quality indices and reducing accidents and diseases caused in the work activities.

The government of Mexico gave the opportunity to foreign or national industries, regarding the NOM-035-STPS-2018 standard, to meet the requirements for November 2020 and have a harmonious working environment, where most companies in the Mexicali city do not have it. This is a prototype analysis for other types of industries and the development of training courses for personnel from companies in this arid zone and science specialists, both those who provide services to companies, as well as research and teaching personnel from educational institutions in AS universities and technological educative institutions.

The ergonomic interventions were applied obtained favorable results were observed, reducing the cases of medical attention for work activities and both the quality and productivity had a considerable increase. Also, in this investigation the cognitive ergonomic interventions were used, being applied to evaluate the behavior of workers in the industrial processes analyzed, obtaining very favorable results, both in increased productivity and quality, and in the reduction of accidents and health symptoms. The training interventions were utilized to be evaluated by the human resources department with specialized people in fundamental manufacturing operations, and it can be observed that new recruited workers were selected and those who relocated for self-improvement or by necessity, with optimal operating performance.

This led to increased productivity and quality, in addition to a reduction in accidents or health symptoms. The multifaceted ergonomic interventions were applied with specialized personnel in the operations of the manufacturing area to support to improve their operational performance by generating confidence and a very favorable self-esteem to increase productivity and quality and decrease accidents and health symptoms in work activities. For this reason, in this research, objectives were achieved that supported improving working conditions and saving the costs, with proposals and the use of continuous improvement tools that reflected rates of high levels of productivity and decreased errors.

References

Ahram T, Jang R (2014) Advances in physical ergonomics and human factors. Book (AHFE conference 2014 [5th international conference on applied human factors and ergonomics]). 20 Volume Set: proceedings of the AHFE conference 19–23 July 2014, p 403

Alexander M, Kusleika R, Walkenbach J (2016) Excel 2019 Bible-the comprehensive tutorial resource. Ed. Wiley. ISBN: 978-1-119-06751-1.1152

Álvarez E, Moya Fernandez PJ, Blanco Encomienda FJ, Muñoz FJ (2015) Methodological insights for industrial quality control management: the impact of various estimators of the standard deviation on the process capability index. J King Saud Univ Sci 27(3):271–277. https://doi.org/10.1016/j.jksus.2015.02.002

Aven T (2015) Risk assessment and risk management: review of recent advances on their foundation. Eur J Oper Res, December 2015. Science Direct Ed. 253(1):1–13. https://doi.org/10.1016/j.ejor.2015.12.023

Aven T, Zio E (2013) Model output uncertainty in risk assessment. Int J Perform Eng 9(5):475–486. https://doi.org/10.23940/ijpe.13.5.p475.mag

Axelsson JRC (2000) Quality and ergonomics management: toward an emerging integrated paradigm. Proceedings of the human factors and ergonomics society annual meeting, 44(12):467–470. https://doi.org/10.1177/154193120004401208

Baugh G (2011) Tissues manual for fashion designers [In Spanish]. Parramon. ISBN 10: 8434238314/ ISBN 13: 9788434238312

Besseris G (2014) Robust process capability performance: an interpretation of key indices from nonparametric view point. TQM J 26(5):445–462. ISSN: 1754-2731

Bevan S (2015) Economic impact of musculoskeletal disorders (MSDs) on work in Europe. Best Pract Res Clin Rheumatol 29(3):356–373. https://doi.org/10.1016/j.berh.2015.08.002

Carrasco Martínez AC (2010) Ergonomic study at work station PT0780 from company S-MEX, S.A. DE C.V [In Spanish]. Industrial engineering theses. Universidad Tecnológica de la Mixteca. Huajuapan de León, Oaxaca, Mexico. October 2010

Chandra A, Chandna P, Deswal S, Kumar R (2009) Ergonomics in the office environment: a review. Proc World Acad Sci Eng Technol 51: 913–919. Conference Paper March 2009. Conference Enviro Energy 2009

Da Costa B, Vieira E (2010) Risk factors for work-related musculoskeletal disorders: a systematic review of recent longitudinal studies [In Spanish]. Am J Ind Med 53(3):285–323. https://doi.org/10.1002/ajim.20750

DOF-Mexico (2019) Diario Oficial de la Federación. Mexico

Drury CG (2010) Global quality: linking ergonomics and production. Int J Prod Res 38(2):4007–4018. https://doi.org/10.1080/00207540050204876

García M, Sánchez Lite A, Sebastián Pérez MA (2011) Ergonomic applications as a improvement tool of the project activity for production processes [In Spanish]. Proceedings of the15th international conference on project engineering, Huesca, Spain, July 2011, pp 2041–2057

González BA, Adenso Díaz B, González Torre P (2003) Ergonomic performance and quality relationship: An empirical evidence case. Int J Ind Ergon 31(1):3–40. https://doi.org/10.1016/S0169-8141(02)00116-6

Harshal TP (2017) Historical milestones of ergonomics: from ancient human to modern human. J Ergon 7(4). ISSN: 2165-7556. https://doi.org/10.4172/2165-7556.1000e169

Hencken EV (2010) Textiles. Concepts and principles, book textiles. 3rd edition. ISBN: 9781563678448. 532

INDEX-Mexicali (2019) Anuario de la industria maquiladora de Mexicali

Imada AS (2007) Book chapter. The value of participation in ergonomics. Book meting diversity in ergonomics. Science Direct. Elsevier Ed. 91–98. https://doi.org/10.1016/B978-008045373-6/50006-4

ISO 9001:2015 (2015) Quality management systems—Requirements

ISO 14001:2015 (2015) Environmental management systems—Requirements with guidance for use

ISO/TC159/SC1. General ergonomics principles

ISO/TC159/SC3. Anthropometry and biomechanics

ISO/TC159/SC4. Ergonomics of human-system interaction

ISO/TC159/SC5. Ergonomics of the physical environment

Karthik T, Gopalakrishnan D (2014) Book Chapter, environmental analysis of textile value chain: An overview. Book. Roadmap to sustainable textiles and clothing. Ed. Springer, pp 153–188

Khan F, Rathnayaka S, Ahmed S (2015) Methods and models in process safety and risk management: past, present and future, Process safety and environmental protection. Science Direct Ed. 98. 116–147. https://doi.org/10.1016/j.psep.2015.07.005

Ley Federal del Trabajo (2019) Official Journal of the Federation—Regulations of working conditions STPS-SSA [In Spanish]

López Torres VG, Marín Vargas ME, Alcalá Álvarez MC (2012) Ergonomics and productivity: related variables to competitiveness of maquiladora plants [In Spanish]. Ingeniería Industrial. Actualidad y Nuevas Tendencias. 9, Julio-Diciembre 2012, pp 17–32. Universidad de Carabobo Carabobo, Venezuela, ISSN: 1856-8327

López Torres VG, Marín E, Zarate Cornejo RE (2011) Work risks in a Maquiladora Plant: A longitudinal analysis [In Spanish]. Revista Int Adm Finan 3(2):103–115

Magrab E, Azarm S, Balachandran B, Duncan J, Herold K, Walsh G (2011) An engineer's guide to MATLAB, 3e: with applications from mechanical, aerospace, electrical, and civil engineering. Ed. Prentice Hal. ISBN: 978-0-13-199110-1. 846

McIntyre JE (2004) Synthetic fibres. Woodhead Publishing. 1rd. Edition. ISBN: 9781855735880. 308. Hardcover ISBN: 9781855735880. eBook ISBN: 9781845690427

Monroe KW (2000) Workplace Risk factors and occupational musculoskeletal disorders. Part 2: a review of biomechanical and psychophysical research on risk factors associated with upper extremity disorders. AIHAJ—Am Ind Hyg Assoc J 61(2). https://doi.org/10.1080/15298660008984532

Norma NOM-172-SEMARNAT-2019 (2019) Guidelines for the collection and communication of the air quality index and health risks [In Spanish]

Norma NOM-011-STPS-2001 (2001) Health and safety conditions in workplaces where noise is generated [In Spanish]

Norma NOM-015-STPS-2001 (2001) High or low thermal conditions—safety and hygiene conditions [In Spanish]

Norma NOM-024-STPS-2001 (2001) Vibrations—conditions of health and safety at workplaces [In Spanish]

Norma NOM-025-STPS-2008 (2008) Lighting conditions in workplaces [In Spanish]

OHSAS 18:000 (OHSAS 18:000-2007) Occupational Health and Safety Management [In Spanish]

Oron GT, Hancock PA (2017) Book chapter. From ergonomics to hedonomics: trends in human factors and technology—the role of hedonomics revisited. Book. In Emotions and affect in human factors and human-computer interaction. Ed. Academic Press. Science Direct, pp 185–194. https://doi.org/10.1016/B978-0-12-801851-4.00007-0

Ozalp V, Oguzhan E (2004) Dimensions of the relationship between ergonomics and quality in manufacturing: a Review. Conference YAEM 2004. Gaziantep-Adana. Turkey. 16–18 June 2004, 404–406

Ozalp V, Oguzhan E (2008) Quality improvement through ergonomics methodology: conceptual framework and an application. Prod Qual Manage J 3(3):311–324

Punnett L, Wegman DH (2004) Work-related musculoskeletal disorders: the epidemiologic evidence and the debate. J Electromyogr Kinesiol 14(1):13–23

Udale J (2014) Diseño textil: Tissues and techniques, fashion design manuals [In Spanish]. Editorial Gustavo Gili. ISBN/EAN. 2ª edición actualizada. Versión en Español. 175, ISBN 8425222699-9788425222696

Watson G (2004) The legacy of Ishikawa. Qual Prog Acad J 37(4):4

Chapter 12
Ergonomic Intervention in a Colombian Meat Processing Plant Using the ERIN Method

Yordán Rodríguez, Elizabeth Pérez, Maria Alejandra Trujillo-Murillo, and Maria Camila Salazar-Marín

Abstract Work-related musculoskeletal disorders remain a major concern for the industrial sector, despite preventive efforts made globally over the past decades. The objective of this chapter is to present an ergonomic intervention process carried out in a Colombian meat processing plant, using the Individual Risk Assessment (ERIN) method. The evaluation using the ERIN method allowed the identification of which aspects of the task under evaluation should be intervened. The intervention proposals were mainly engineering, were modeled in 3D, and focused on reducing exposure to physical risk factors. The impact of the proposals was projected before their implementation, using the ERIN method. As a result, the level of exposure to risk factors for work-related musculoskeletal disorders is estimated to be reduced from Very High (Actual Total Risk = 38) to Medium (Projected Total Risk = 24). It is expected that this case study will serve as a reference on how ergonomic interventions can be carried out in the industry.

Keywords Ergonomic assessment · Ergonomics in manufacturing · Ergonomic risk · Musculoskeletal disorders · Workplace intervention

12.1 Introduction

For several decades, statistics on occupational diseases in different regions of the world have shown that work-related musculoskeletal disorders (WMSDs) are among the diseases with the highest prevalence and incidence (Bernard and Putz-Anderson 1997; Baldwin 2004; Punnett and Wegman 2004; Eurofound 2019) and that they have a high negative impact from a socio-economic point of view (Baldwin 2004; Piedrahita 2006). Some authors even argue that the situation in industrially developing countries is more unfavorable (O'Neill 2000; Torres et al. 2011). At the same

Y. Rodríguez (✉)
National School of Public Health, Universidad de Antioquia, Medellín, Colombia
e-mail: yordan.rodriguez@udea.edu.co

E. Pérez · M. A. Trujillo-Murillo · M. C. Salazar-Marín
School of Industrial Engineering, Universidad Pontificia Bolivariana, Medellín, Colombia

A. Realyvásquez Vargas et al. (eds.), *New Perspectives on Applied Industrial Ergonomics*,
https://doi.org/10.1007/978-3-030-73468-8_12

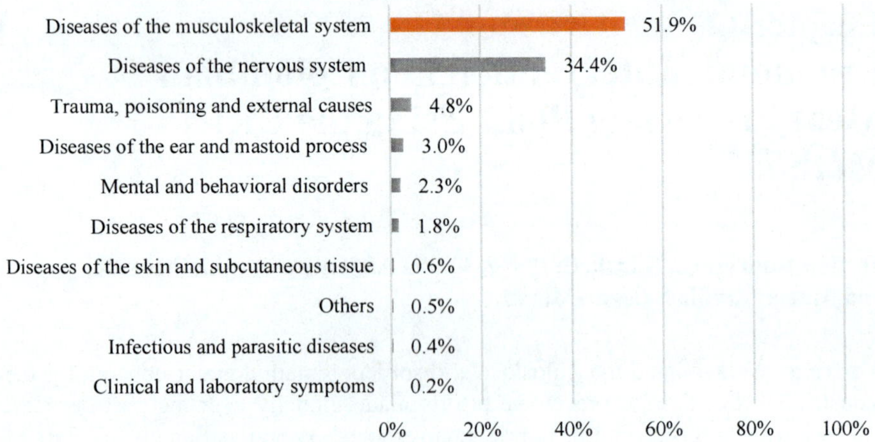

Fig. 12.1 Occupational diseases reported in Colombia by the SGRL during the period 2015–2017. *Source* Based on Castillo and Bravo (2019)

time, it is recognized that Ergonomics, as a scientific discipline and profession, has a relevant role in the prevention of these diseases (Dempsey 2007).

In Colombia, according to the Executive Report of the Second National Survey of Occupational Safety and Health Conditions published by the General System of Occupational Risks (SGRL), WMSDs represented between 87 and 90% of the total occupational diseases reported during the period 2009–2012 (MINTRABAJO 2013). More recent statistics show that WMSDs continue to be the main cause of occupational disease in Colombia (Castillo and Bravo 2019). Figure 12.1 shows that WMSDs account for more than 50% of the illnesses treated by the SGRL during the period 2015–2017.

In correspondence with the increase of WMSD statistics at a global level, professionals, researchers, and institutions have increased efforts for their prevention. However, the positive effects of these actions have not had the impact expected by society (Wells 2009). This leads us to ask the following question: Why have we not been as effective in preventing WMSDs?

The following are some reasons given by Richard Wells, a well-known researcher on the subject, that answer this question (Wells 2009):

- Risk assessments have focused mainly on physical risk factors, overlooking organizational factors that impact on the workplace.
- The weaknesses presented in the assessment process have led to the implementation of ineffective measures.
- The concentration of efforts in conducting etiological studies, instead of conducting more studies on interventions.
- The existing gaps between the knowledge generated and its adequate application in prevention activities.

- The lack of documentation of the effectiveness of evaluation methods and techniques for making changes in the workplace.
- Insufficient attention to training requirements for effective use of available evaluation tools and methods.
- Insufficient information on the effectiveness of interventions in musculoskeletal health or in reducing levels of exposure to risk factors.

Another reason is the imbalance between the actions and resources used to evaluate (diagnose) and those destined to intervene (transform) working conditions (Rodríguez 2018, 2019). In the practice of industrial ergonomics, it is common that preventive actions advance only to the assessment stage and no interventions are made to reduce exposure to WMSD risk factors. Some professionals have called this phenomenon as over-diagnosis of working conditions, which leads to limited expected results in terms of prevention. If we add to this situation that the number of practitioners with adequate training in Ergonomics in Latin America is insufficient, it warns of the need and importance of establishing strategies aimed at facilitating the implementation of ergonomic interventions and increasing the number of professionals who can carry them out.

One strategy that has been promoted by the World Health Organization (WHO), the International Ergonomics Association (IEA), and the International Labor Organization (ILO) is the development of easy-to-use ergonomic methods and tools for non-expert personnel to carry out job and work activity assessments (ILO and IEA 2010; Occhipinti and Colombini 2016). These tools would facilitate the process of identifying the aspects that need to be intervened in the short term, achieving an immediate impact. In addition, the number of people who could initiate preventive activities in the workplace would be increased. In practice, most workplace interventions do not require detailed evaluation and can be carried out immediately and at low cost (Corlett 2003). The approach taken by the ISO ergonomics standards to evaluate manual load handling activities (e.g., pulling, pushing, lifting, and handling light loads at high frequency), depending on the level of detail, demonstrates the convenience of starting the evaluation process with simpler methods (Occhipinti and Colombini 2016).

In line with this strategy, the ERIN method was developed (Rodríguez et al. 2013). The ERIN method was created so that non-experts in ergonomics can carry out massive evaluations of jobs and work tasks and thus identify which aspects should be intervened to reduce the levels of exposure to risk factors of WMSDs.

This chapter shows how an ergonomic intervention was carried out in a workplace in a Colombian meat processing plant using the ERIN method.

12.2 Individual Risk Assessment: ERIN

ERIN is an observational method for assessing exposure to risk factors of WMSDs in static and dynamic tasks. It can be used for the design and redesign of tasks and workplaces (Rodríguez et al. 2013; Rodríguez 2019). It should be highlighted that ERIN is a method that has gradually increased its use and acceptance by ergonomics practitioners in several parts of the world and mainly in Latin America.

ERIN assesses and quantifies the risk of seven variables:

1. Interaction of posture and movement frequency of the trunk.
2. Interaction of posture and frequency of movement of the shoulder/arm.
3. Interaction of posture and frequency of movement of the wrist.
4. Interaction of posture and frequency of movement of the neck.
5. Rhythm: the relationship between the speed of work and the effective duration of the task.
6. Intensity of effort: the relationship between the perceived effort in performing the task and the frequency of effort.
7. Self-assessment: stress perceived by the worker when carrying out the task evaluated.

The total risk is obtained by adding up the risk of the seven variables. With this value of the total risk, the level of risk established in the ERIN method is determined (Table 12.1).

ERIN can be described as an easy to use, fast learning and low-cost method. This method was developed on the premise that it could be used by people with little experience and who, with minimal training in the use of the method, could begin to make ergonomic assessments of tasks and workstations.

The short time required for the assessment is an advantage of the method if a large number of tasks are to be assessed on a large scale. Another advantage of the method is that it allows the impact of intervention proposals to be quantified by variable.

ERIN is a method that has been proven to have acceptable levels of reliability and validity. At the same time, according to its users, it presents high levels of usability (Rodríguez 2018).

To facilitate the use of the ERIN method, a free mobile application has been developed and is available on Google Play Store in Spanish. This application facilitates the

Table 12.1 ERIN: Risk levels and recommended action

Zone	Total risk	Risk level	Action
Green	6–14	Low	No changes are required
Yellow	15–24	Medium	Further investigation is needed and changes may be required
Orange	25–34	High	Investigation and changes are required soon
Red	≥ 35	Very high	Investigation and changes are required immediately

Source Rodríguez (2019)

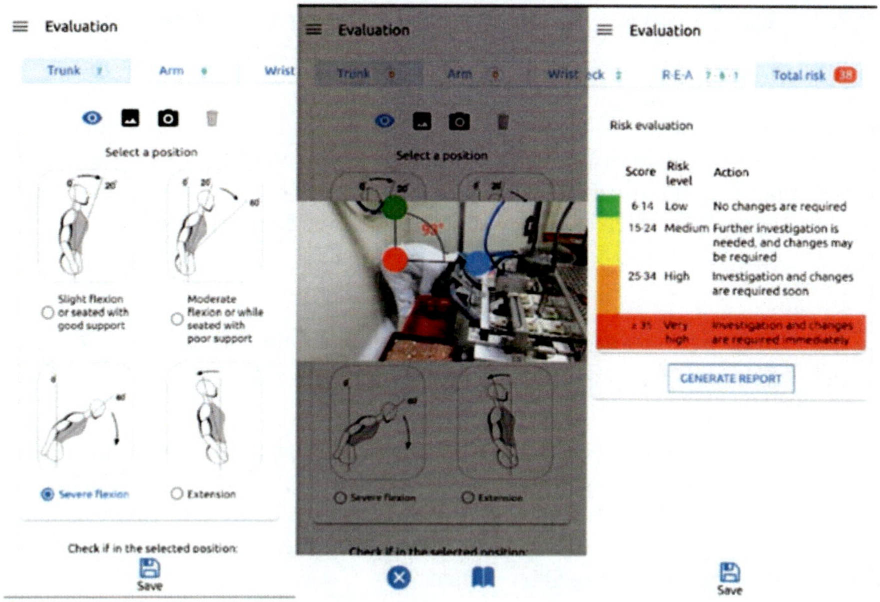

Fig. 12.2 ERIN 1.0 mobile application

assessment and estimation of body angles, among other functionalities. Figure 12.2 shows some images of this application.

12.3 Ergonomic Intervention Process Using the ERIN Method

The evaluation and intervention of the task were carried out following the five-step process proposed by the author of the method (Rodríguez 2018).

12.3.1 Step 1: Select the Workstations/Tasks to Be Assessed

Before starting the ergonomic assessment, an order of priority should be established according to the most critical tasks in the company. This allows the resources available for the intervention process to be used in a rational and organized way. For the selection of the critical tasks, the statistics of WMSDs, the opinion of workers and supervisors, the musculoskeletal symptoms, the importance of the task in the production process should be considered. The expertise of the person conducting the study is a factor that favors the correct selection.

12.3.2 Step 2: Assess the Task Using the ERIN Method

The tasks selected in the previous step are assessed using the ERIN method. The evaluation can be done using the ERIN worksheet or the available mobile application. The steps for evaluation are described in the header of the ERIN worksheet. It is possible that during the assessment process other risk factors not assessed with ERIN may be identified, in which case they should be recorded and analyzed later.

12.3.3 Step 3: Propose Solutions Aimed at Lowering ERIN's Scores

Proposals for intervention should aim to lower the risk scores of the most critical variables because of the assessment with the ERIN method. The expected impact of the proposals on the variables should be measured and evaluated with the ERIN method. In this step, the participation of workers involved in the change is recommended. The cost of the proposals should be estimated.

12.3.4 Step 4: Implement the Solutions

Some intervention proposals do not require significant costs and bring many benefits to the productive system (e.g., changes in work methods that decrease cycle time). But this is not always possible. High implementation costs and inadequate justification of the proposals from a productive and health point of view can become barriers to the implementation of solutions. The ergonomist must search, as far as possible, for the most economical solutions to favor their implementation in the company.

12.3.5 Step 5: Re-evaluate Using ERIN

Once the intervention proposals have been implemented, the task should be re-evaluated using the ERIN method. This step is performed with the aim of verifying the effectiveness of the proposals in terms of reducing the level of exposure to WMSD risk factors, according to ERIN.

12.4 Ergonomic Intervention in a Meat Processing Plant Using the ERIN Method

The processing plant under study belongs to the manufacturing sector in Colombia. This economic sector reports the second-highest rate of occupational disease in the country (Castillo and Bravo 2019). Figure 12.3 shows the behavior of occupational diseases in this sector where more than 50% are WMSDs. This demonstrates the importance and need for ergonomic interventions aimed at the prevention of WMSDs in this sector.

12.4.1 Description of Finished Product Packaging Task

The ergonomic intervention focused on the task of packaging finished products. This task is part of the vacuum packaging workstation and is executed during the last phase of the production process. This task was selected because the workers reported muscle pain and fatigue in their feet, back, wrists, and fingers. Additionally, they complained about the low temperatures, the repetitiveness of the task and its uninterrupted execution during the working day.

The task is to inspect and place the products coming out of the vacuum packing machine into baskets for later storage. The products handled are meat packages weighing 500 g. Each full container has 30 packs and weighs approximately 15 kg. The ambient temperature in the workplace is 1 °C. The established cycle time is 40 s, and the task is carried out for 8 h effectively, which is equivalent to 720 cycles during the working day. The activities that conform the cycle are: (1) take with both hands the meat packages that come out of the vacuum packing machine, (2) place 30 meat packages inside a plastic container, (3) move the full container to a temporary

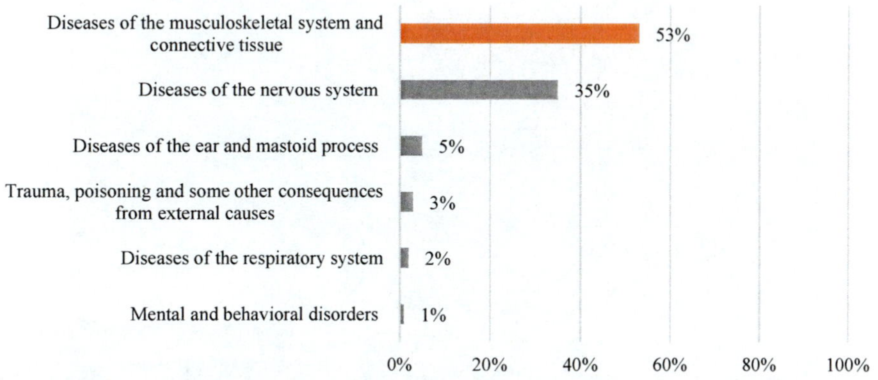

Fig. 12.3 Occupational diseases in the manufacturing sector in Colombia during 2017. *Source* Based on Castillo and Bravo (2019)

storage area located two meters away, and (4) place an empty container near the vacuum packing machine.

12.4.2 *Finished Product Packaging Task Assessment Using the ERIN Method*

The task assessment was carried out using the mobile application of the ERIN method. Figures 12.4–12.7 show the critical postures selected to evaluate the postural load of the trunk, shoulder/arm, wrist, and neck. The body angles were estimated using the ERIN mobile application.

For the evaluation of the trunk, the moment when the worker places the container full of products on the floor (lowest level) was selected as a critical posture. Figure 12.4 shows that the trunk is in "severe flexion," over 90°, which corresponds to a score of (3) for postural loading. With a trunk movement frequency of "more than 10 times per minute," which is equivalent to "very frequent," the risk score for the trunk variable gave a value of (7).

The right shoulder/arm was evaluated. Figure 12.5 shows that critical posture occurs when the worker takes the empty container from the highest point of the container pile. At this point, the worker performs a "severe flexion" of the shoulder/arm of more than 90°, corresponding a score of (3), to which is added the adjustment (+1) for having the arm abducted. With a postural load score of (4) and a "very frequent" movement of frequency, the shoulder/arm variable score was (9).

The right wrist was assessed. As shown in Fig. 12.6, critical posture occurs when the worker accommodates and presses down on the packs inside the container. In

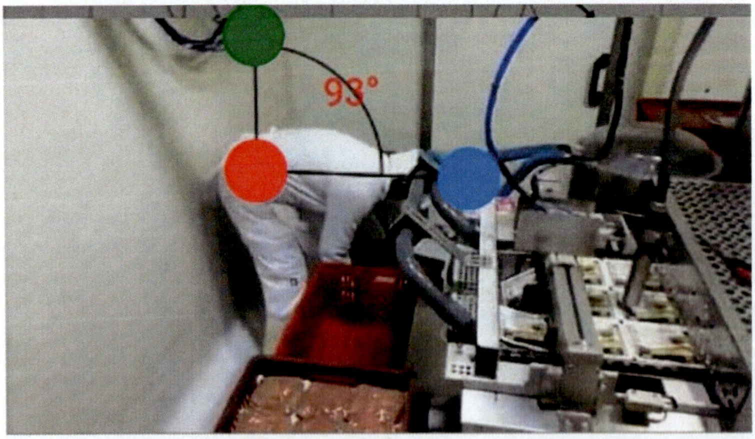

Fig. 12.4 Critical posture of the trunk

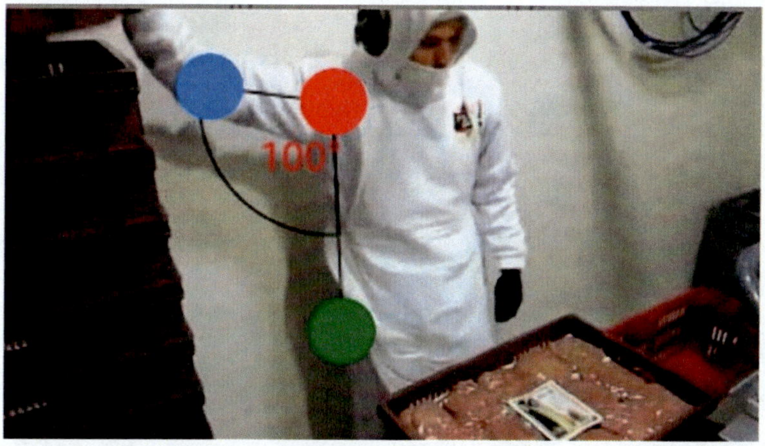

Fig. 12.5 Critical posture of the shoulder/arm

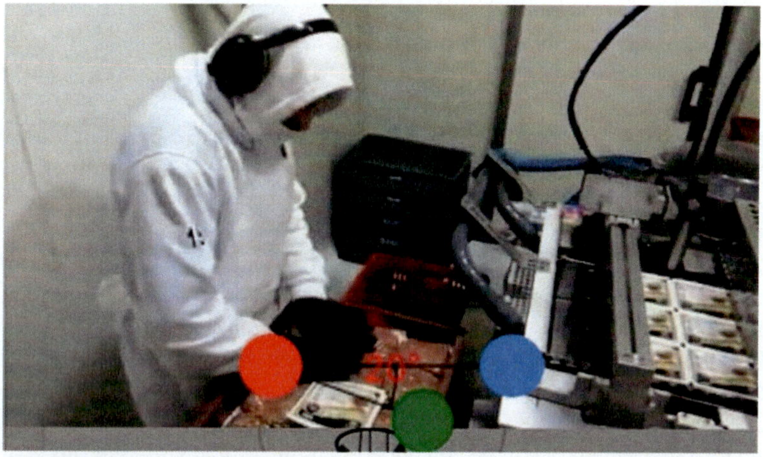

Fig. 12.6 Critical posture of the wrist

this posture, an "extension" angle on the wrist was identified as less than 45° (1). The adjustments corresponding to wrist pronation (+1) and holding the packs in the hand for more than 50% of the cycle time (+1) were added, resulting in a score of (3) for postural loading. The frequency of wrist movement was classified as "very frequent," so the variable's risk score was (6).

In the case of the neck, as shown in Fig. 12.7, the critical posture was a "severe flexion" of more than 20°, occurring when the worker accommodates the packages in the container. Thus, the postural load score was (2). The frequency of neck movement was classified in the "occasionally" category. The score for the interaction of posture and neck movement frequency was (2).

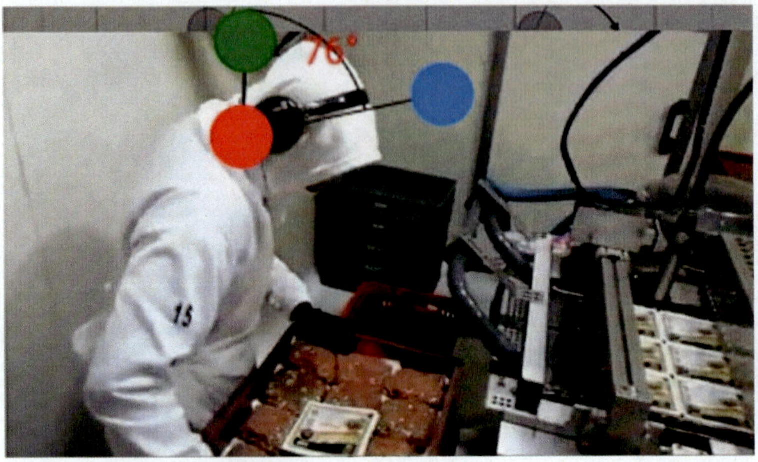

Fig. 12.7 Critical posture of the neck

Rhythm considers the effective duration of the task, in this case, 8 h; and the speed at which the work is performed, for this case classified as "fast." Therefore, the rhythm variable gave a score of (7).

The effort in this task was identified as the action of the worker in placing the 500-g meat packages in the container. The worker classified this effort as "slight" and his movement frequency was "more than 10 efforts per minute." Therefore, the result of the intensity of the effort was a score of (6).

In the self-assessment variable, a score of (1) was obtained, since the worker rated his task as "mildly stressful," due to the high repetitiveness of the task and the precision required to locate the finished products within the container.

The risk scores per variable were summed, giving a total risk of 38. This result corresponds to a "Very High" risk level, indicating that changes are required immediately according to the ERIN method.

From this assessment, it can be concluded that the critical variables, according to the ERIN method, were the shoulder/arm (risk score (9), maximum possible value (9)), the wrist (risk score (6), maximum possible value (6)), the rhythm (risk score (7), maximum possible value (7)) and the trunk (risk score (7), maximum possible value (9)). Therefore, intervention proposals should aim at decreasing the risk score on these variables.

Figure 12.8 presents the results of the task evaluation using the ERIN method worksheet.

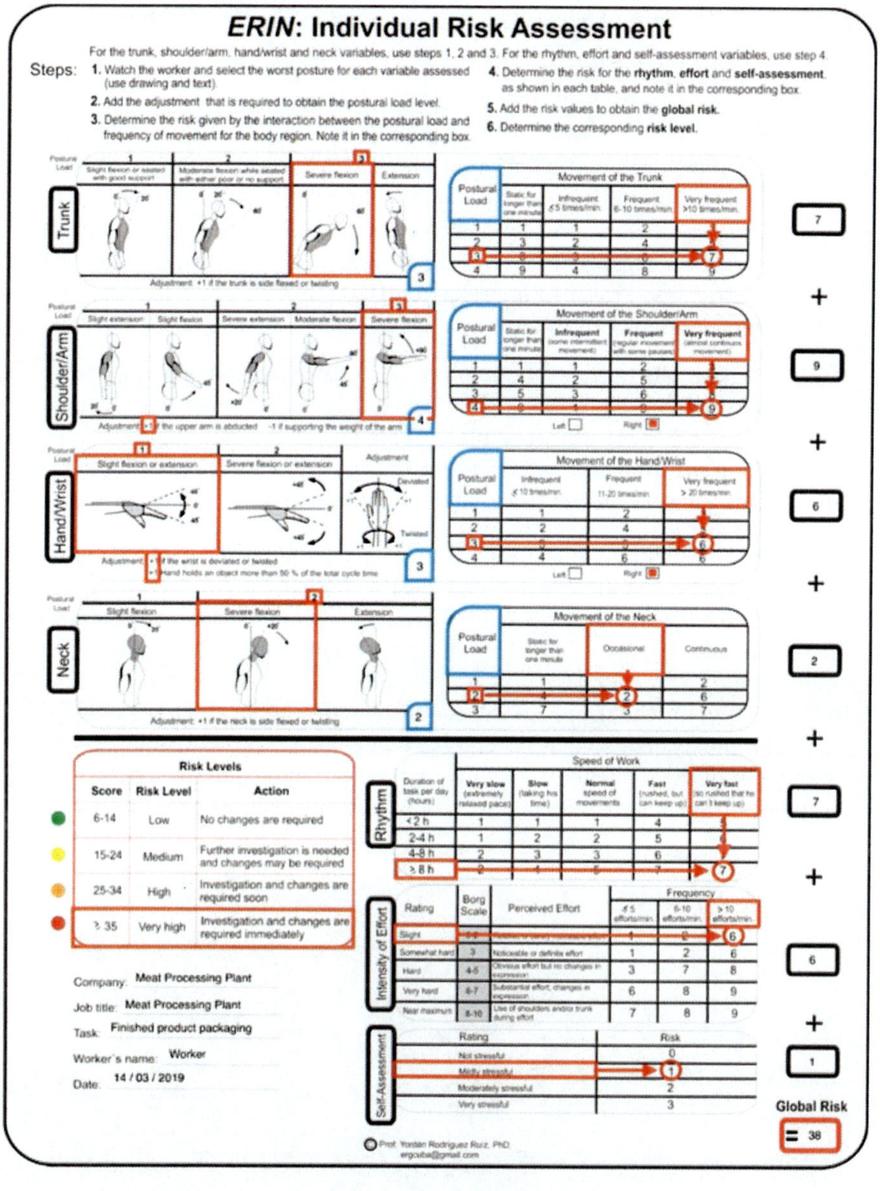

Fig. 12.8 ERIN worksheet. Results of the finished product packaging task evaluation

12.4.3 Proposals of Ergonomic Intervention in the Finished Product Packaging Task

The guidelines of the ISO 14738 standard (ISO 14738 2002) on anthropometric requirements for the design of workstations at machinery were considered in order to make proposals for intervention. The source of anthropometric data used for the design was the study of anthropometric parameters of the Colombian labor population (Estrada et al. 1995). The proposals were projected using the software Sketchup in its professional version Pro-2019 in its version 18.0.1697

5. The intervention proposals are explained below.

- **Manual scissor lift table cart**: the proposal consists of a transport system that incorporates an adjustable lifting platform so that the container is located at the same level as the worker's work plane. This proposal contributes to reducing the flexion of the trunk and facilitates the transport of the containers in the area, reducing the effort of pushing and pulling. Figure 12.9 shows the proposal.

- **Height of the working plane**: the proposal is to incorporate an inclined plane and a system of rollers at the outlet of the vacuum packing machine. This proposal to adjust the working surface would reduce the flexion of the trunk and neck at the time of picking up the finished products. Figure 12.10 shows the proposed modification.
- **Roller system for the supply of the containers**: the proposal is to incorporate an inclined roller system so that the empty containers reach the worker by gravity. The supply of the containers would be carried out by another worker who is in charge of this task in this area. This proposal would eliminate the displacement

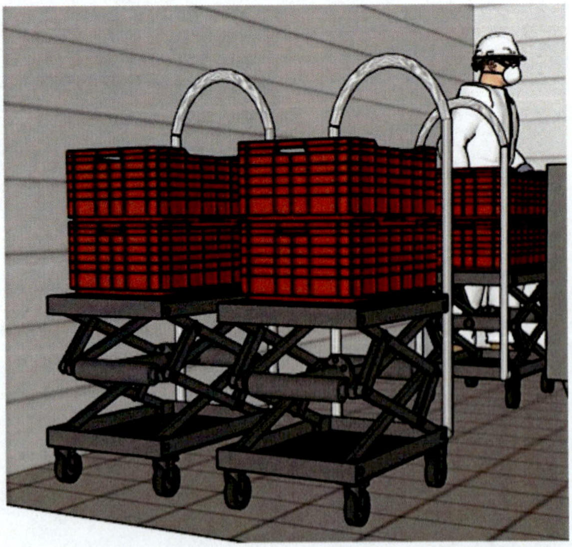

Fig. 12.9 Manual scissor lift table cart

Fig. 12.10 Proposed modification of the work surface

of the worker to reach the containers and take them to the worksite. On the other hand, it would decrease the severity of the postures of the trunk (Fig. 12.4) and arm (Fig. 12.5). The modeled proposal is shown in Fig. 12.11.

- **Changes in work method**: during the evaluation process, it was detected that the worker makes unnecessary pro-supination movements of the wrist when placing the meat packages inside the plastic container. The worker and the production manager were consulted, concluding that this movement could be eliminated from the cycle.

As a result of the elimination of movements and displacements that are part of the current work cycle, it is expected to reduce the score on the variable rhythm, since the work speed would be reduced from "fast" to "normal."

Figure 12.12 shows the integration of the intervention proposals in the work area.

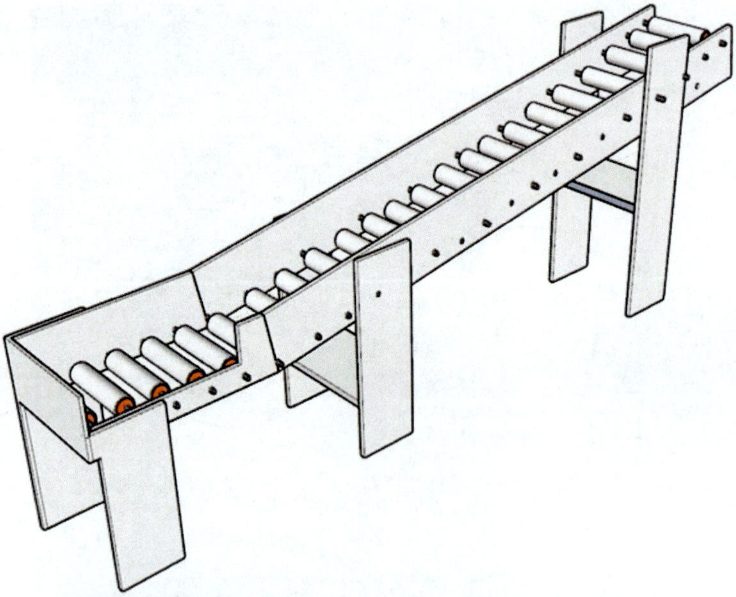

Fig. 12.11 Roller system for the supply of the containers

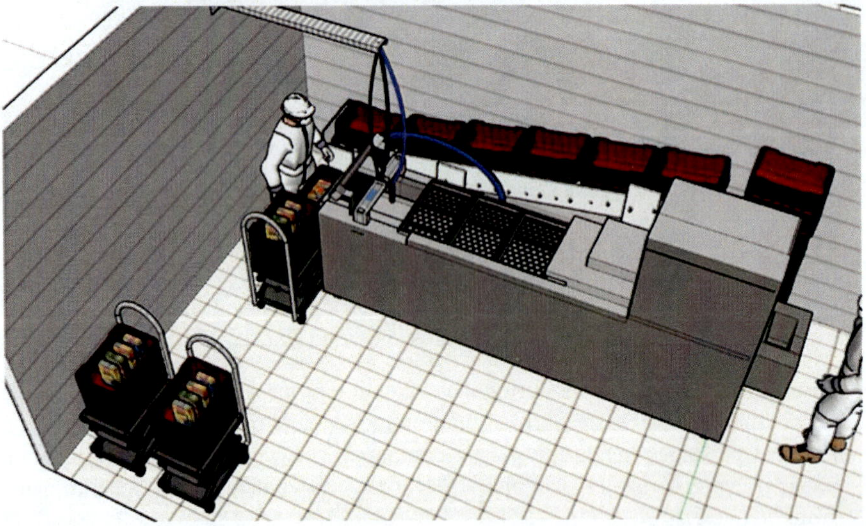

Fig. 12.12 Intervention of the finished product packaging task

Table 12.2 Impact of intervention proposals according to the ERIN method

ERIN variables	Actual	Projected	Impact
Posture and frequency of movement of Trunk	7	3	4
Posture and frequency of Shoulder/Arm movement	9	3	6
Posture and frequency of movement of the wrist	6	5	1
Posture and frequency of movement of Neck	2	1	1
Rhythm	7	5	2
Intensity of effort	6	6	0
Self-assessment	1	1	0
Total risk	38	24	14
Risk level	Very high	Medium	

12.4.4 Impact Assessment of Intervention Proposals

Table 12.2 shows the projected impact on the reduction of exposure with the implementation of the intervention proposals.

As can be seen, the level of risk decreases from Very High (Total Risk = 38) to Medium (Total Risk = 24) taking as a reference the evaluation with the ERIN method, which indicates an improvement in working conditions. In this work, the importance of integrating all the proposals is highlighted, in order to increase their impact on the prevention of WMSDs.

12.5 Discussions

The implementation or planning of an ergonomic intervention involves the reduction or elimination of exposure to WMSD risk factors (Dempsey 2007). Therefore, exposure levels should be measured before and after the intervention. In this work, it was estimated that with the implementation of the proposed measures, a reduction in the level of exposure to risk factors related to WMSD from Very High (Actual Total Risk = 38) to Medium (Projected Total Risk = 24) would be achieved. In the case presented in this chapter, as in many other situations in different industrial sectors, the number of workers directly targeted by an ergonomic intervention is small. This limits the power of the statistical analyses and conclusions obtained, although, from an ergonomic point of view, the mere reduction of exposure to WMSD risk factors could be considered a successful outcome (Dempsey 2007).

In the presentation of ergonomic intervention proposals to decision-makers, it is not recommended to use only health indicators, as they may be unattractive to them since they are not usually experts in ergonomics. For this reason, it is increasingly common to use productive indicators to justify the efficiency of ergonomic interventions. Productivity can be estimated even before the intervention by studying the activities, movements, and times required with the new working methods and means. For example, reductions in cycle times due to the elimination of activities can be estimated (Dempsey 2007; Rodríguez and Pérez 2013). However, the acceptance and implementation of an ergonomic intervention in an organization will largely depend on workers and managers being convinced that WMSD is a real problem (Whysall et al. 2004).

The intervention presented in this study focused mainly on providing engineering solutions for the task of packaging finished products. These types of micro-level interventions are often effective in gaining credibility with decision-makers and workers in enterprises, as these interventions can usually be easily implemented, and their impact can be measured in the short term. However, it should be mentioned that interventions that include multiple factors (e.g., organizational, environmental, individual, etc.) achieve greater impact in the medium and long term (Karsh et al. 2001; Silverstein and Clark 2004).

The proposals for the ergonomics intervention and a general estimate of the implementation costs were presented to the company's management, who expressed the relevance of carrying out the proposals. However, it should be mentioned that, due to the financial situation of the company to date and the fact that production levels have not been affected by the deficiencies found in the design of the task analyzed, it was decided not to establish an implementation plan in the short term. This case study is an example of how economic factors influence organizational decisions, despite showing that ergonomic intervention proposals reduce worker exposure to WMSD risk factors.

It should also be mentioned that the workers in charge of health and safety issues were satisfied with the use of the ERIN method and that, given the usefulness of the method, they have used it to evaluate other workplaces in the company.

This study once again demonstrates the importance of including ergonomic requirements in the initial phases of industrial projects (e.g., purchase of equipment, tools, furniture, layout, etc.), since ergonomic intervention aimed at improving existing working conditions generates higher additional costs and even reduces the possibilities of improvement and its impact on workers' health and productivity (Hendrick 2008).

12.6 Conclusions

The performance of ergonomic interventions is fundamental in the prevention of WMSDs in the industry. Its results should be focused not only on reducing exposure to WMSD risk factors but also on improving indicators of effectiveness and efficiency of production or service processes. In this way, the acceptance by decision-makers is increased, driving its implementation.

This case study shows how, using the ERIN method, aspects of the workplace can be identified that can be transformed and the expected impact on exposure reduction can be projected. It also highlights the usefulness of showing the proposals through

3D modeling techniques, as this facilitates understanding by the different actors involved in the intervention. As stated at the beginning of this chapter, it is important to increase actions and studies in the field of ergonomic interventions aimed at the prevention of WMSDs. A key action is the generation of tools, methods, approaches, procedures, software, and technologies, which are easy to apply and serve as a guide for people with basic training in ergonomics, on how to carry out ergonomic intervention processes in the industry.

It is expected that this case study will serve as a reference on how to carry out ergonomic interventions, which are necessary for our Latin American context.

References

Baldwin ML (2004) Reducing the costs of work-related musculoskeletal disorders: targeting strategies to chronic disability cases. J Electromyogr Kinesiol 14:33–41. https://doi.org/10.1016/j.jelekin.2003.09.013

Bernard BP, Putz-Anderson V (1997) Musculoskeletal disorders and workplace factors. A critical review of epidemiologic evidence for work-related musculoskeletal disorders of the neck, upper extremity, and low back. Department of Health and Human Services, Public Health Service, Centers for Disease Control and Prevention, National Institute for Occupational Safety and Health (NIOSH), U.S.

Castillo SP, Bravo GP (2019) Occupational disease in Colombia 2015–2017 [In Spanish]. Fasecolda, pp 48–55

Corlett N (2003) Rapid Upper Limb Assessment (RULA). In: Karwowski W, Marras WS (eds) The occupational ergonomics handbook. CRC Press, Boca Raton, FL, USA, pp 437–445

Dempsey PG (2007) Effectiveness of ergonomics interventions to prevent musculoskeletal disorders: beware of what you ask. Int J Ind Ergon 37:169–173. https://doi.org/10.1016/j.ergon.2006.10.009

Estrada J, Camacho J, Restrepo M, Parra C (1995) Anthropometric parameters of the Colombian labor population -Acopla-95 [In Spanish]

Eurofound (2019) Working conditions and workers' health. Publications Office of the European Union, Luxembourg

Hendrick HW (2008) Applying ergonomics to systems: some documented "lessons learned." Appl Ergon 39:418–426. https://doi.org/10.1016/j.apergo.2008.02.006

ILO, IEA (2010) Ergonomic checkpoints: practical and easy-to-implement solutions for improving safety, health and working conditions, 2nd edn. International Labour Office in collaboration with the International Ergonomics Association, Geneva

ISO 14738 (2002) ISO 14738:2002 Safety of machinery—anthropometric requirements for the design of workstations at machinery

Karsh B-T, Moro FBP, Smith MJ (2001) The efficacy of workplace ergonomic interventions to control musculoskeletal disorders: a critical analysis of the peer-reviewed literature. Theor Issues Ergon Sci 2:23–96. https://doi.org/10.1080/14639220152644533

MINTRABAJO (2013) II National Survey of Safety and Health Conditions at Work in the Colombian Labor Risk System [In Spanish]. Bogotá, Colombia

Occhipinti E, Colombini D (2016) A toolkit for the analysis of biomechanical overload and prevention of WMSDs: criteria, procedures and tool selection in a step-by-step approach. Int J Ind Ergon 52:18–28. https://doi.org/10.1016/j.ergon.2015.08.001

O'Neill DH (2000) Ergonomics in industrially developing countries: does its application differ from that in industrially advanced countries? Appl Ergon 31:631–640. https://doi.org/10.1016/S0003-6870(00)00033-8

Piedrahita H (2006) Costs of work-related musculoskeletal disorders (MSDs) in developing countries: Colombia case. Int J Occup Saf Ergon 12:379–386. https://doi.org/10.1080/10803548.2006.11076696

Punnett L, Wegman DH (2004) Work-related musculoskeletal disorders: the epidemiologic evidence and the debate. J Electromyogr Kinesiol 14:13–23. https://doi.org/10.1016/j.jelekin.2003.09.015

Rodríguez Y (2018) Individual risk assessment (ERIN): method for the assessment of workplace risks for work-related musculoskeletal disorders. In: Handbook of research on ergonomics and product design, IGI Global, pp 1–27

Rodríguez Y (2019) ERIN: a practical tool for assessing exposure to risks factors for work-related musculoskeletal disorders. In: Bagnara S, Tartaglia R, Albolino S et al (eds) Proceedings of the 20th congress of the international ergonomics association (IEA 2018), Springer International Publishing, Cham, pp 369–379

Rodríguez Y, Pérez E (2013) Ergonomic improvements. Good effects for the organization. Rev Téc Ing Univ Zulia 36:183–192

Rodríguez Y, Viña S, Montero R (2013) ERIN: a practical tool for assessing work-related musculoskeletal disorders. Occup Ergon 11:59–73. https://doi.org/10.3233/OER-130210

Silverstein B, Clark R (2004) Interventions to reduce work-related musculoskeletal disorders. J Electromyogr Kinesiol 14:135–152. https://doi.org/10.1016/j.jelekin.2003.09.023

Torres Y, Rodríguez Y, Viña S (2011) Preventing work-related musculoskeletal disorders in Cuba, an industrially developing country. Work 38:301–306. https://doi.org/10.3233/WOR-2011-1133

Wells R (2009) Why have we not solved the MSD problem? Work 34:117–121. https://doi.org/10.3233/WOR-2009-0937

Whysall ZJ, Haslam RA, Haslam C (2004) Processes, barriers, and outcomes described by ergonomics consultants in preventing work-related musculoskeletal disorders. Appl Ergon 35:343–351. https://doi.org/10.1016/j.apergo.2004.03.001

Chapter 13
Postural and Fatigue Analyses for Ergonomic Workstations Design as an Integrated Approach to Sustainable Workplaces

Arturo Realyvásquez Vargas, Aide Aracely Maldonado-Macías, and Jorge Luis García-Alcaraz

Abstract Deficient workstation designs are an important risk factor of suffering musculoskeletal disorders (MSDs) among operators in manufacturing companies. Postural and fatigue analyses can provide valuable information for the ergonomic design of workstations, as well as increasing the sustainability of manufacturing companies. The objective of this research is to propose a novel integrated approach by using postural and fatigue analyses for the ergonomic design and improvement of workstations as a strategy to increase sustainability of manufacturing companies. The proposed methodology comprises three stages: (1) workstations prioritization, (2) postural assessment and fatigue analyses using Rapid Upper Limb Assessment (RULA) and Muscle Fatigue Analysis (MFA), as well as (3) intervention and control stage. To validate this integrated approach, a case study was conducted analyzing the manual tasks from a group of employees in two production lines in a manufacturing company. As main results, 16 workstations were analyzed using the proposed methodology. In three of them the risk level for MSDs was found as moderate and reduced after ergonomic interventions. Absenteeism decreased by 88% and the production rate was increased in 31.35% after four weeks of the ergonomic design implementation. As a result, the integration of postural and fatigue analyses may lead to the ergonomic design for workstations and their improvement. Consequently, this approach can facilitate the increase of sustainability in manufacturing companies

A. Realyvásquez Vargas (✉)
Department of Industrial Engineering, Tecnológico Nacional de México/I.T. Tijuana, Tijuana, Baja California, Mexico
e-mail: arturo.realyvazquez@tectijuana.edu.mx

A. A. Maldonado-Macías · J. L. García-Alcaraz
Department of Industrial and Manufacturing Engineering, Autonomous University of Ciudad Juárez, Ave. del Charro 450 Norte. Col. Partido Romero, 32310 Ciudad Juárez, Chihuahua, Mexico
e-mail: amaldona@uacj.mx

J. L. García-Alcaraz
I.T. Ciudad Juárez, Tecnológico Nacional de México, Ciudad Juárez, Chihuahua, Mexico
e-mail: jorge.garcia@uacj.mx

A. Realyvásquez Vargas et al. (eds.), *New Perspectives on Applied Industrial Ergonomics*,
https://doi.org/10.1007/978-3-030-73468-8_13

and employees' well-being by making a better use of the human resources, reducing absenteeism, and increasing production rates and performance.

Keywords Postural analysis · RULA · Muscle Fatigue Analysis (MFA) · Workstation design · Sustainability

13.1 Introduction

According to the International Ergonomics Association (IEA), ergonomics is the scientific discipline that helps understand the interactions between human beings and other system elements (e.g., technology, tools, environment, tasks, organizational elements) to optimize human well-being and the overall system performance (International Ergonomics Association (IEA) 2018). The main goal of ergonomics is to preserve employees' health, safety, and comfort (Meyer et al. 2017) in order to improve work performance (Dul and Neumann 2009) by minimizing or eliminating occupational sources of ergonomic risk, such as environmental, psychosocial, or posture-related factors, to mention a few. In addition, environmental risk factors include lighting (Nava et al. 2015; Omidiandost et al. 2015), noise (Nava et al. 2015; Lee et al. 2016), and temperature (Nava et al. 2015; Califano et al. 2017), whereas psychosocial risk factors in the workplace include mental workload and job demands, among others (Pereira 2014). Finally, unnecessary and forced postures (i.e., postures that involved forceful exertions) are also a source of ergonomic hazards at work (Nava et al. 2015; Shirzaei et al. 2015).

A concept that has been closely related to ergonomics is sustainability. According to several authors, sustainability can be described as longevity, meaning "the longer a system can be maintained, the more suitable it is" (Lew et al. 2016; Marchese et al. 2018; Roostaie et al. 2019). However, the most cited definition regarding sustainability, which is retrieved from the 1987 Brundtland Commission Report, describes sustainable development as a "development that fulfills current needs without compromising the ability of future generations to meet their own needs" (Roostaie et al. 2019). In job activities, odd and forced postures can have a negative effect on the sustainability of companies, since they are a cause of musculoskeletal disorders (MSDs) in employees in any type of industrial sector (Sarkar et al. 2015; Ng et al. 2016; Nourollahi et al. 2018). These postures are caused by a poor workstation design, and they can be eased with ergonomic workstations design (Singh and Singh 2019). Regarding MSDs, they have a significant direct effect on employees' performance (Odebiyi et al. 2016; Madadizadeh et al. 2017) and absenteeism (Madadizadeh et al. 2017). Therefore, MSDs can lead to companies to have an inappropriate optimal use of the human resource, and then to decrease their sustainability level and their competitive advantage. According to Hitka et al. (2019), human resource plays a strategic role in the company development. It represents a prerequisite for a successful performance in companies and its further development. Also, Hitka et al. (2019) mention that, currently, in order to stay sustainably successful in the market, enterprises are aware of the fact that material, financial, and information resources must be implemented and connected with human resources. Therefore, odd and forced postures will be reduced or deleted by applying an ergonomic design that can help companies to have a better implementation of the human resource to increase their sustainability level.

In Mexico, manufacturing is a key industry sector, since it includes 604,250 manufacturing companies (Instituto Nacional de Estadística Geografía e Informática (INEGI) 2020a), and generate more than 2,280,504 direct employments (Realyvásquez et al. 2016). In Mexico, the hours worked in the manufacturing industry add up to an average of 521,065,000 per month, which requires an employee to work 11 h per day on average (Instituto Nacional de Estadística Geografía e Informática (INEGI) 2020b). This provides an idea the amount of physical fatigue that can be presented in an employee. Although ergonomics has been promoted and implemented across different industries and sectors (de Freitas Penteado et al. 2012; Higgins et al. 2017; Hanafie et al. 2018; Maldonado-Macías et al. 2018; Bergsten et al. 2018), occupational risk factors are still seen in several job tasks and workstations, at least in the manufacturing industry. Ergonomic studies conducted in Mexico demonstrate that Mexican manufacturing companies have a lack of appropriate ergonomic work conditions, consequently, they are a source of numerous occupational hazards (Hernández et al. 2012; Maldonado-Macías et al. 2013, 2015; Hernández-Arellano et al. 2015). For instance, a study conducted in a computer manufacturing company that is established in Ciudad Juárez found that the task of repairing computers is likely to cause MSDs (Maldonado-Macías et al. 2015). Other authors (Hernández-Arellano et al. 2015) explored the relationship between workload and fatigue among Mexican assembly operators and found that manipulated weight ranging from 3 to 14 kg contributed to employee exhaustion and to a poor performance. Finally, it has been reported that the tasks performed by advanced manufacturing technology operators demanded both physical and mental efforts that together caused a rapid fatigue (Hernández et al. 2012).

Based on these backgrounds, the present research aims to propose an integrated methodology based on postural and fatigue analyses and workstation designs to build sustainable workplaces and increase the sustainability of manufacturing companies. In order to show the impact of the proposed methodology, a case study conducted in a Mexican manufacturing company is presented. It can be stated that the sustainability level of a company has decreased, since human resource has had an unoptimized performance. In this case study, workstations are assessed and redesigned to decrease the levels of ergonomic risks at least to low, therefore increase the sustainability level. Also, ergonomic improvement actions in the identified moderate or high risk in workstations are then implemented to decrease the levels of ergonomic risks.

Moreover, the research methodology is based on the critical realism (CR) approach. According to Easton (2010b), CR states that a single case study research method is enough to generalize theoretical and empirical findings, providing a new, rigorous, and coherent philosophical position that helps develop the theoretical and research process. Also, this author mentions that CR is a coherent, rigorous, and a new philosophical position that not only substantiates single case research as a research method, but also provides helpful implications for both the theoretical development and research process. Similarly, Tsang (2014) states that CR highlights the impacts of a case study on the theoretical process, empirical generalization, and theoretical evidence. Additionally, Tsang presents the fallibility of knowledge, who establishes that all developed theory requires being subjected to empirical evidence and evaluations; in that sense, case studies are appropriate research strategies to illustrate and analyze proposed theories. Therefore, only one case study is enough to generalize results (Easton 2010a).

According to that position, the main objective of this research is to demonstrate the impact of the proposed integrated approach on the sustainability level of manufacturing companies using a single case study. In fact, this study attempts to validate that the integrated approach of postural analysis and fatigue analyses can lead to ergonomic workstations design and significantly reduce the MSD risk level and absenteeism of employees while increasing their performance.

Finally, the last part of the paper is organized as follows: Subsections 13.1.1 and 13.1.2 provide theoretical information about the RULA and MFA methods applied in this research. Section 13.2 addresses a description of materials and methods that are implemented in the present case study; Sect. 13.3 shows the findings that were obtained; and finally, Sect. 13.4 presents the conclusions and industrial implications regarding the proposed methodology.

13.1.1 RULA Method

The Rapid Upper Limb Assessment (RULA) method assesses concrete (i.e., specific and individual, not global) postures, which is conducted by direct observation while employees perform their job (Singh et al. 2012), or by video or photo observation (Kathiravan and Gunarani 2018). Specifically, RULA evaluates the number of movements performed by the employee to complete a task, static muscle work, as well as the use of force (Escalante 2009), since they are the main causes of MSDs (Rodríguez-Ruíz and Guevara-Velasco 2011). To perform the assessment, RULA divides the body into two groups of segments: (A) arm, forearm, and wrist; while (B) neck, trunk, and legs (McAtamney and Nigel Corlett 1993; Escalante 2009). Both left and right sides of the body are assessed independently from each other (Escalante 2009; Diego-Mas 2015) using a worksheet that includes figures of working postures for each body segment and a scoring scale. The final score provided by the RULA method ranges from 1 to 7, which is proportional to the risk involved in the task. In other words, a high score in a posture indicates greater possibilities of disorders at a musculoskeletal level.

Furthermore, RULA results guide evaluators for further actions (Escalante 2009) and help organizations focus on decreasing occupational hazards. Table 13.1 displays RULA traditional scale of occupational risk levels and their corresponding corrective actions (Escalante 2009). Note that a low level of risk corresponds to a RULA score of 1–2. In this case, the assessed posture is acceptable and does not require to be

Table 13.1 Risk and action levels in RULA (McAtamney and Nigel Corlett 1993)

Risk level	RULA score	RULA risk indicator	Action level
1	1–2	Minimum	Posture is acceptable and changes are not required
2	3–4	Low	In-depth research is required, it is possible to make changes
3	5–6	Moderate	It is required to redesign the task; it is necessary to carry out research activities
4	7	High	Immediate changes are required

changed. However, when a body posture is scored 7 by the RULA method, it denotes a high level of MSD risk. Therefore, ergonomic improvement actions must be implemented to minimize these risks. In long and short terms, RULA assessments and their corresponding corrective actions (i.e., ergonomic improvements) benefit companies and employees equally.

13.1.2 Rodgers's MFA Method

The Suzanne Rodgers or Muscle Fatigue Analysis (MFA) method was designed by Rodgers and Williams to characterize the employee discomfort in automobile assembly lines and production tasks (Ma et al. 2009). The goal of the MFA is to assess muscle fatigue caused by efforts from odd body postures and manual material handling (Damayanti and Tesavrita 2018). The MFA divides the body into seven regions: (1) neck, (2) shoulders, (3) back, (4) arms/elbow, (5) wrists/hands/fingers, (6) legs/knees, and (7) ankles/feet/toes (Damayanti and Tesavrita 2018). Each body section is assessed according to three aspects: effort level, effort duration, and effort frequency.

To perform the assessment, the MFA method uses a discrete and ordinal four-point scale, where 1 is used to indicate a light effort (or < 6 seconds duration and 1 < effort/minute of frequency), 2 for a moderate effort (or 6–20 seconds duration and 1–5 efforts/minute of frequency), 3 for a heavy effort (or 20–30 seconds duration and > 5–15 efforts/minute of frequency), and 4 to indicate an effort that cannot be performed by most people (or 30 seconds duration and > 15 efforts/minute of frequency) (Stanton et al. 2005). The combination of the three aspects (effort level-duration-frequency) is used to determine the priority to change the score of a task. Tasks with a high score should be analyzed and redesigned to reduce risks of MSDs (Stanton et al. 2005; Ma et al. 2009).

13.2 Methodology

13.2.1 Materials

The materials used in this research for assessments and analyses are the following:

- A video camera with 20.9 million of effective megapixels with a sensor size of 23.5 mm × 15.7 mm. It is touch-sensitive device with 170 degrees viewing angle (Nikon).
- A computer with a touch screen of 15.6 inches and an Intel Core i5-7200U processor. Also, it has a RAM memory of 8 gigabytes and a hard disc of 2 terabytes, as well as functions like Bluetooth, wi-fi, DVD access, and Windows 10 operative system (Amazon).
- An ergonomic assessment software (RULA, Rodgers MFA): In the case of the RULA method, the software consists of an application that was developed by the Society of Ergonomists of the Mexico Civil Association (SEMAC). Similarly, the software from the Rodgers MFA method consists of a program in Microsoft Excel® developed by SEMAC.
- Microsoft Excel® software program: This is the most used and known program for the creation and management of spreadsheets around the world. It allows to perform mathematical operations from easy additions and subtractions to complex trigonometric functions (MEDIAactive 2010a). It also allows the creation of professional-looking graphics from registered data (MEDIAactive 2010b), and it is supported by Windows 10.
- A Flexometer: The flexometer used in this study is formed by a metal tape with a scale graduated in meters, centimeters, and millimeters. This tape is stuck on itself due to a spiral spring that pulls it into a housing. The flexometer used in this study can measure objects or space distances ranging from 20 cm to more than 5 m (López-Cañero 2016).

13.2.2 Methods

As depicted in Fig. 13.1, the methodology entails several procedures and comprises three stages: (1) workstations prioritization, (2) postural assessment and fatigue analysis, and (3) intervention and control of occupational risks. In turn, each stage comprises a series of more specific activities labeled as steps.

13.2.2.1 Stage 1: Workstations Prioritization

Step 1a. Overview of production areas. In order to perform this step, a team is created, which must be integrated by senior managers that know the manufacturing process, such as Ergonomists, product engineers, quality engineers, plant engineers,

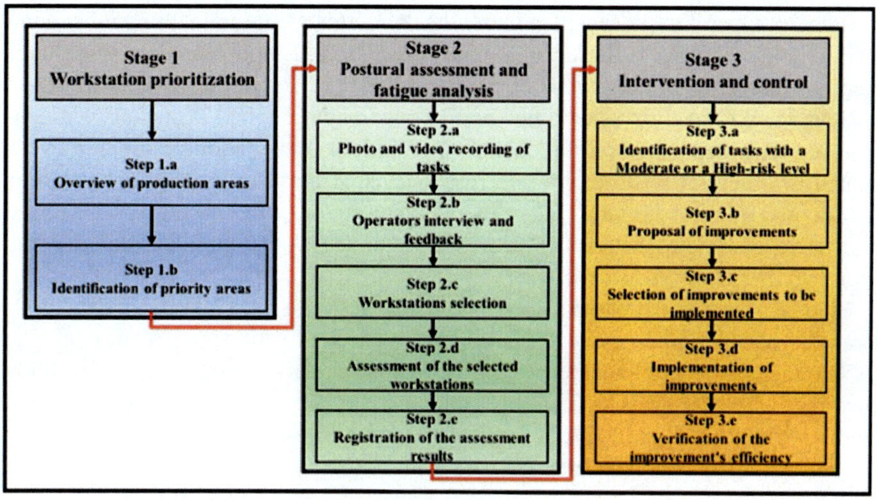

Fig. 13.1 Methodology stages and steps

and production engineers, to mention few. It is recommended that the team members receive a course about ergonomics, occupational risk factors, and evaluation methods.

In this step, the team has to visit a plant to see the company firsthand and detect, by direct observation, potential occupational risk factors in the workstations, such as repetitive movements, odd postures, manual material handling, and environmental conditions. Sixteen workstations and tasks are observed.

Step 1b. Identification of priority areas. In this step, the tasks and workstations are labeled according to the potential level of risk (i.e., low, moderate, high) noticed during the plant visit. The labeling consists of notes made in an Excel file that contained the names of the workstations. The potential risk level is based on the movements and unnecessary movements performed by employees that are observed by the team.

13.2.2.2 Stage 2: Postural Assessment and Fatigue Analysis

Step 2a. Photo and video recording of tasks. In this step, videos are recorded, and photographs are taken only of the workstations that are labeled with sources of moderate, or high ergonomic risks. Videos and photographs are not recorded neither taken of those workstations with minimum or low sources of ergonomic risks. Employees can be expected to participate in the ergonomic assessments; however, it is recommended to develop and provide a consent format where they are notified about the project. The goal of this step is to collect as much data as possible on the different positions of the body segments involved in each task. The video camera is used to record all videos and take all the photographs as well. The videos comprise at least five work cycles, and the photos are taken from both sides (left and right).

Step 2b. Operators interview and feedback. In this step, the team members and plant employees discuss the occupational risk factors. Employees then share their opinions on the potential sources of risks that they can find in their workstations.

Step 2c. Workstation selection. The recorded/photographed tasks and workstations are analyzed based on the deviation of body postures adopted by employees related to a neutral posture, and the frequency that these uncomfortable postures are adopted or their duration. Later, the workstations worthy of being assessed are selected, according to the level of priority ergonomic risk that was considered by the team members.

Step 2d. Assessment of the selected workstations. The most appropriate methods to assess occupational risk levels in the selected workstations are identified. The RULA method is implemented to assess ergonomic risk levels caused by body postures, whereas levels of muscular fatigue risks are measured using the MFA method. The RULA and MFA are two of the most popular methods for ergonomic risk measurement and assessment in a workplace (Plantard et al. 2017; Sharan et al. 2019).

Based on cost, time, and employees' training, changes are made only in those workstations/tasks that require improvements immediately, i.e., only when the risk indicators are moderate or high, both with RULA and MFA. In addition, for each workstation, and based on the critical realism approach, only one employee is assessed (Easton 2010b; Maldonado-Macías et al. 2015). After workstations are redesigned, employees with similar physical characteristics are assigned to a specific workstation to avoid risk factors related to body postures.

Step 2e. Registration of assessment results. In this step, the team members compile the data and register the assessment results in a worksheet template provided by the company, which is used for recording ergonomic assessment data.

13.2.2.3 Stage 3: Intervention and Control of Occupational Risks

Step 3a. Identification of tasks with moderate or high-risk level. This step involves using the assessment results obtained with RULA and MFA to identify tasks with moderate or high levels of occupational risks. In this step, the risk levels are not found based on the experience of the ergonomist and other team members, therefore, the level of risk in the tasks can be different from the one detected in step 1b. The goal of this step is to prioritize ergonomic redesigns in tasks and workstations that need the greatest attention. Tasks and workstations with a higher risk indicator have a higher priority.

Step 3b. Proposal of improvements. In this step, meetings are scheduled with the team members to brainstorm and plan potential ergonomic improvement proposals, which can minimize ergonomic risks in the tasks and workstations selected in the previous step. Such proposals can include workstations and tasks redesigns, job rotations of employees, or assignments of the same tasks to employees with similar physical characteristics to avoid odd body postures.

Step 3c. Selection of improvements to be implemented. The suitable improvement proposals are chosen to implement them in the selected tasks and workstations. Once the team members select the proposals, they inform their corresponding departments on this selection and the changes to workstations or tasks. These departments must know this information because they perform management tasks (equipment maintenance, products quality inspection, to name a few) on such workstations.

Step 3d. Implementation of improvements. The selected improvement proposals are implemented by the team members; therefore, selected tasks and workstations are redesigned. Team members from different company departments are involved in the implementation of the ergonomic improvement actions.

Step 3e. Verification of improvements efficiency. The final step involves assessing the redesigned tasks and workstations with the same methods implemented before the redesign (i.e., RULA and MFA). The goal is to determine whether the implemented ergonomic improvements contribute to decreasing levels of ergonomic risks or not.

13.3 Results of the Case Study

The case study took place in a Mexican manufacturing company. The company is established in Tijuana, Mexico in which electronic and automotive products are manufactured, and operates 24 h per day, seven days a week. Regarding its workforce, the company employs 3,000 workers across four shifts. In this company, employees of 16 workstations in two production lines—Line A and Line B—are exposed to odd postures who have reported a feeling of pain in different parts of the body, consequently, absenteeism has increased.

Moreover, the production level has decreased, to the extent that the company has a difficulty to meet customers' demand. The manufacturing process begins with the task of Clean Housing/Fan/Router (Fan). In this task, the employee takes an aluminum housing from a roller rack. The housing weights 100 g, it is placed in a machine to clean it. Then, in the tasks of Clean Housing/Fan/Router (Router), the employee electronically analyzes a card called PCB, and then places it in a rack. In the task of Manual Thermal Grease, the employee places a rubber in an aluminum component with a manual applicator, which already has the programmed points where the rubber must be applied. Later, the component is placed in a worktable. In the workstation of Post Test, the product functionality is tested. In order to perform this, the product is placed in a rack, which through a lever moves to connect the product to the current and to a computer. Regarding the task of Run In, the employee scans a piece of 100 g, and places it in a test cabinet with several shelves. In the tasks of Inspection 1 and Packing 1, the electronics board is inspected and packed, respectively.

In addition, the manufacturing process continues with the task of Secure Cover, where the employee unpacks the electronics board and places an aluminum cover using a drill. Later, in the task of Plenum, the employee places some components in the electronics board using a drill. Some other components are placed in the board in the tasks of Stick Lead 1 and 2 by using a small pallet and rails. Then, in the Pallet Return task, the employee removes the pallet from the electronics board and inspects it. Also, the employee marks it down. Next, in the Touch Up task, the employee inspects the electronics board and welds some components based on the marks down. In the task of Information and Communication Technology (ICT), an electronics board is inspected by using a machine. To do this, the operator opens the machine, places the electronics board in it, and closes the machine. Finally, in the task of Inspection and Packaging, the employee inspects the electronics board with a special light and packs it in a box.

13.3.1 Results of Stage 1

A team formed by an Ergonomist, a product engineer, a quality engineer, a plant engineer, and a production engineer received a course and training by SEMAC. The team members were all Industrial Engineers. The course was focused on detection and assessment of ergonomic risks, including training on ergonomic methods. Also, the course included some examples of applied ergonomics, specifically about workstations and tasks redesign, in other similar case studies. It is necessary to mention that, before the course, only the Ergonomist had knowledge regarding ergonomics. Table 13.2 shows the backgrounds of each team member.

In this case study, sixteen workstations and tasks were observed to detect potential occupational risk factors.

According to the results of the first ergonomic assessment (i.e., by direct observation), 16 tasks (i.e., all the tasks) across production lines A and B (Table 13.3) were sources of low, moderate, or high ergonomic risk, according to the RULA scale (see Table 13.1).

Most of the risk factors identified in these workstations involved body postures, such as twisted or inclined trunks, extended arms, or flexed necks.

Table 13.2 Backgrounds of the team's members

Member	Seniority in the company (years)	Department
Ergonomist	1	Continuous Improvement
Product Engineer	3	Engineering
Quality Engineer	5	Quality
Plant Engineer	3	Maintenance
Production Engineer	2	Production

Table 13.3 Assessed workstations

Production Lines and Workstations	
Line A	Line B
Clean Housing/Fan/Router (Fan)	Stick Lead 1
Clean Housing/Fan/Router (Router)	Stick Lead 2
Manual Thermal Grease	Pallet Return
Post Test	Selective
Run In	Touch Up
Inspection 1	Information and Communication Technology (ICT)
Packing 1	Inspection and Packaging
Secure Cover	
Plenum	

13.3.2 Results of Stage 2

13.3.2.1 Assessment of the Production Line A

Nine workstations were assessed in the production line A. The following paragraphs thoroughly discuss the assessment results.

Clean Housing/Fan/Router (Fan). The occupational risk factors detected in this workstation involved two body postures. Both postures had a score of 4, which according to the RULA scale, a low level of MSD risk is indicated. Therefore, improvement actions were not implemented in this task. Additionally, in both postures, the neck and the trunk had the highest sores. In the first posture, both body parts scored 3, whereas in the second posture, the neck scored 3, and the trunk scored 4. Such results imply that it might be necessary to readjust the height of the worktables, yet immediate ergonomic improvements were not considered to be necessary.

Clean Housing/Fan/Router (Router). The risk factor identified in this workstation also involved a body posture adopted by the employee. The Clean Housing/Fan/Router (Router) workstation scored 3 in the overall RULA assessment, which denotes a low level of MSD risk, according to the RULA scale. In addition, the highest individual scores were reported in the body part: trunk, therefore, implying that the height of the worktable must be modified. As in the previous case, immediate ergonomic improvement actions were not considered to be necessary because the risk level was low.

Manual Thermal Grease. For this workstation, a final RULA score of 3 was reported, which denotes a low level of MSD risk, according to the RULA scale. Regarding the individual body part scores, the left lower arm scored 3, thus implying that in the future (not immediately) it is necessary to redesign the workstation to help the employee to easily reach the materials and remove unnecessary efforts.

Post Test. This workstation reported a final RULA score of 4 that denotes a low risk of MSD, according to the RULA scale. Immediate ergonomic improvement actions were not necessary in this workstation, yet both the neck and the wrist scored 3. Moreover, the employee observed stated that sometimes a pain in the neck is presented, but not in the wrist.

Run In. Figure 13.2 depicts the assessed posture and workstation along with the results from the RULA assessment. As can be seen, the workstation had an overall score of 5 that denotes a moderate level of MSD risk, according to the RULA scale. Four body parts—the upper arm, the wrist, the neck, and the trunk—had the highest individual scores. Ergonomic improvement actions (i.e., to place the scanners in a higher position) were considered for this workstation, focusing on the previous body parts.

Inspection 1. This workstation had an overall score of 4, with a particularly high score in the neck (i.e., 3). Immediate ergonomic improvement actions in this workstation were not considered necessary.

Packaging 1. This workstation had an overall score of 2, which denotes a minimum level of MSD risk according to the RULA scale. Therefore, immediate ergonomic improvements were considered unnecessary in this workstation.

Secure Cover. This workstation was assessed using Suzanne Rodgers's method (i.e., the MFA). The assessment results indicated that the highest level of risk—a low level—could be found in the shoulders, the arms, and the elbows. Therefore, immediate ergonomic improvements were not considered for this workstation.

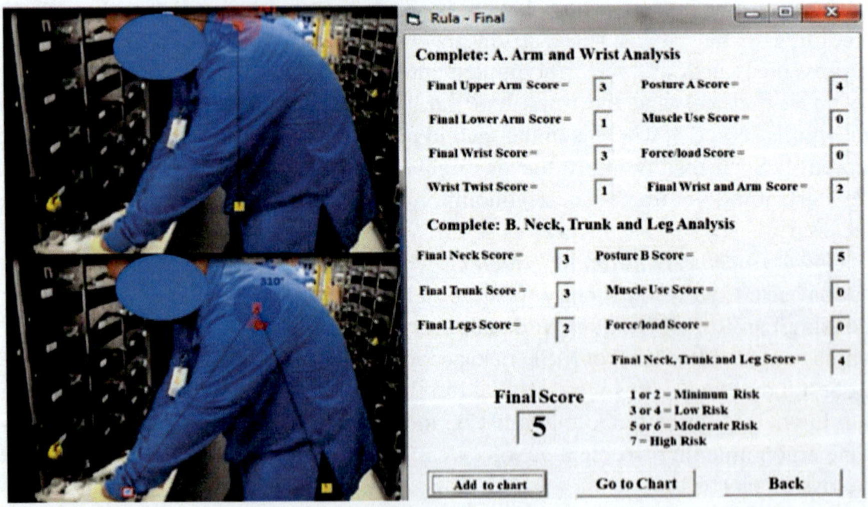

Fig. 13.2 Ergonomic assessment of the Run-In workstation

Plenum. The *Plenum* workstation was assessed though the RULA method. The overall score of workstations was 4, thus denoting a low level of MSD risk, according to the RULA scale and a moderate level of risk according to the company scale. In other words, ergonomic improvements were not necessary in this workstation.

In conclusion, merely one workstation in the production line A—i.e., Run-In workstation—had a high level of ergonomic risk, according to the RULA scale.

13.3.2.2 Assessment of the Production Line B

Seven workstations were assessed in the production line B. The following paragraphs thoroughly discuss the findings.

Stick Lead 1. For this workstation, the overall RULA score was 5, thus denoting a moderate level of MSD risk according to the RULA scale. As Fig. 13.3 displays, the body parts with the highest scores included the upper arm, the neck, and the trunk. Therefore, immediate ergonomic improvement actions were considered necessary, with a focus on these body parts.

Stick Lead 2. This workstation reported the highest overall RULA score, therefore, denoting a moderate level of MSD risk, according to the RULA scale. As depicted in Fig. 13.4, the most remarkable sources of risk included the upper arm and the trunk, with individual scores of 4 and 3, respectively. Therefore, immediate ergonomic improvement actions in the *Stick Lead 2* workstation sought to reduce these scores.

Pallet Return, Selective, and *Touch Up*. In these three workstations, the overall RULA score was 4, thus denoting a low level of MSD risks. Immediate ergonomic interventions were considered unnecessary; consequently, special attention must be given to body parts such as the lower arm and the neck, which reported the highest

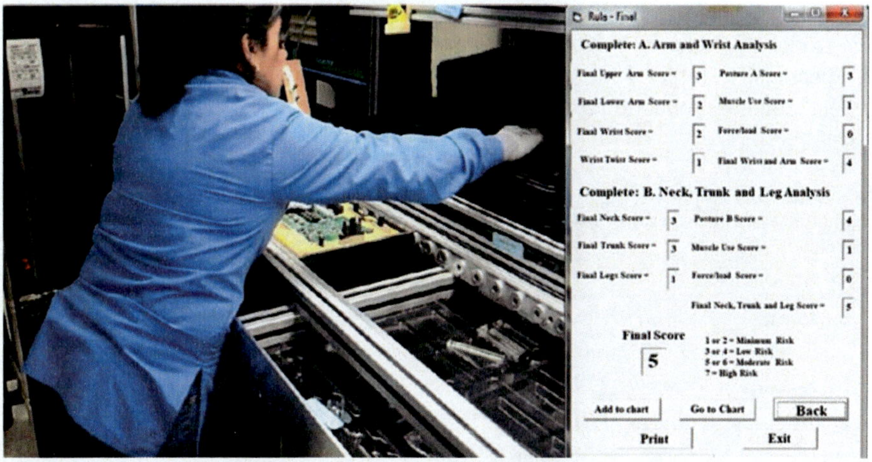

Fig. 13.3 Ergonomic assessment of the Stick Lead 1 workstation

Fig. 13.4 Ergonomic assessment of the Stick Lead 2 workstation

individual scores. In other words, if the company aims to implement ergonomic improvements in these workstations, they must be focused on these body parts.

Information and Communication Technology (ICT). The MFA was implemented to assess this workstation. The results indicated that the most noticeable level of risk; a low level, which could be found in the shoulders, arms, and elbows. Therefore, no immediate ergonomic improvements were necessary in this workstation.

Inspection and Packaging. This workstation was assessed through the RULA method. The overall score was 4, thus denoting a low level of MSD risk. Then, immediate ergonomic interventions were not necessary in this workstation; however, individual RULA scores were high in the upper arm and the neck. Such results imply that if the company is focused on implementing ergonomic improvement actions in this workstation, these should be centered in these body parts.

Specifically, Tables 13.4 and 13.5 display a summary of the assessment results collected in the production lines A and B, respectively.

Table 13.4 Summary of ergonomic assessment results in Production Line A

Workstation	RULA Risk Level
Packaging 1	Minimum
Clean Housing/Fan/Router (Fan)	Low
Clean Housing/Fan/Router (Router)	Low
Manual Thermal Grease	Low
Post Test	Low
Inspection 1	Low
Plenum	Low
Run In	Moderate
Secure Cover	Low MFA

Table 13.5 Summary of ergonomic assessment results in Production Line B

Workstation	RULA Risk Level
Pallet Run	Low
Selective	Low
Touch Up	Low
Inspection and Packaging	Low
Stick Lead 1	Moderate
Stick Lead 2	Moderate
Information and Communication Technology (ICT)	Low MFA

In general, three workstations showed moderate levels of ergonomic risks, according to the RULA scale. The following section discusses the corrective actions applied to redesign these workstations, and consequently decrease the levels of ergonomic risks.

13.3.3 Results of Stage 3

13.3.3.1 Improvements in the Production Line A

An ergonomic improvement action was implemented in the Run-In workstation to decrease the level of ergonomic risks in the production line A. Originally, in the Run-In workstation, the employee had to flex the trunk to perform the scanning task. Hence, the solution was to place the scanners at a higher position. To this end, the height at which an average-height employee (i.e., 167 cm) has the elbows positioned at 90° when scanning the components was taken as a reference. The new position of the scanners was 25.4 cm higher from their original position, which placed them at the same height as the worktables. This modification prevents the operator from constantly lifting and putting a component down during its transportation. Figure 13.5 shows the ergonomic assessment of the Run-In workstation once the improvement action was implemented. As can be observed, the level of risk decreased from 5 to 4, which is a low risk level according to the RULA scale. In fact, immediate ergonomic improvements were not necessary in this workstation.

13.3.3.2 Improvements in the Production Line B

Improvement actions were implemented in two workstations—Stick Lead 1 and Stick Lead 2—from the production line B. Due that these are consecutive workstations, they have the same structure and require the employees to perform the same tasks, the same ergonomic improvement action was implemented in both of them. According to the assessment results, the leading risk factor in Stick Lead 1 and Stick Lead 2 was

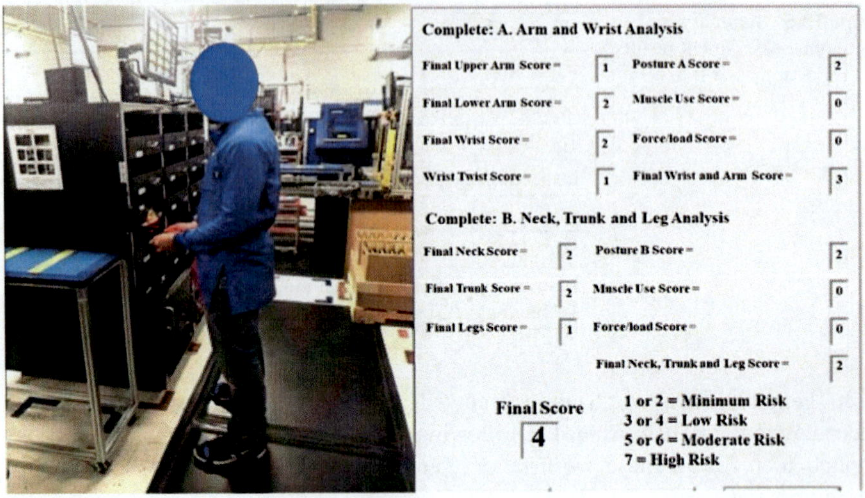

Fig. 13.5 Ergonomic assessment of the Run-In workstation following ergonomic improvements

the posture of the trunk when the employees reach for the components. Originally, the reach distance was 70.1 cm, yet it was sought to be reduced to prevent employees from being harmed. To this end, the channel of the plastic bins was removed, since the employees do not use the bins. Consequently, the reach distance point decreased by 15 cm; that is from 70.1 cm to 55.1 cm. This new reach distance point was consistent with the average reach distance of an average-height employee in the company. In fact, the redesigned workstation became suitable to the shortest employee in the company, whose reach distance was calculated with the flexometer to be 55 cm. Figure 13.6 illustrates Stick Lead 1 and Stick Lead 2 workstations before and after the improvement.

Once the improvement action was implemented, the employees were no longer required to flex their trunk, and the level of risk in Stick Lead 1 and Stick Lead 2 workstations decreased from moderate to low (Fig. 13.7), according to the RULA scale.

After the ergonomic workstations design, employees with similar anthropometric characteristics were assigned to each workstation.

13.3.3.3 Verification of Improvements

Absenteeism was reduced at 88%, after implementing the ergonomic workstation design, whereas the production level increased in more than 14,000 products per week, which represented an increase of 31.35%. Figure 13.8 shows the evolution of the production before and after the ergonomic improvements were implemented. Therefore, ergonomic improvements work efficiently.

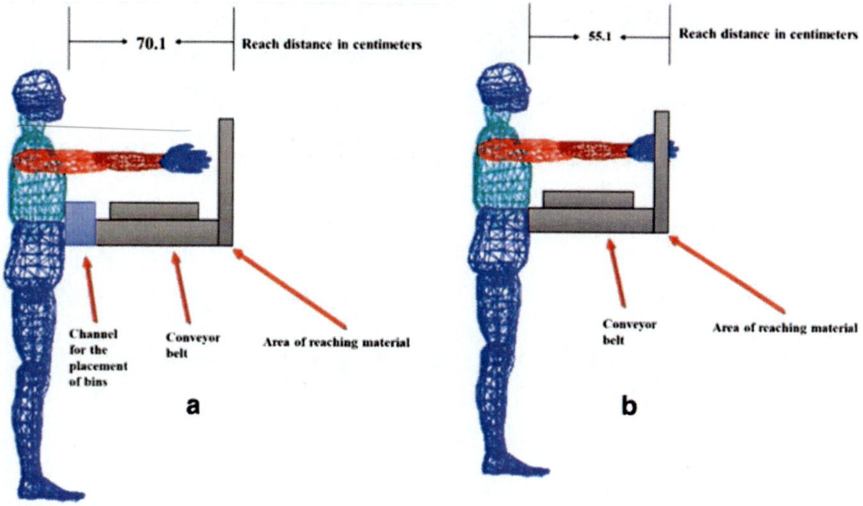

Fig. 13.6 Reach distance point for the Stick Lead workstation: (**a**) before the improvement, and (**b**) after the improvement

13.4 Discussion

In Mexico, the negative effects of odd body postures are frequently neglected in the workplaces. Additionally, the positive effects of ergonomic designs are often neglected and underestimated by employees, engineers, and even managers of manufacturing companies. More precisely, ergonomics is a scientific field that is poorly known in Mexico, as the short- and long-term effects are not considering ergonomics in the workplaces (Munguía-Vega et al. 2019). This study confirms that statement and helps promote ergonomic interventions, in terms of evaluation and design.

Furthermore, the results of this study confirm the efficiency of postural methods such as RULA and MFA to determine risk factors for musculoskeletal disorders among a variety of workstations and applications in a Mexican context. Similar results were found by Escalante (2009), as well as by Rodríguez-Ruíz and Guevara-Velasco (2011), who applied methods like RULA, ERIN, and LEST to develop procedures to recognize, assess, and determine risk factors and promote ergonomic interventions. Currently, Mexico is applying new regulations such as the Mexican Official Standard-036 (i.e., Official Mexican Norm-036, NOM-036-STPS) for evaluating ergonomic risks among industries and companies in an effort to provide better work conditions for operators in manufacturing industries (Diario Oficial de la Federación 2018). This effort will help disseminate knowledge about ergonomics, methodologies, and benefits from ergonomics in the workplace.

Additionally, the proposed integrated methodology integrating postural analysis and fatigue analyses for the ergonomic workstations design offers advantages for manufacturing industries, since analyses can be provided by a more prescriptive

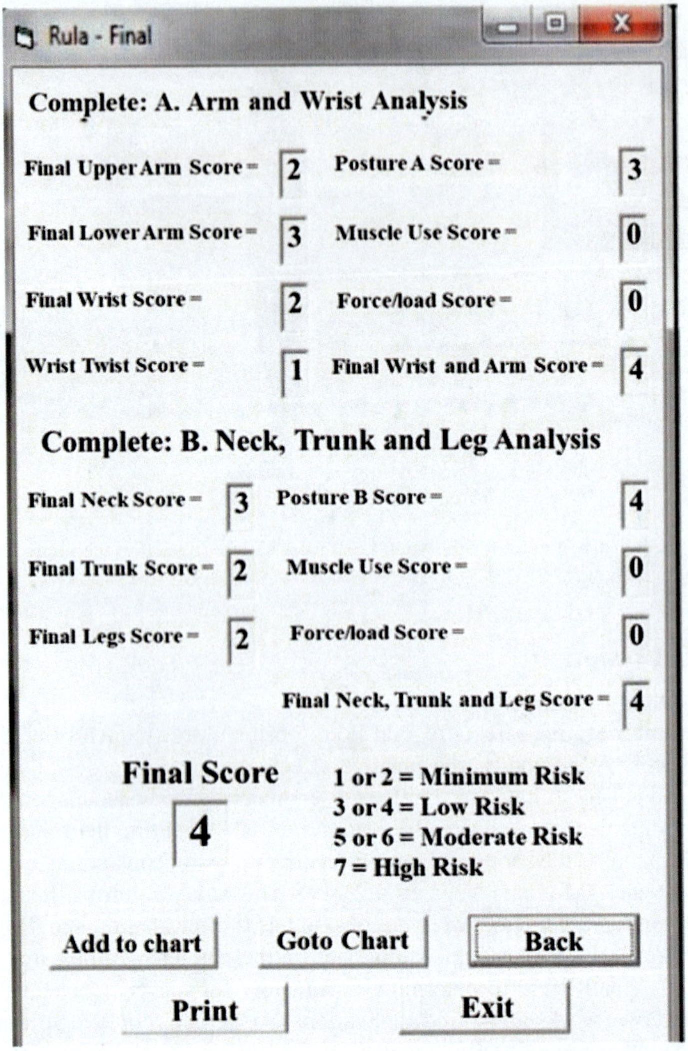

Fig. 13.7 Ergonomic assessment of the Stick Lead workstation following ergonomic improvements

approach for the recognition, discrimination, and assessment of ergonomic risks. In addition, in the last stage, risk levels can be determined more clearly for establishing priorities, proposing, and implementing improvements. In this sense, the proposed methodology is considered suitable for being applied along with recognized improvement philosophies such as Kaizen, which is an appropriate strategy for a better competitiveness in the manufacturing industry (Vieira et al. 2012).

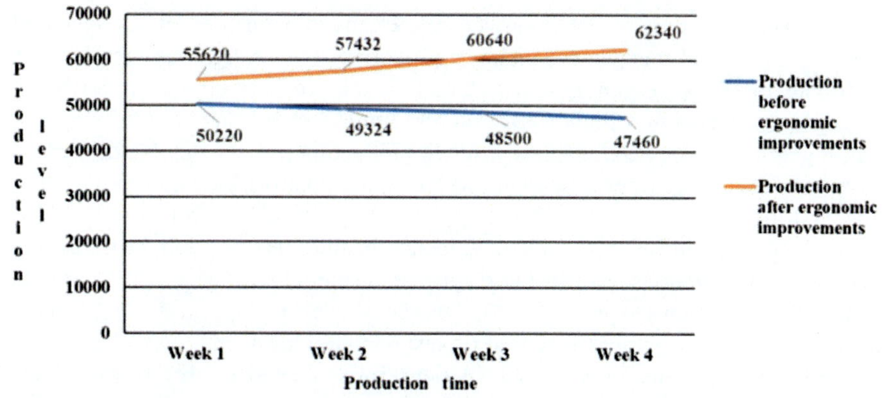

Fig. 13.8 Verification of ergonomic improvements

13.5 Study Limitations

Some limitations are addressed in the present study. First, for each workstation analyzed, only one operator was included in the analysis. Moreover, workstation redesigns were performed without a previous anthropometric study of the employees. Consequently, new workstation designs may not be comfortable or appropriate for all employees. However, there are some ergonomic studies that included only one operator, such as the study conducted by Maldonado-Macías et al. (2015), which in this research this fact is supported by the critical realism approach (Easton 2010a, b). The company was required to assign employees with similar physical characteristics for each workstation to avoid that they will adopt odd postures.

13.6 Conclusions and Recommendations

This research proposed and validated an integrated approach (postural analysis and fatigue analyses for ergonomic workstations design) to develop sustainable workplaces. A case study was carried out in a Mexican manufacturing company, where the integrated approach was successfully implemented in high-risk workstations. The objective of this research was accomplished, since the proposed integrated approach helped reduce, at first, the levels of ergonomic risks in several workstations. Consequently, absenteeism decreased in more than 80%, production increased up to 30%, and a better use and performance of human resources was obtained. Therefore, the manufacturing company increased its level of sustainability. These results indicate that the postural analysis, fatigue analyses, and ergonomic workstations design as an integrated approach, help manufacturing companies to have a better use and performance of human resources, therefore increasing the level of sustainability.

Moreover, according to the improvement actions implemented, it is concluded that small, fast, and inexpensive changes in workstations design can significantly contribute to reduce ergonomic risk levels in the workplace. For instance, by placing the scanners at a higher position and removing the channel of the plastic bins, ergonomic risks were addressed in two production lines of the manufacturing company. Also, employees now claim to feel more comfortable when doing their jobs, and the company has to enhance performance aspects, such as productivity, quality, and competitiveness, such as evidence in the literature demonstrates that ergonomic interventions lead to these type of results (Vieira et al. 2012; Hamja et al. 2019; Pakhomova et al. 2020). All these consequences support the fact that the manufacturing company has increased its sustainability level.

Finally, through the improvement actions implemented in this research, the company seeks to comply with the ergonomic requirements from the NOM-036-STPS of ergonomics. As a final recommendation, it is suggested that the company replicates the research study in its other production plants and departments, perform short-term ergonomic improvements in the other workstations analyzed, and adopt ergonomics as a work philosophy.

Acknowledgements Authors want to thank the manufacturing company for allowing to conduct this research in its facilities among its employees. Also, thanks to the Technological Instituto of Tijuana for providing the equipment, materials, and facilities that were necessary to conduct this project. Finally, authors want to thank SEMAC for providing training and software to develop the present research project.

References

Amazon HP 15.6-inch Notebook PC [In Spanish]. https://www.amazon.com/-/es/portátil-HP-Dual-Core-Bluetooth-altavoces/dp/B077HZQKJQ/ref=sr_1_7?qid=1574825789&refinements=p_89%3AHP&s=pc&sr=1-7. Accessed 26 Nov 2019

Bergsten EL, Mathiassen SE, Larsson J, Kwak L (2018) Implementation of an ergonomics intervention in a Swedish flight baggage handling company—a process evaluation. PLoS ONE 13:1–17. https://doi.org/10.1371/journal.pone.0191760

Califano R, Naddeo A, Vink P (2017) The effect of human-mattress interface's temperature on perceived thermal comfort. Appl Ergon 58:334–341. https://doi.org/10.1016/j.apergo.2016.07.012

Damayanti KA, Tesavrita C (2018) Fatigue measurement of elderly workers in small and medium enterprises. Rev Integr Bus Econ Res 7:144–151

de Freitas Penteado EV, de França MG, de Brito Ramalhoto AM, de Oliveira AM, Machado BR, Genipapeiro JA (2012) Implementation of ergonomics in a service unit: challenges and advances. Work 41:2633–2636. https://doi.org/10.3233/WOR-2012-0654-2633

Diario Oficial de la Federación (2018) Official Mexican Standard NOM-036-1-STPS-2018, Ergonomic risk factors at work-Identification, analysis, prevention and control. Part 1: Manual handling of loads [In Spanish]. https://diariooficial.gob.mx/nota_detalle.php?codigo=5544579&fecha=23/11/2018. Accessed 17 Dec 2019

Diego-Mas J (2015) Postural evaluation using the RULA method [In Spanish]. https://www.ergonautas.upv.es/metodos/rula/rula-ayuda.php. Accessed 11 Oct 2016

Dul J, Neumann W (2009) Ergonomics contributions to company strategies. Appl Ergon 40:745–752. https://doi.org/10.1016/j.apergo.2008.07.001

Easton G (2010) One case study is enough. Lancaster, The Department of Marketing

Easton G (2010) Critical realism in case study research. Ind Mark Manag 39:118–128. https://doi.org/10.1016/j.indmarman.2008.06.004

Escalante M (2009) Evaluación Ergonómica de Puestos de Trabajo. In: Seventh LACCEI Latin American and Caribbean Conference for Engineering and Technology. 7th Latin American and Caribbean Conference for Engineering and Technology. San Cristobal, pp 1–7

Hamja A, Maalouf M, Hasle P (2019) Assessing the effects of lean on occupational health and safety in the Ready-Made Garment industry. Work 64:385–395. https://doi.org/10.3233/WOR-192982

Hanafie A, Haslindah A, Saripuddin M, Yunus A (2018) Implementation of Ergonomics in the Management of Crop Yields Using Combine Harvesters. Int J Adv Sci Res Eng 4. https://doi.org/10.31695/IJASRE.2018.32786

Hernández-Arellano JL, Castillo J, Serratos-Perez N, García-Alcaraz JL (2015) Relationship between Workload and Fatigue among Mexican Assembly Operators. Int J Phys Med Rehabil 3. https://doi.org/10.4172/2329-9096.1000315

Hernández J, Brunette M, Ibarra G, García-Alcaraz J (2012) Fatigue factors in operators of semi-automated machinery in Mexico [In Spanish]. Ing Ind 30:11–28

Higgins NA, Talone AB, Fraulini NW, Smither JA (2017) Human factors and ergonomics assessment of food pantry work: a case study. Work 56:455–462. https://doi.org/10.3233/WOR-172511

Hitka M, Kucharčíková A, Štarchoň P et al (2019) Knowledge and human capital as sustainable competitive advantage in human resource management. Sustainability 11:4985–5003. https://doi.org/10.3390/su11184985

Instituto Nacional de Estadística Geografía e Informática (INEGI) (2020a) National Statistical Directory of Economic Units [in Spanish]. https://www3.inegi.org.mx/sistemas/mapa/denue/. Accessed 5 July 2020

Instituto Nacional de Estadística Geografía e Informática (INEGI) (2020b) Manufacturing Statistics and Export Maquiladora [in Spanish]. In: Manuf. Stat. Export Maquiladora [in Spanish]. https://www.inegi.org.mx/temas/manufacturasexp/

International Ergonomics Association (IEA) (2018) Definition and Domains of Ergonomics. https://iea.cc/what-is-ergonomics/. Accessed 2 Nov 2018

Kathiravan S, Gunarani GI (2018) Ergonomic Performance Assessment (EPA) Using RULA and REBA for Residential Construction in Tamil Nadu. Int J Civ Eng Technol 9:836–843

Lee PJ, Lee BK, Jeon JY et al (2016) Impact of noise on self-rated job satisfaction and health in open-plan offices: a structural equation modelling approach. Ergonomics 59:222–234. https://doi.org/10.1080/00140139.2015.1066877

Lew AA, Ng PT, Ni C (Nickel), Wu T (Emily) (2016) Community sustainability and resilience: similarities, differences and indicators. Tour Geogr 18:18–27. https://doi.org/10.1080/14616688.2015.1122664

López-Cañero J (2016) Assembly of air conditioning equipment [in Spanish], 1st edn. Paraninfo, Madrid

Ma L, Chablat D, Bennis F, Zhang W (2009) A new simple dynamic muscle fatigue model and its validation. Int J Ind Ergon 39:211–220. https://doi.org/10.1016/J.ERGON.2008.04.004

Madadizadeh F, Vali L, Rafiei S, Akbarnejad Z (2017) Risk factors associated with musculoskeletal disorders of the neck and shoulder in the personnel of Kerman University of Medical Sciences. Electron Physician 9:4341–4348. https://doi.org/10.19082/4341

Maldonado-Macías A, Alvarado A, García JL, Balderrama C (2013) Intuitionistic fuzzy TOPSIS for ergonomic compatibility evaluation of advanced manufacturing technology. Int J Adv Manuf Technol 70:2283–2292. https://doi.org/10.1007/s00170-013-5444-5

Maldonado-Macías A, Realyvásquez A, Hernández J, García-Alcaraz JL (2015) Ergonomic assessment for the task of repairing computers in a manufacturing company: a case study. Work 52:393–405. https://doi.org/10.3233/WOR-152118

Maldonado-Macías AA, Alferez-Padron CR, García-Alcaraz JL, Avelar-Sosa L (2018) Knowledge Management and Ergonomics Implementation in Manufacturing Systems: Development and Validation of a Questionnaire for Critical Success Factors. In: Baporikar N (ed) Knowledge Integration Strategies for Entrepreneurship and Sustainability, 1st edn. IGI-Global, Hershey, Pensilvania, pp 188–213

Marchese D, Reynolds E, Bates ME, Morgan H, Clark SS, Linkov I (2018) Resilience and sustainability: similarities and differences in environmental management applications. Sci Total Environ 613–614:1275–1283. https://doi.org/10.1016/j.scitotenv.2017.09.086

McAtamney L, Nigel Corlett E (1993) RULA: a survey method for the investigation of work-related upper limb disorders. Appl Ergon 24:91–99. https://doi.org/10.1016/0003-6870(93)90080-S

MEDIAactive (2010a) Learn Excel 2010 with 100 practical exercises [in Spanish], 1st edn. Marcombo, Barcelona

MEDIAactive (2010b) Excel 2010 Manual [in Spanish], 1st edn. Marcombo, Barcelona

Meyer F, Eweje G, Tappin D (2017) Ergonomics as a tool to improve the sustainability of the workforce. Work 57:339–350. https://doi.org/10.3233/WOR-172563

Munguía-Vega NE, Flores-Barboa VS, Zepeda-Quintana DS, Velazquez-Contraras LE (2019) Assessing the effectiveness of integrating ergonomics and sustainability: a case study of a Mexican maquiladora. Int J Occup Saf Ergon 25:587–596. https://doi.org/10.1080/10803548.2017.1419589

Nava R, Castro J, Rojas L, Gómez M (2015) Evaluación ergonómica de los Puestos de Trabajo del Área Administrativa. REDIELUZ 3:27–35

Ng A, Hayes MJ, Polster A (2016) Musculoskeletal disorders and working posture among dental and oral health students. Healtcare 4:13–27. https://doi.org/10.3390/healthcare4010013

Nikon Nikon D500| DSLR de lentes intercambiables de Nikon. In: Especificaciones. https://www.nikon.com.mx/nikon-products/product/dslr-cameras/d500.html#tab-ProductDetail-ProductTabs-TechSpecs. Accessed 26 Nov 2019

Nourollahi M, Afshari D, Dianat I (2018) Awkward trunk postures and their relationship with low back pain in hospital nurses. Work 59:317–323. https://doi.org/10.3233/WOR-182683

Odebiyi DO, Akanle OT, Akinbo SR, Balogun SA (2016) Prevalence and Impact of Work-Related Musculoskeletal Disorders on Job Performance of Call Center Operators in Nigeria. Int J Occup Env Med 7:98–106. https://doi.org/10.15171/ijoem.2016.622

Omidiandost A, Sohrabi Y, Poursadeghiyan M, Yarmohammadi H, Mosavi A (2015) Evaluation of general and local lighting as an environmental ergonomics factor in different parts of a hospital in the city of Kermanshah in 2015. Tech J Eng Appl Sci 5:255–259

Pakhomova A, Salnikova Y, Namestnikova L, Marahovskaya I (2020) The use of ergonomic methods for ensuring the competitiveness of business structures. In: Kantola JI, Nazir S (eds) Advances in Human Factors, Business Management and Leadership. AHFE 2019. Advances in Intelligent Systems and Computing. Springer Verlag, Washington, DC, pp 361–367

Pereira F (2014) Mental workload, task demand and driving performance: what relation? Procedia—Soc Behav Sci 162:310–319. https://doi.org/10.1016/j.sbspro.2014.12.212

Plantard P, Shum HPH, Le Pierres AS, Multon F (2017) Validation of an ergonomic assessment method using Kinect data in real workplace conditions. Appl Ergon 65:562–569. https://doi.org/10.1016/j.apergo.2016.10.015

Realyvásquez A, García-Alcaraz JL, Blanco-Fernández J (2016) Development and validation of a macroergonomic compatibility questionnaire [in Spanish]. Contaduría Y Adm 61:478–498. https://doi.org/10.1016/j.cya.2016.04.002

Rodríguez-Ruíz Y, Guevara-Velasco C (2011) Assessment of Workstations Using ERIN and RULA Ergonomic Tools [in Spanish]. Ing Ind 32:19–27

Roostaie S, Nawari N, Kibert CJ (2019) Sustainability and resilience: a review of definitions, relationships, and their integration into a combined building assessment framework. Build Environ 154:132–144. https://doi.org/10.1016/j.buildenv.2019.02.042

Sarkar K, Dev S, Das T et al (2015) Examination of postures and frequency of musculoskeletal disorders among manual workers in Calcutta, India. Int J Occup Environ Health 22:151–158. https://doi.org/10.1080/10773525.2016.1189682

Sharan D, Jose JA, Rajkumar JS (2019) How to perform an ergonomic workplace analysis? In: Bagnara S, Tartaglia R, Albolino S, et al. (eds) Proceedings of the 20th Congress of the International Ergonomics Association (IEA 2018). IEA 2018. Advances in Intelligent Systems and Computing. Springer Verlag, pp 7–11

Shirzaei M, Mirzaei R, Khaje-Alizade A, Mohammadi M (2015) Evaluation of ergonomic factors and postures that cause muscle pains in dentistry students' bodies. J Clin Exp Dent 7:e414–e418. https://doi.org/10.4317/jced.51909

Singh H, Singh LP (2019) Musculoskeletal disorders among insurance office employees: a case study. Work 64:153–160. https://doi.org/10.3233/WOR-192978

Singh J, Lal H, Kocher G (2012) Musculoskeletal Disorder Risk Assessment in Small Scale forging Industry by using RULA Method. Int J Eng Adv Technol 1:513–518

Stanton NA, Hedge A, Brookhuis K et al (2005) Handbook of Human Factors and Ergonomics Methods. CRC Press, Boca Raton, FL

Tsang EWK (2014) Case studies and generalization in information systems research: a critical realist perspective. J Strateg Inf Syst 23:174–186. https://doi.org/10.1016/j.jsis.2013.09.002

Vieira L, Balbinotti G, Varasquin A, Gontijo L (2012) Ergonomics and Kaizen as strategies for competitiveness: a theoretical and practical in an automotive industry. Work 41:1756–1762. https://doi.org/10.3233/WOR-2012-0381-1756